Throughout the book, we will see changes in colour. Cooler colours introduce speculations on the material, technology, and programming; and warmer colours indicate a focus on the user. From speculations on omnipotent technology to personal freedoms and desires, from objects to subjects and back, and from the individual to the collective, this book oscillates between these two sets of conditions defined by materiality and user experience.

Throughout the book, we will see changes in colour. Cooler colours introduce speculations on the material, technology, and programming; and warmer colours indicate a focus on the user. From speculations on omnipotent technology to personal freedoms and desires, from objects to subjects and back, and from the individual to the collective, this book oscillates between these two sets of conditions defined by materiality and user experience.

Barba

The Why Factory: Winy Maas with Ulf Hackauf, Adrien Ravon, Transformer studio TU Delft and Wieland and Grouwens

Imagine a composite of molecules. Its structure can be steered and altered in real time.
It forms a material that can change shape, shrink and expand to do almost everything.

The material can be solid and can form walls and ceilings. These walls and ceilings can be enclosed but can also open up if we'd like, to allow for more light or a better view. The walls can even open themselves further, to create doors. Doors that can then twist and turn to become a stair . . .

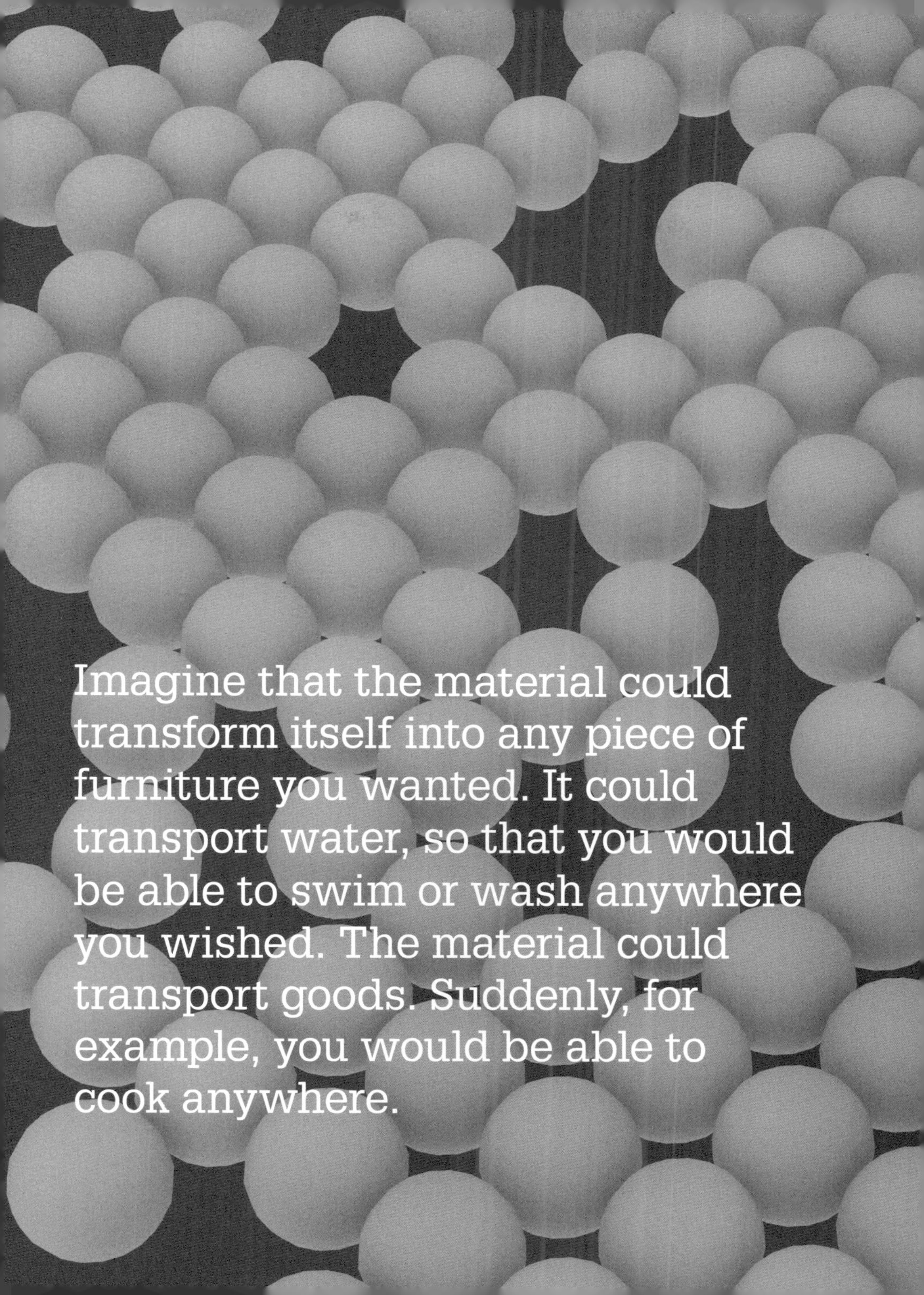

Imagine that the material could transform itself into any piece of furniture you wanted. It could transport water, so that you would be able to swim or wash anywhere you wished. The material could transport goods. Suddenly, for example, you would be able to cook anywhere.

And the material could conduct electricity so that you could have light and power anywhere and at any time. The material could clean up and move waste . . .

You could direct the material to do all of this. With your hand, with your body. The material would sense you . . .

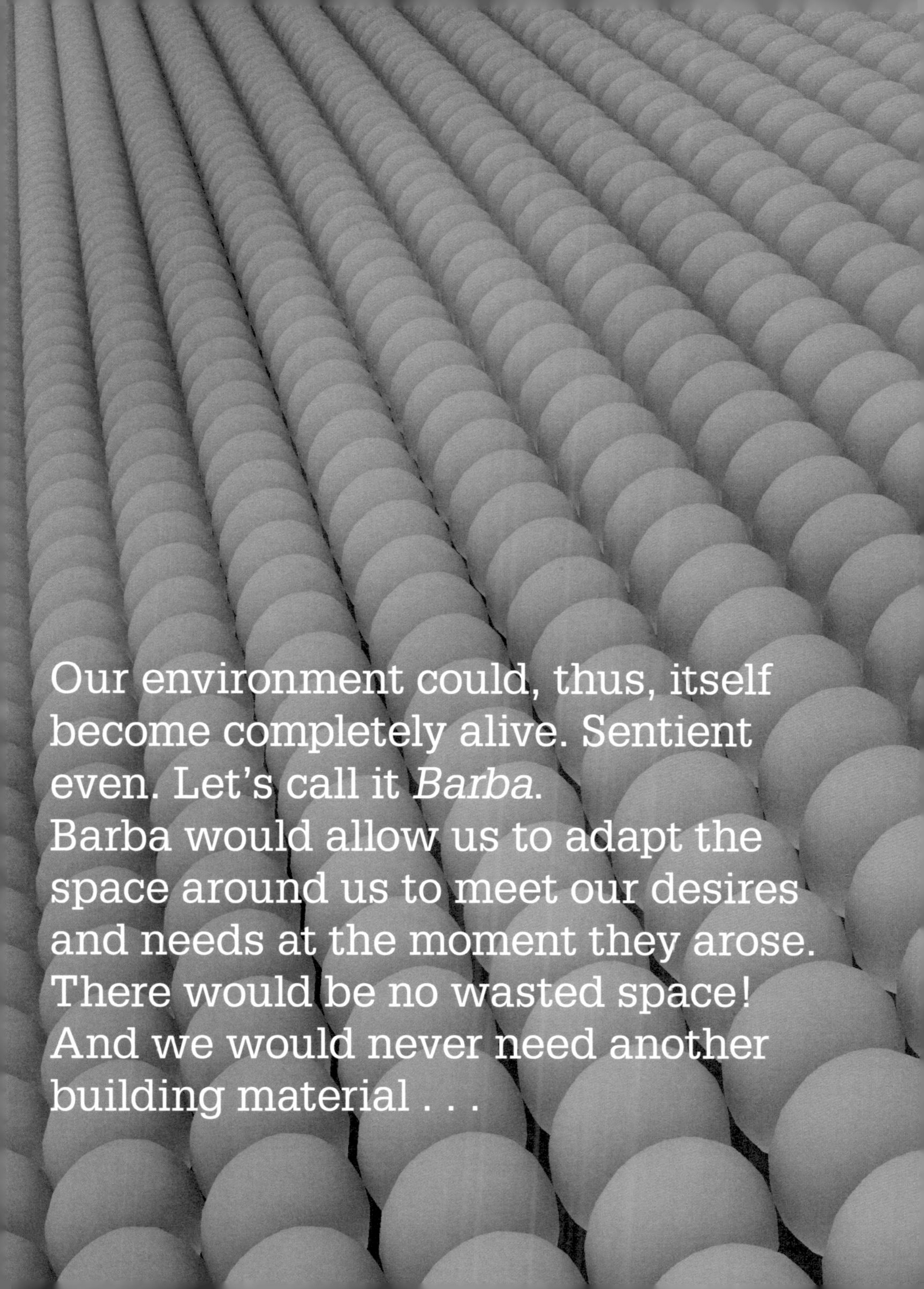

Our environment could, thus, itself become completely alive. Sentient even. Let's call it *Barba*.
Barba would allow us to adapt the space around us to meet our desires and needs at the moment they arose. There would be no wasted space! And we would never need another building material . . .

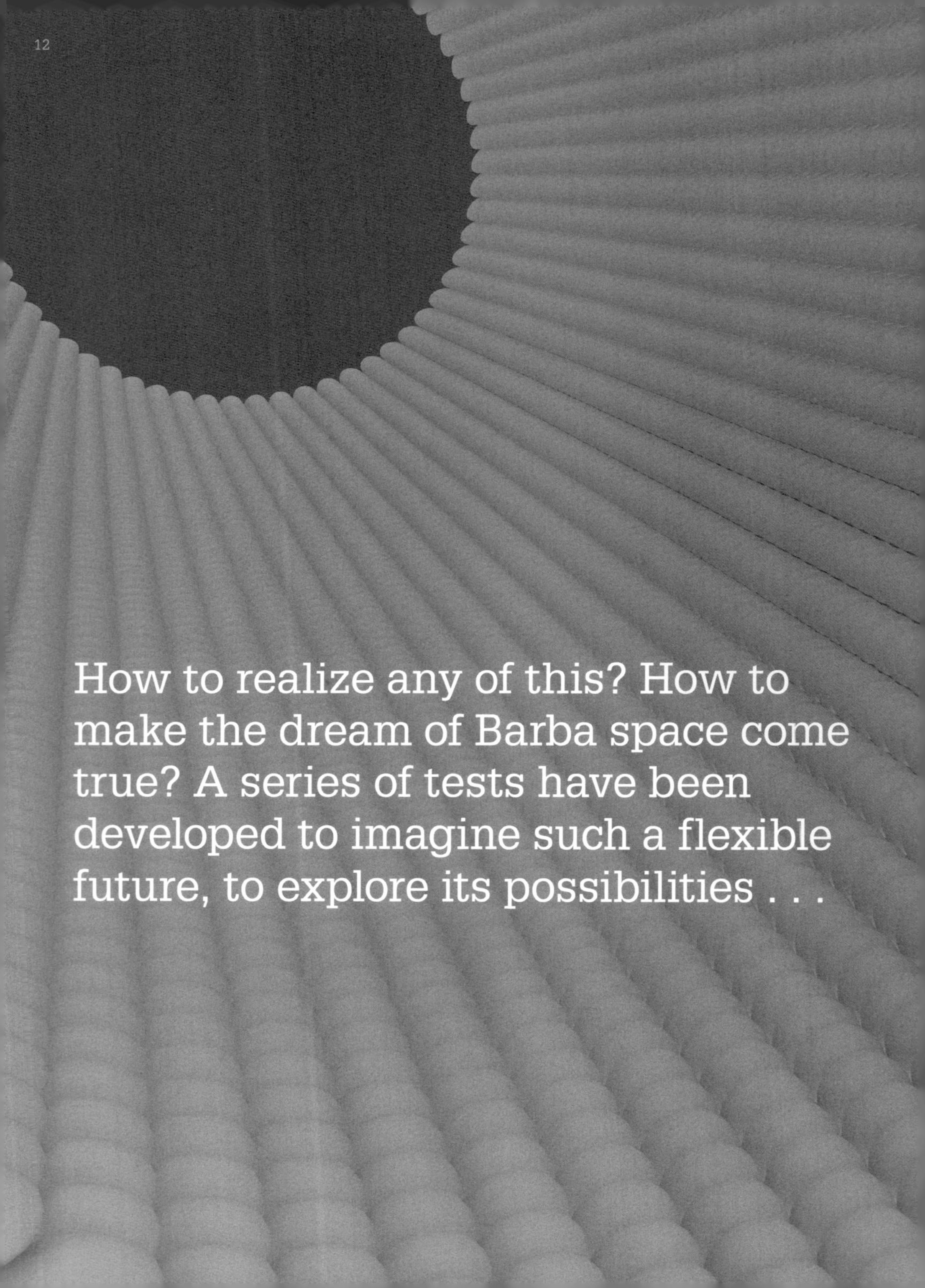

How to realize any of this? How to make the dream of Barba space come true? A series of tests have been developed to imagine such a flexible future, to explore its possibilities . . .

But this speculation itself raises many questions. What will architecture be under these new conditions? Will it lose its classical definition? With this new adaptability, will it even be possible to define architecture at all? And what happens to identity? In a morphing, flexible space, would we still know where we were — or who we were?

And what about property?
If everything were always in flux, would we still be able to take stock of what we have or had? What would real estate be, then? A surreal estate? What would 'value' mean?

Beautiful and horrendous thoughts come alive. Thoughts of a future and of a new way of living that are closer to us now then you might ever have thought . . .

Barba, our dream material, has a very powerful potential for the future of cities. If buildings could grow when needed and shrink when empty, cities could be used much more efficiently. Think of all the office buildings left empty at night or on the weekends; Barba would allow them to deflate and give way to other buildings, to parks or to nature.

Imagine what this might mean for an apartment. At night, only a bedroom would be there. During dinner, the bathroom and bedroom would be folded away. And when you take a bath, the bathroom would be the only room extended to its necessary size. All of this could dispose of a tremendous amount of wasted space. Cities could be much more open and much more green. Or they could be much denser, with a much reduced footprint.

Wasted Space

The Why Factory: Robotic City and Adaptive City studios, TU Delft

We would like to estimate how much space we could gain by building with Barba and thereby working towards the development of a fully adaptable environment. A simple, visual case study indicates the general scale of unused space. We follow an average couple during a regular day. They get up and have breakfast together. One person leaves to work in an office while the other works at home. The couple eats dinner together and spends the evening at home. Throughout the day, we monitor which rooms are used and which remain empty.

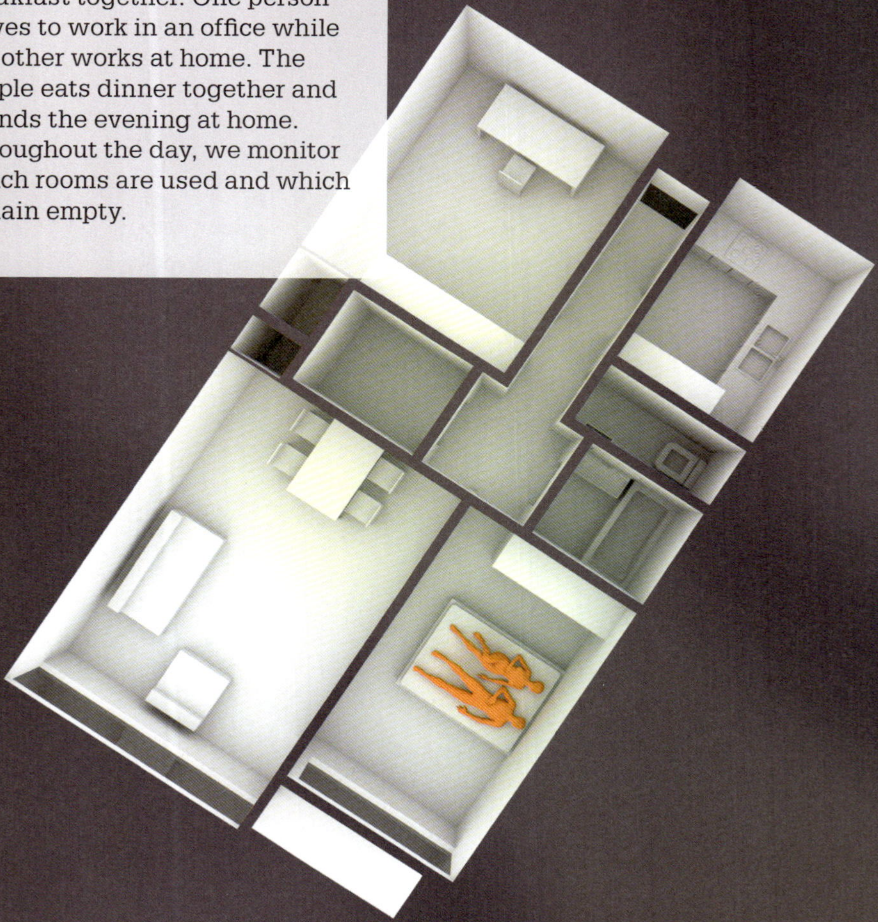

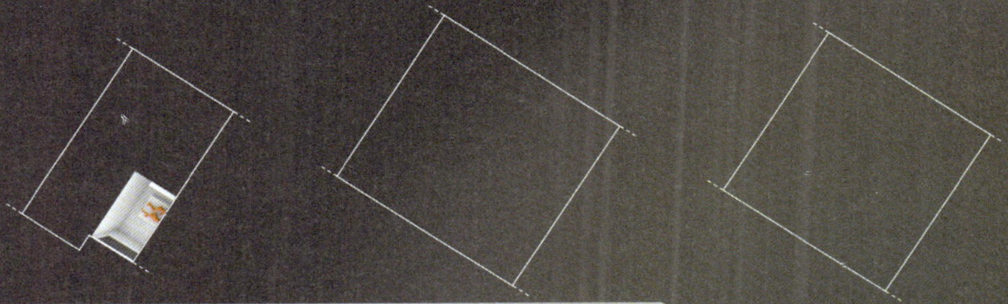

06:15 We begin our observation in the morning. Our subjects are asleep.

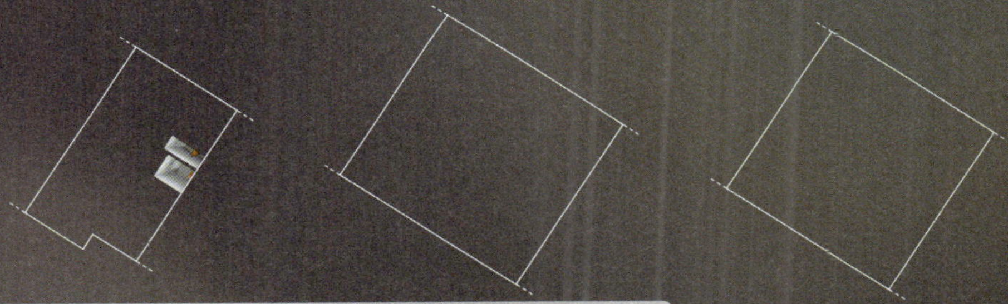

06:30 A moment later, the alarm goes off and the morning routine starts.

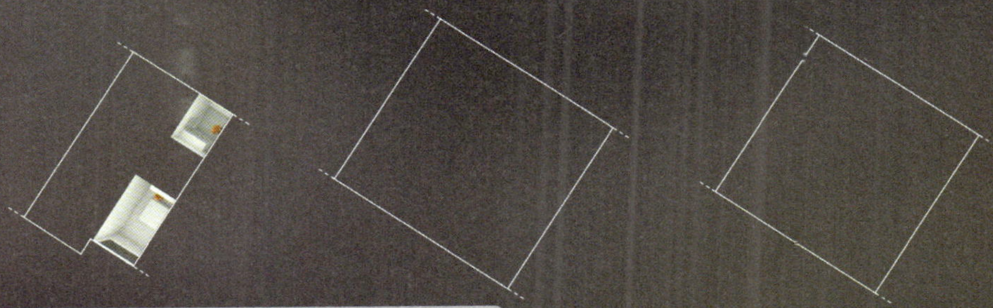

06:45 The couple gets dressed and prepares breakfast.

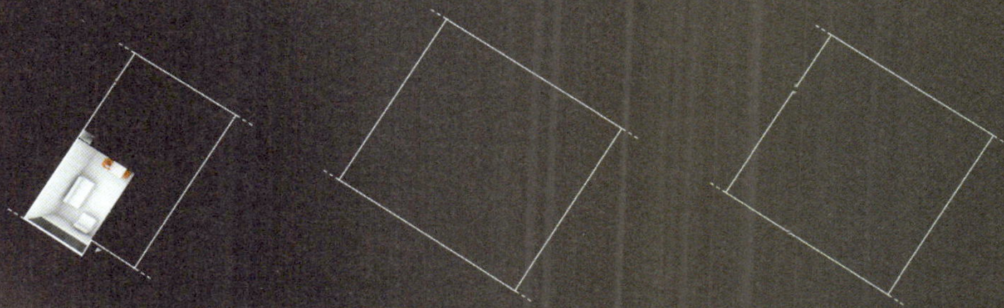

07:00 Our subjects eat breakfast together.

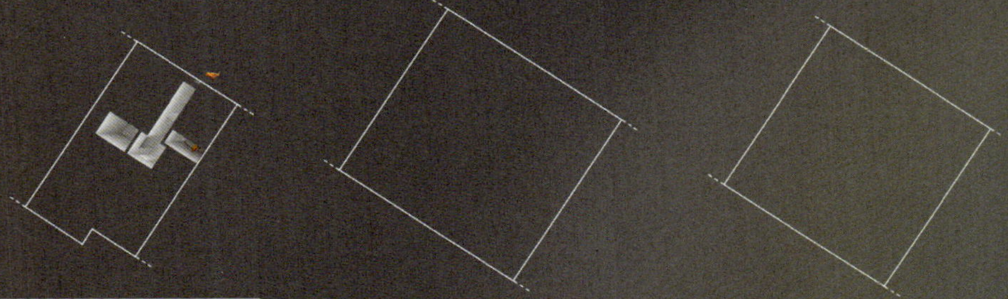

07:15 One leaves the home.

07:30 While one person walks to the office, the other has time for a short workout.

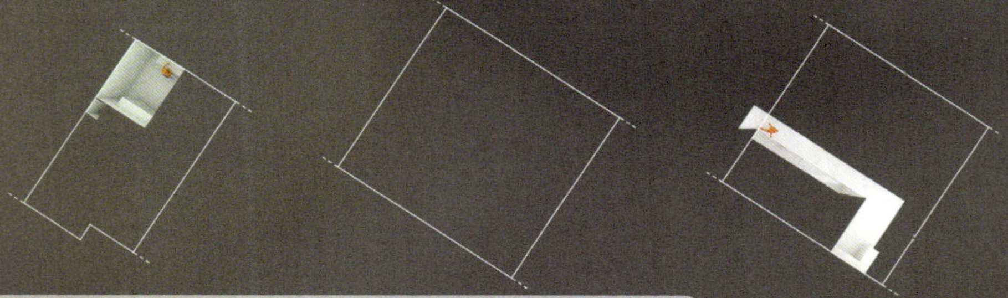

08:00 Work starts with another coffee, at the office and in the study at home.

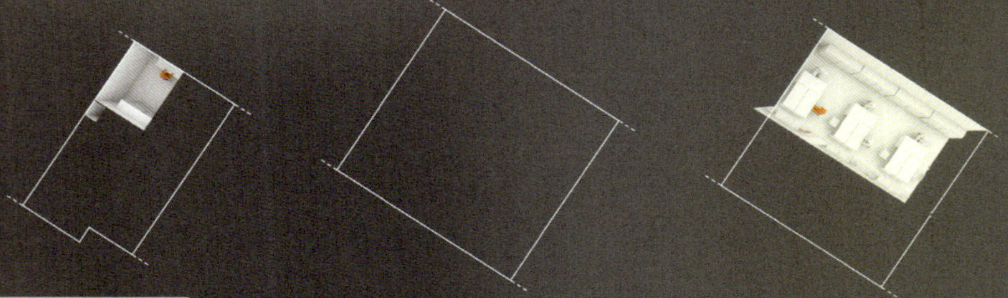

08:45 E-mailing . . .

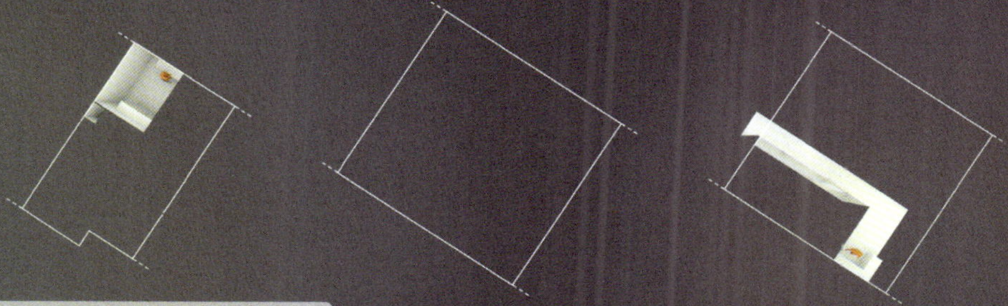

09:30 . . . and another coffee break.

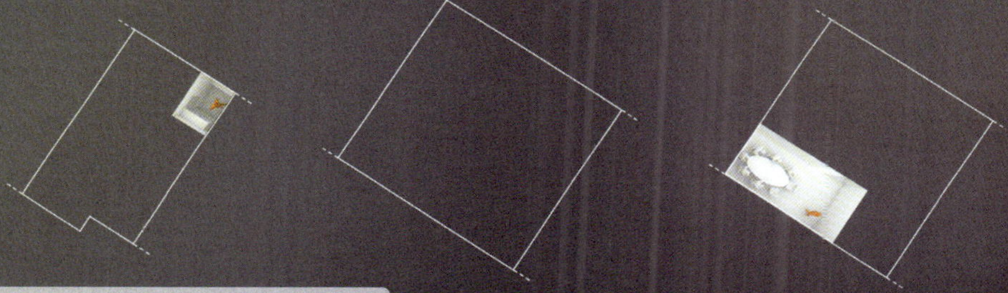

10:30 At the office, a project is presented.

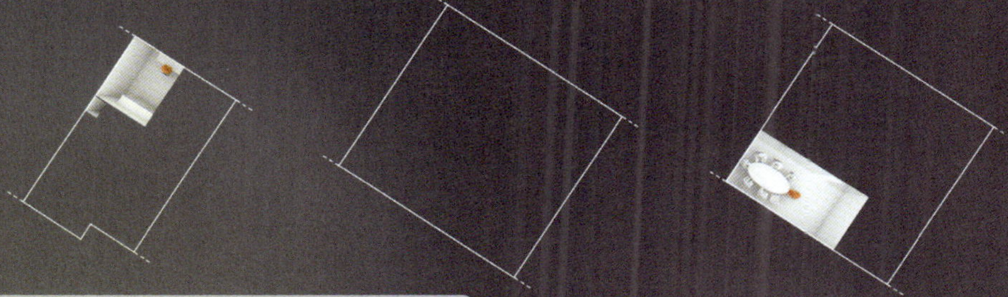

11:00 The presentation is followed by a discussion.

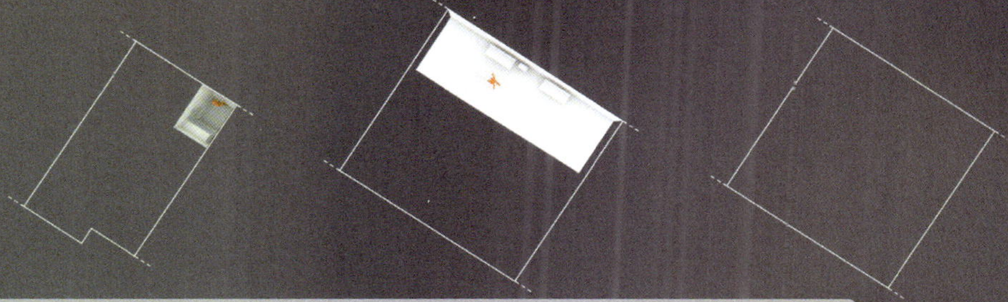

12:30 While one prepares lunch in the kitchen, the other leaves the office to pick something up.

13:00 Coffee after lunch: on the balcony and on a sunny bench outside.

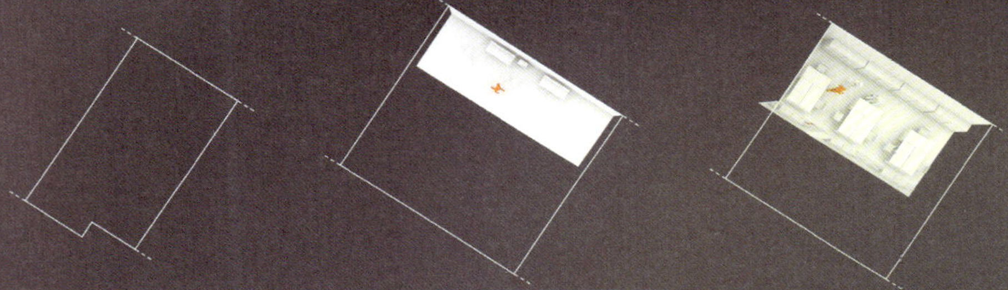

17:15 At the office, a last chat and then grocery shopping on the way home.

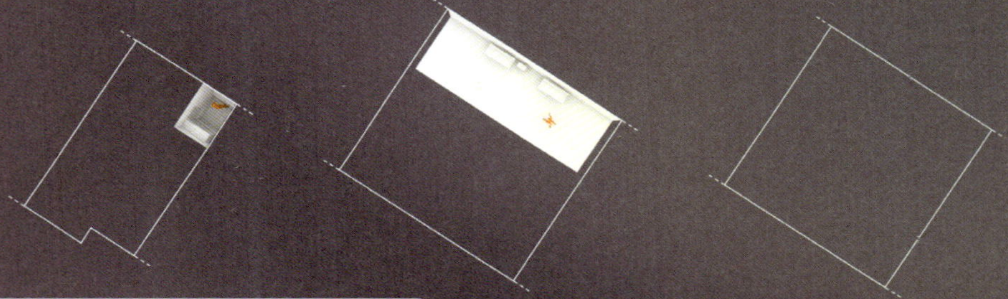

17:30 Preparing dinner at home, first alone . . .

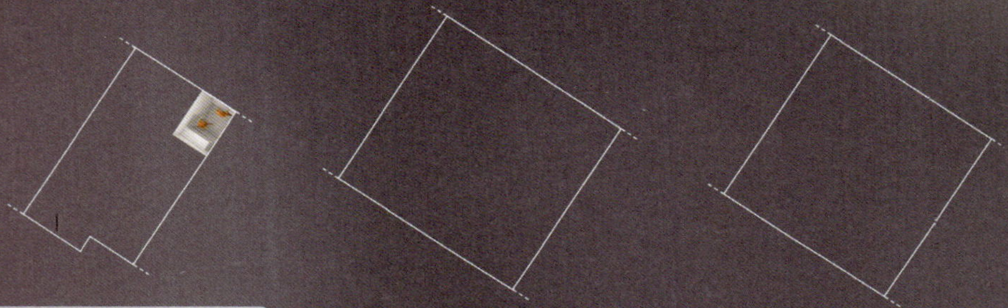

17:15 . . . and then together.

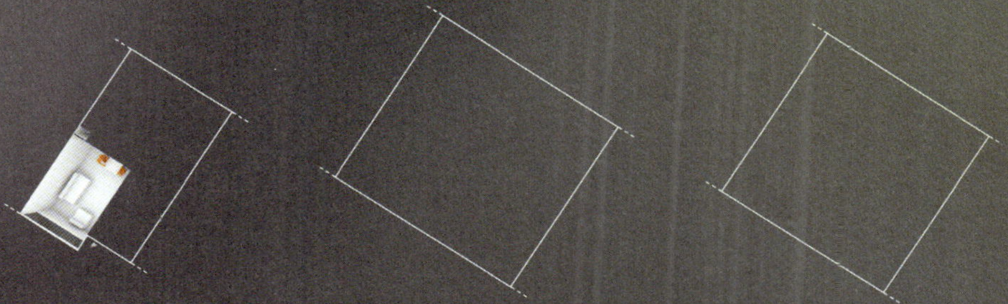

18:15 Dinner together.

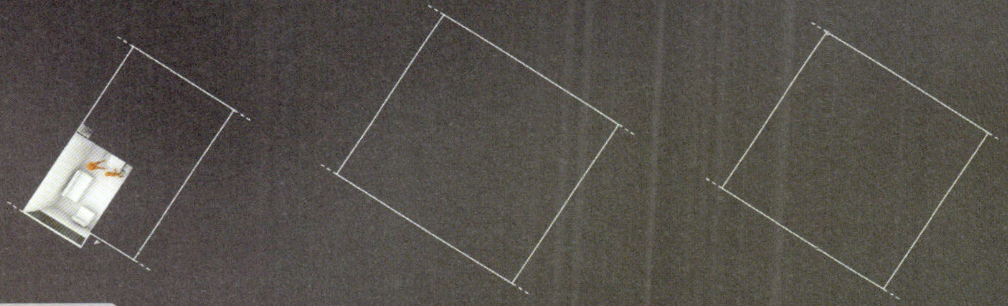

18:45 . . .

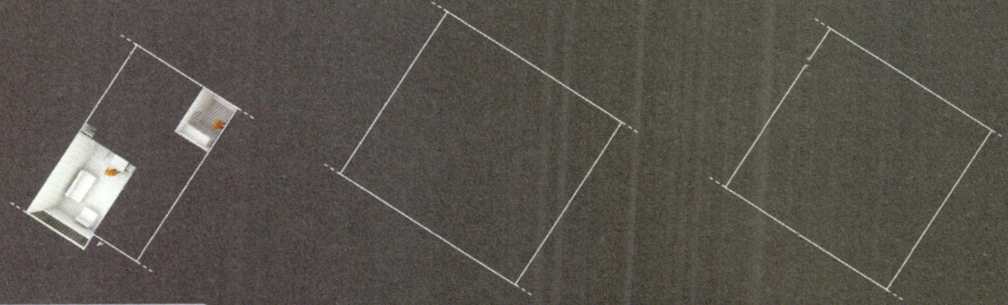

19:00 Cleaning up.

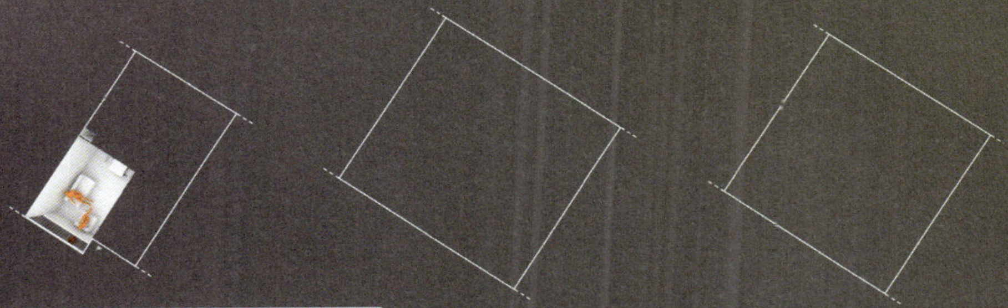

20:00 Time to watch the evening news.

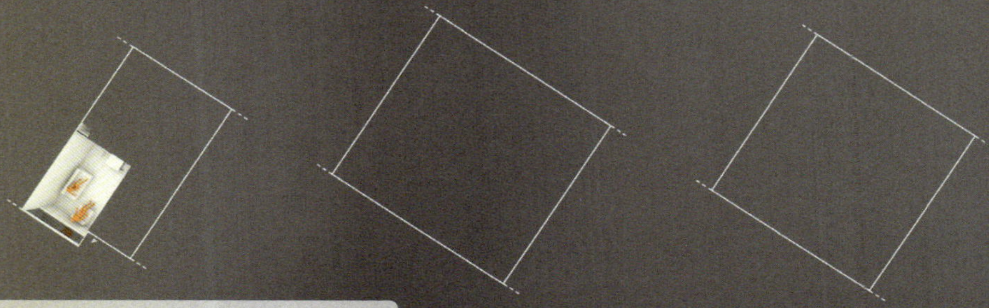

21:00 Getting more comfortable for a movie.

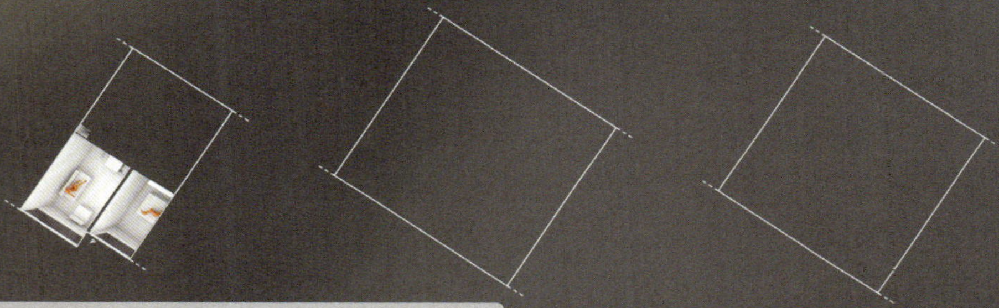

22:30 While one watches sports, the other reads a book.

23:30 The evening routine . . .

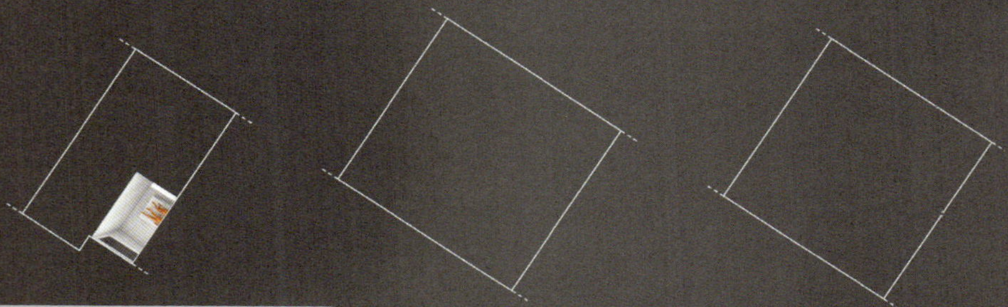

00:30 . . . and both are asleep again.

In this visual case study, only a small portion of each of the two buildings described is in use at any given moment. In order to actually measure occupancy, we estimate how much space is used at all times of the day. We calculate this as time multiplied by space and measure it in square-metre-hours, or m^2h. We consider the apartment, the street space and the office separately.

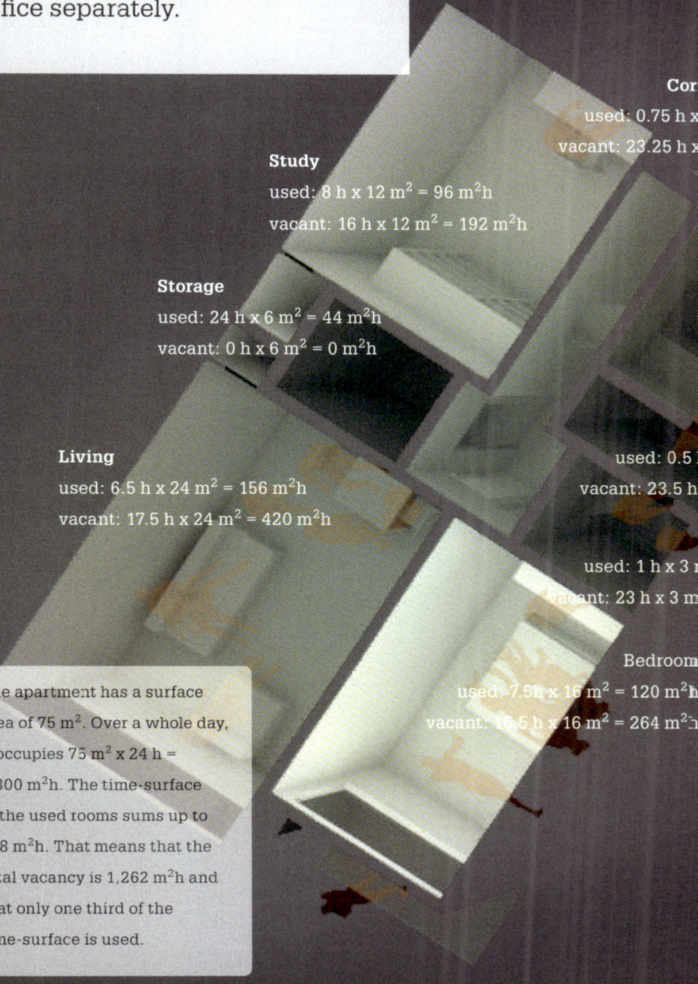

The apartment has a surface area of 75 m^2. Over a whole day, it occupies 75 m^2 x 24 h = 1,800 m^2h. The time-surface of the used rooms sums up to 538 m^2h. That means that the total vacancy is 1,262 m^2h and that only one third of the time-surface is used.

This first graphic evaluation indicates how much vacant space an average apartment and office may have on a regular day. Over 24 hours, approximately two-thirds of the apartment and three-quarters of the office could be qualified as dead space. These calculations are rough. Peaks are not taken into account. The office, for example, is not used at night and could give way to other urban spaces.

Our case study is not suitable for calculating the space efficiency of the street space, as there are many users other than those two we've monitored and described.

At times when most people in the city are asleep, for example, there may be not as much need for that kind of workspace. Still, the ratio of vacancy to used space is impressive. We found it intriguing that at least half of the apartment is unused at any given moment of the day, without compromising on any comfort. Imagine Barba in full use and what such a flexible material could mean for the efficiency of cities.
Urban footprints could shrink, giving way to agriculture and nature . . .

The office has a surface area of 100 m². Over the course of a day, it occupies 100 m² x 24 h = 2,400 m²h. The time-surface of the used rooms adds up to 430 m²h. This means that the total vacancy is 1,970 m²h and that only one third of the time-surface is used.

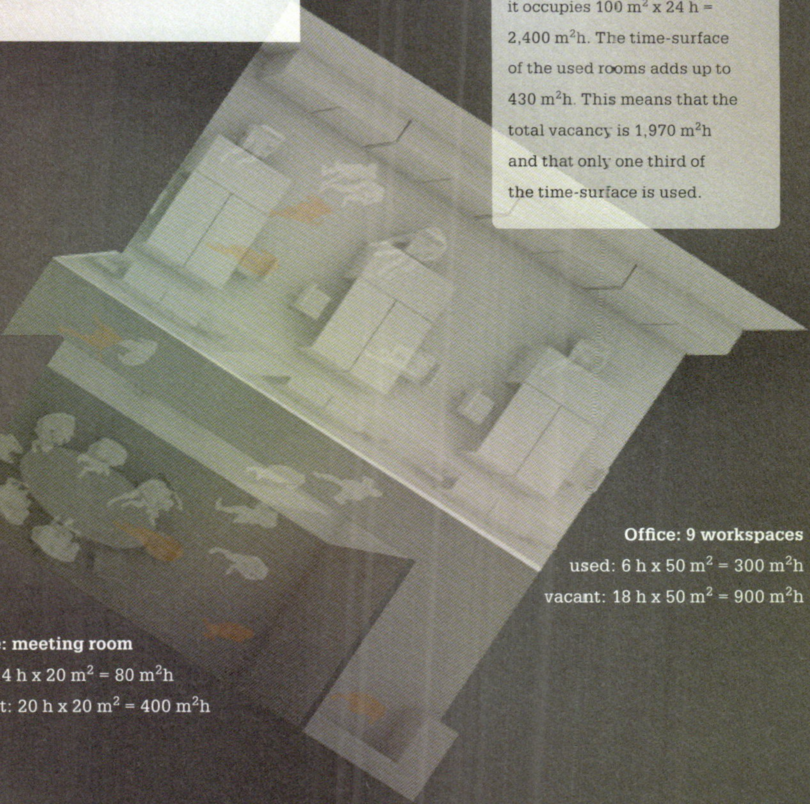

Office: 9 workspaces
used: 6 h x 50 m² = 300 m²h
vacant: 18 h x 50 m² = 900 m²h

Office: meeting room
used: 4 h x 20 m² = 80 m²h
vacant: 20 h x 20 m² = 400 m²h

Can we imagine living in an environment that would constantly change according to our needs and desires? We'll illustrate this Barba world, step by step. We'll start with the user himself and with how Barba reads his body. When the user wakes up, Barba gently pushes him up and follows him wherever he walks. Barba continuously adapts to his movements. Breath enlarges the space. Our gestures direct Barba.

In a series of logical steps, our story moves from the user to his room to the home. With more people involved, space has to be negotiated and the negotiation will create a dynamic, nearly sculptural composition of volumes. Eventually, these dynamic volumes lead to a living village that constantly calibrates the balance between social collectiveness and individual desire.

Life in Barba

The Why Factory and ETH Zürich: Adrien Ravon and the Transformer at ETH Zürich Studio

Transformer at ETH Zürich Studio:
Paolo Birrer, Tania Carl, Riccardo Caruso, Melina Cerfeda, Elena Chestnova, Nicholas Ganz, Marcel Hodel, Léone Kündig, Marisa Muscionico, Sebastian Oswald, Kai Peter, Pascal Ruckstuhl, Enrique Ruiz, Paula Schilliger, Vera Schmidt, Henry Stehli, Ying Yi Tan, Yangzom Wujohktsang
Teachers: Winy Maas, Ulf Hackauf, Adrien Ravon, Stefan Wülser

Imagine life in a fully transformable environment . . .

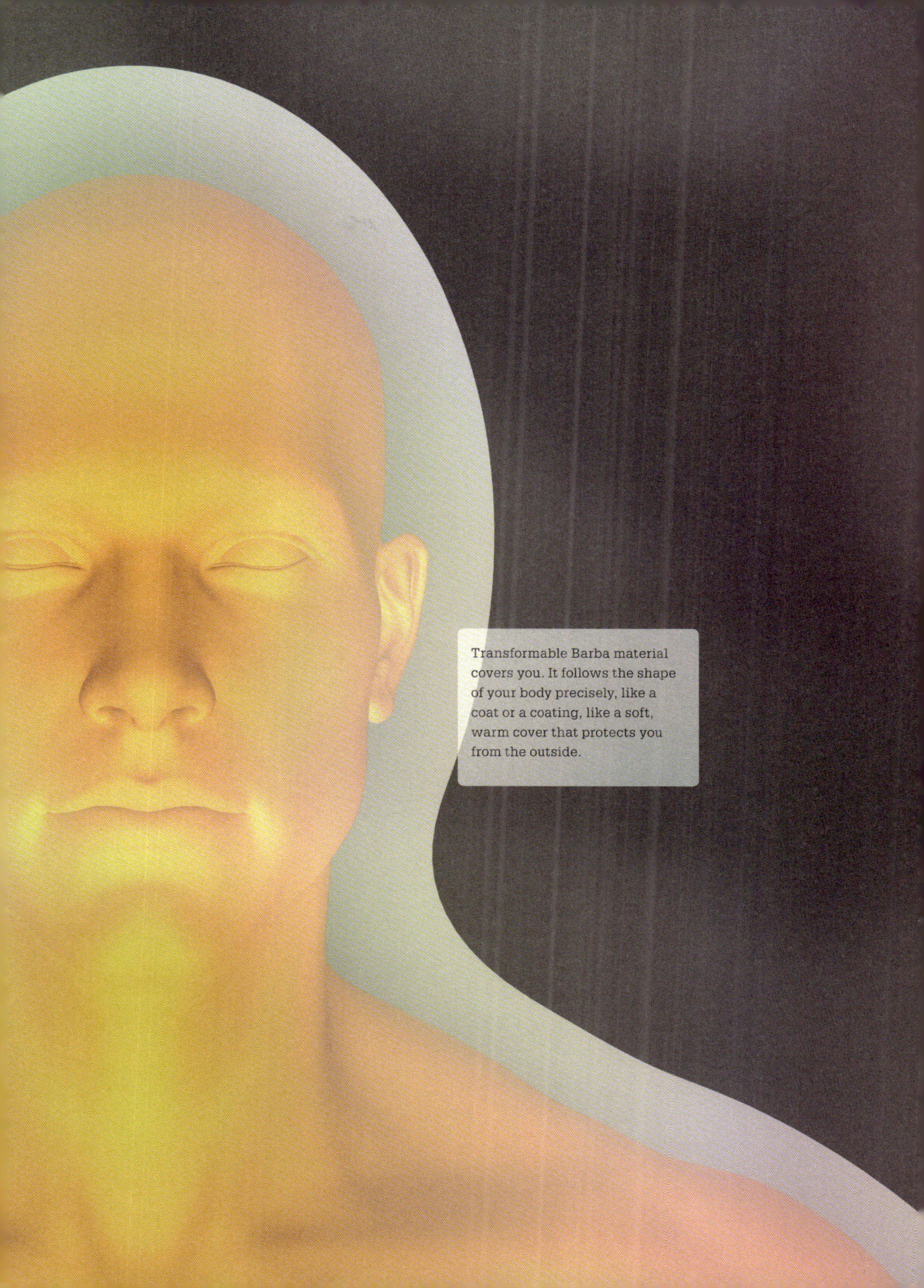

Transformable Barba material covers you. It follows the shape of your body precisely, like a coat or a coating, like a soft, warm cover that protects you from the outside.

. . . in which a flexible and responsive material surrounds and shelters you.

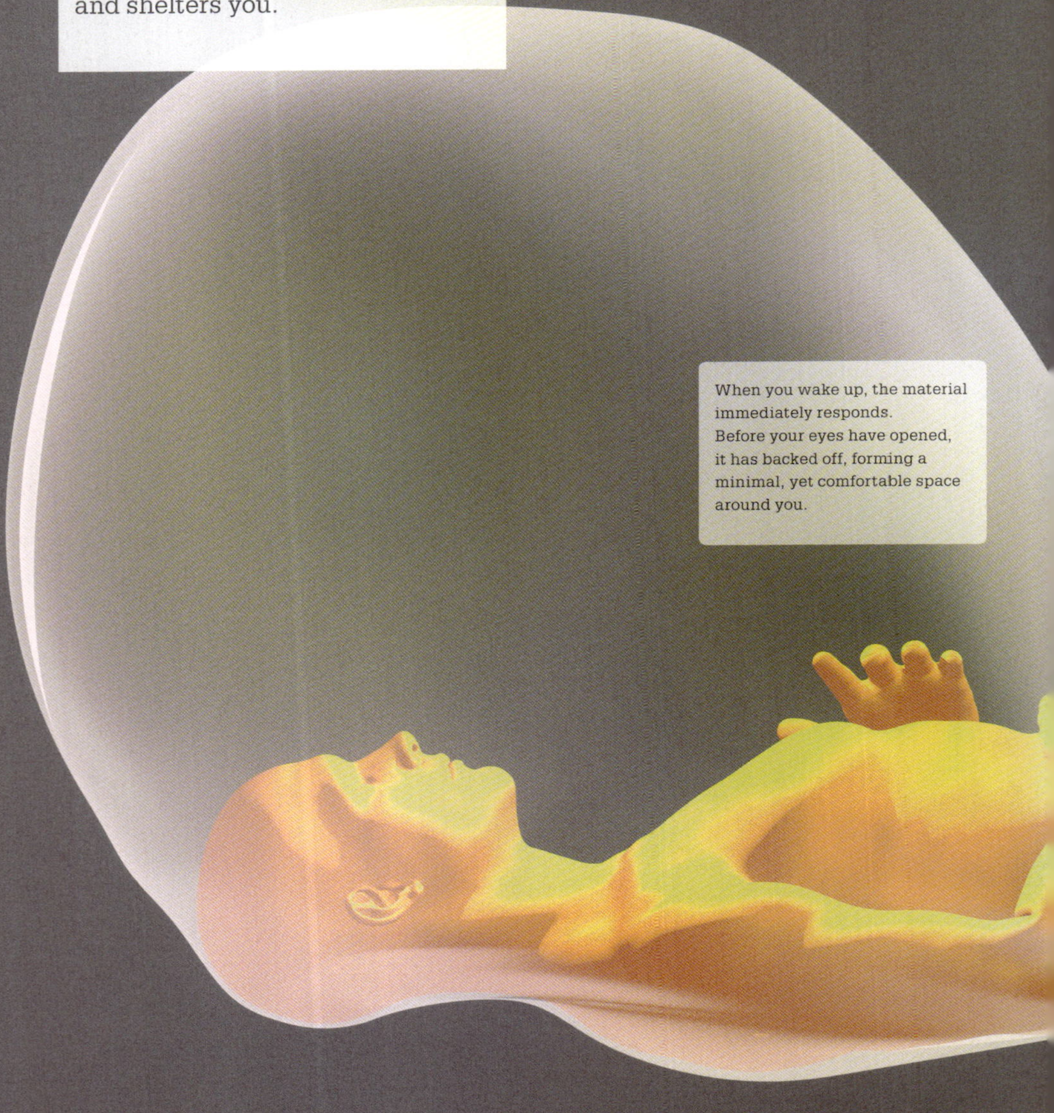

When you wake up, the material immediately responds.
Before your eyes have opened, it has backed off, forming a minimal, yet comfortable space around you.

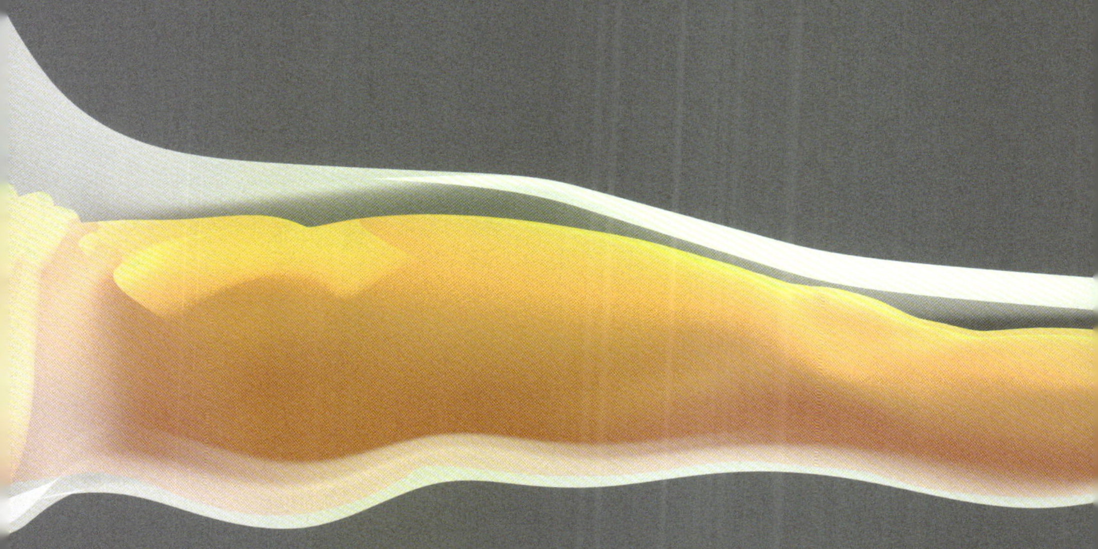

The material follows your body and comes to fit you as would bespoke apparel.

The material gently helps you to get up.

Once rigid, it can push you along, or carry you somewhere.

By default, Barba material is soft. But when needed, it becomes rigid.

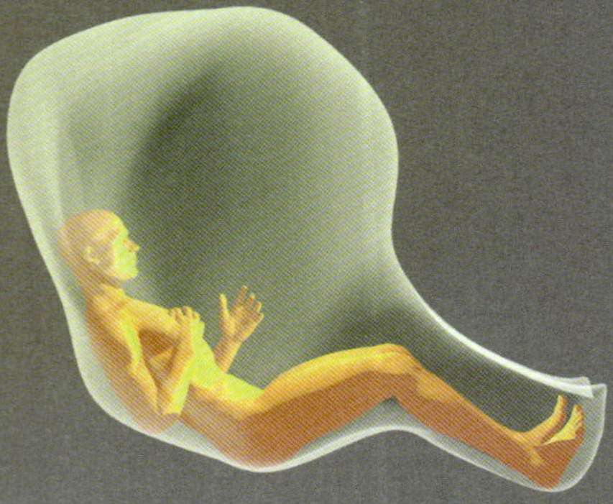

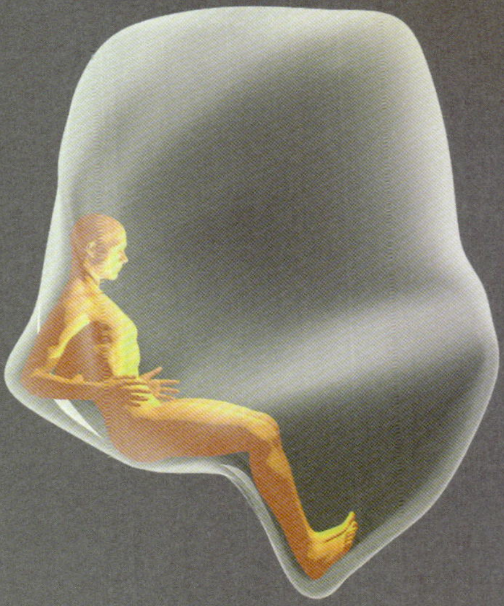

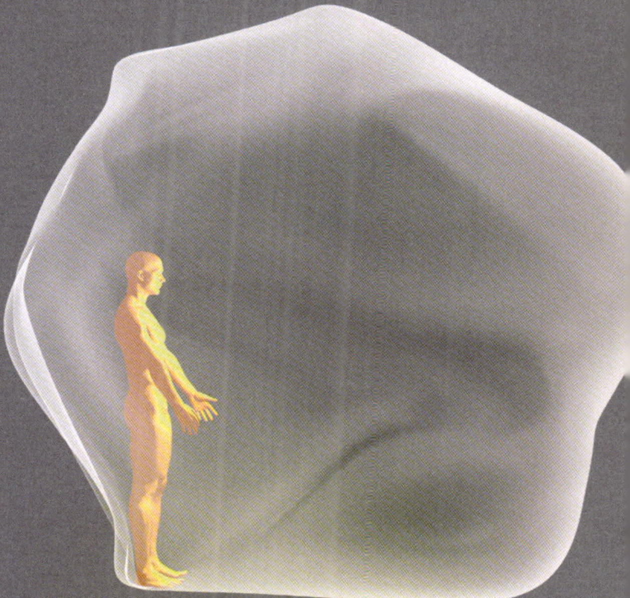

Before you've even touched it, the material anticipates your actions and creates a safe and comfortable buffer zone around you.

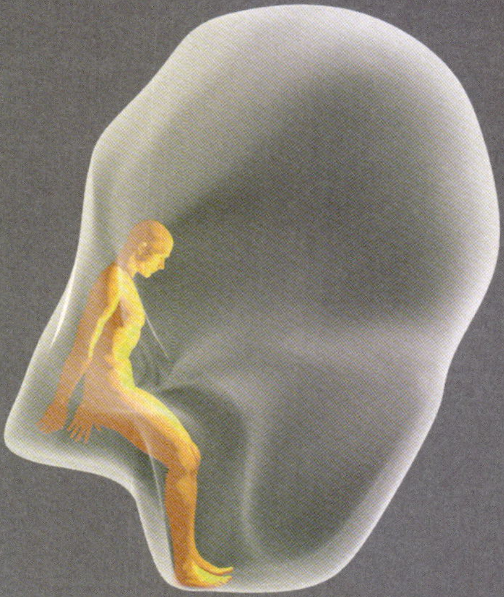

It can help you to stand up. And then retreat and become soft again.

Look around and see how your gaze increases the space around you.

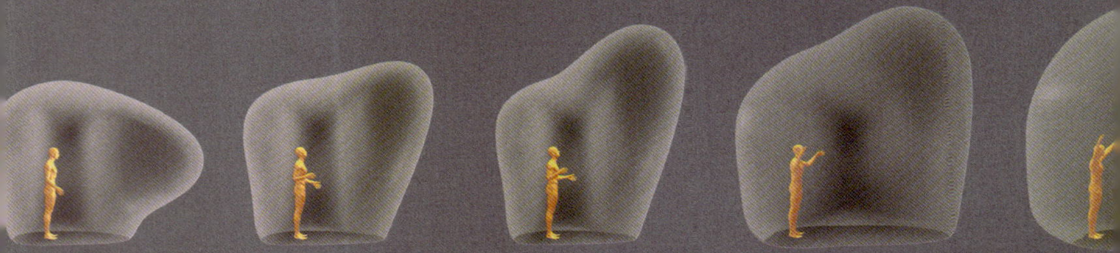

Barba responds to the turn of your head and increases space where you look.

You can further increase the space with your arms.

The Barba space reads your gestures and expands.

Barba space can extend almost endlessly.

Barba is able to anticipate and memorize your movement. Prediction and history become physical.

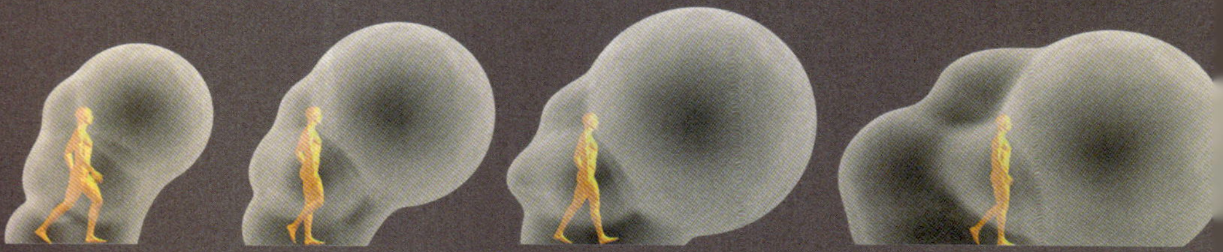

Barba anticipates your movements and extends the space in front of you.

The more quickly you go, the larger the space of anticipation becomes.

The material creates a buffer, for comfort and for safety.

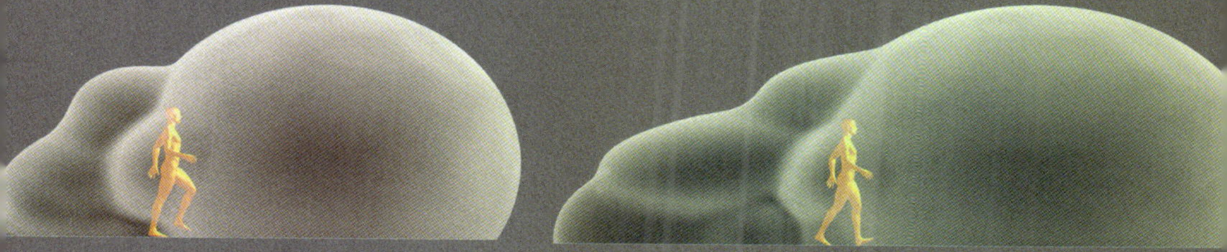

The material keeps track of your past movements, trails your history behind you. This structural inertia temporarily preserves the space you've just vacated, before it smoothly fades away.

And you can open the Barba space; you can look or step outside of it.

Programmed gestures allow you to create and position a window. You can adjust the size with your arms.

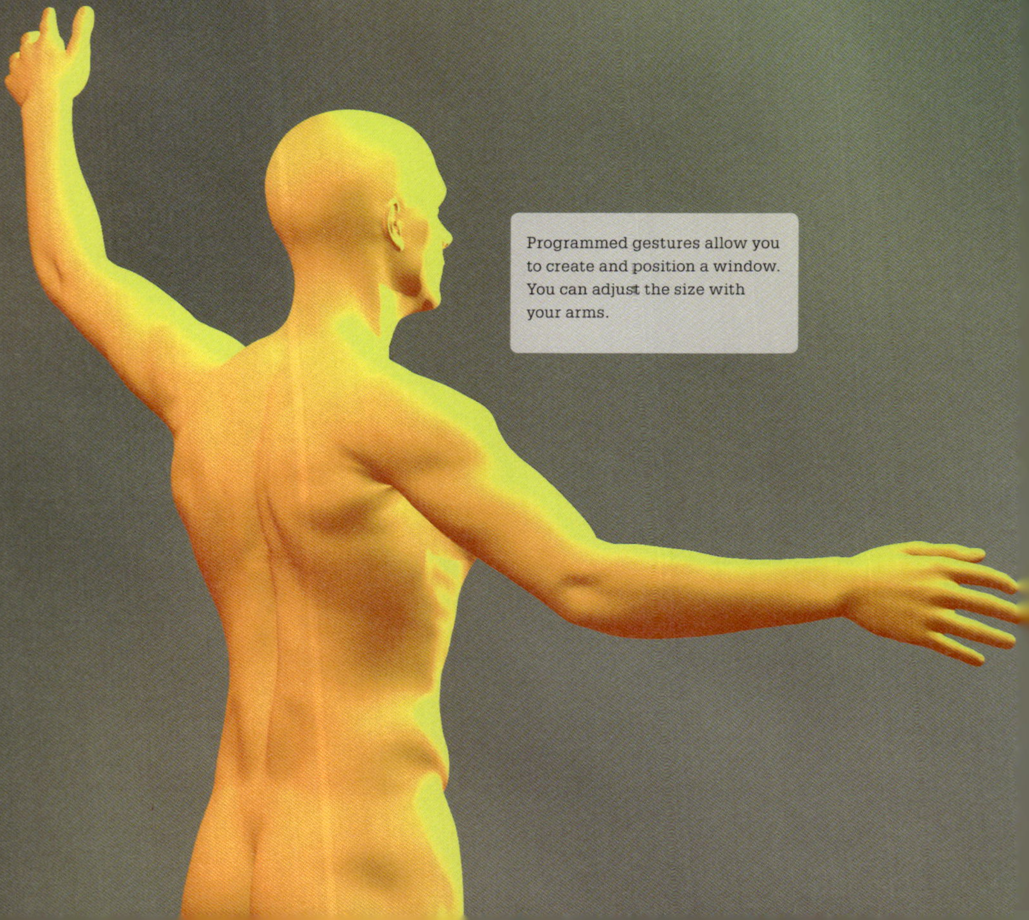

There are different ways
to create a window.

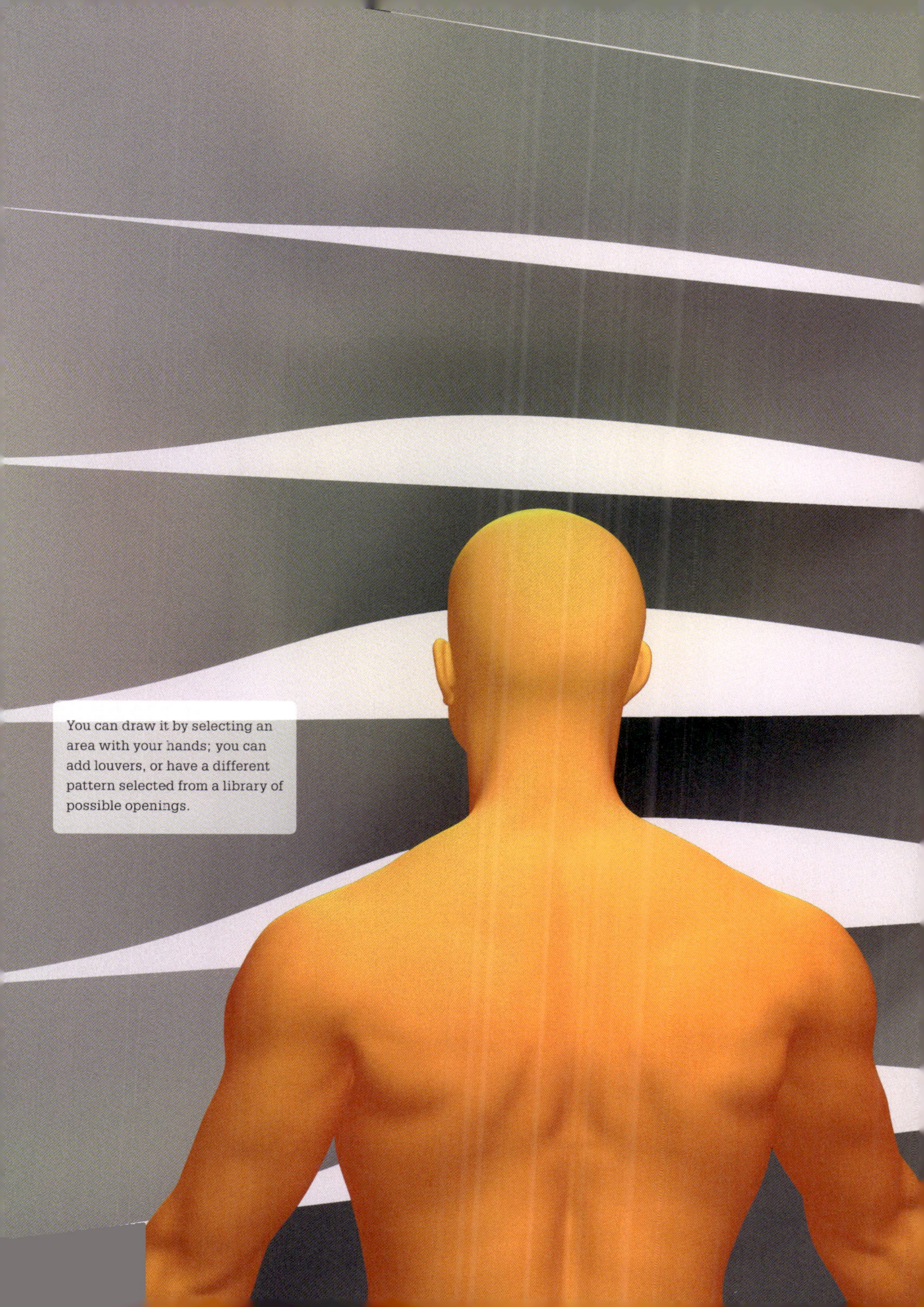

You can draw it by selecting an area with your hands; you can add louvers, or have a different pattern selected from a library of possible openings.

Imagine starting your daily routine in this fully adaptable environment. A washbasin would simply appear where you needed it . . .

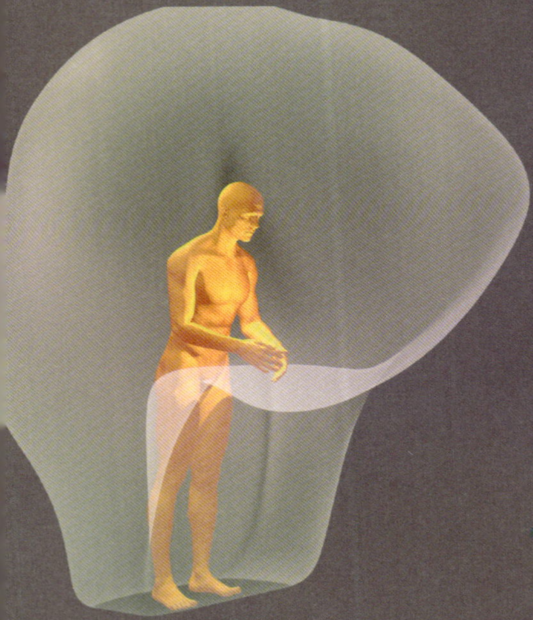

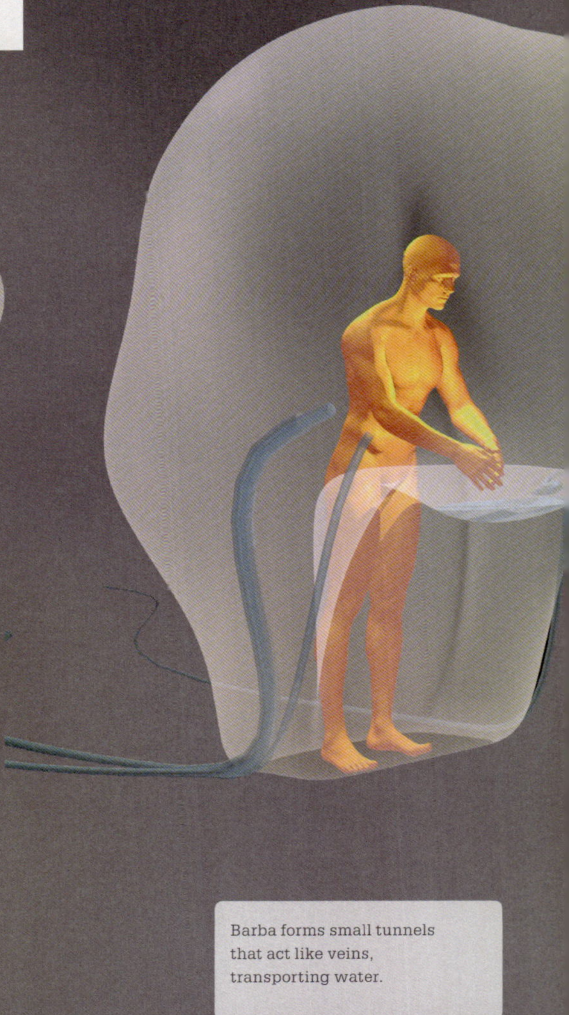

You can mould and shape an ergonomic sink with your hands.

Barba forms small tunnels that act like veins, transporting water.

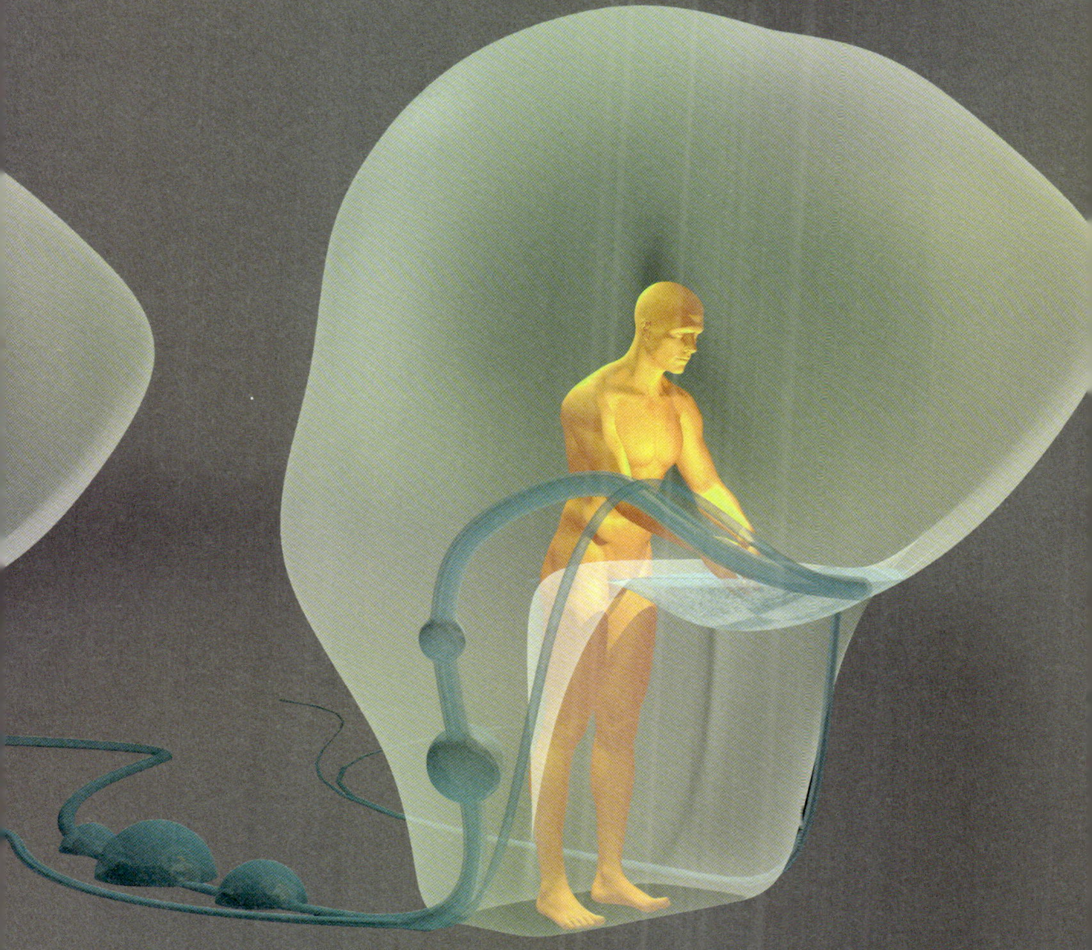

Through peristaltic movement, the veins – not unlike organic organs – pump water towards you.

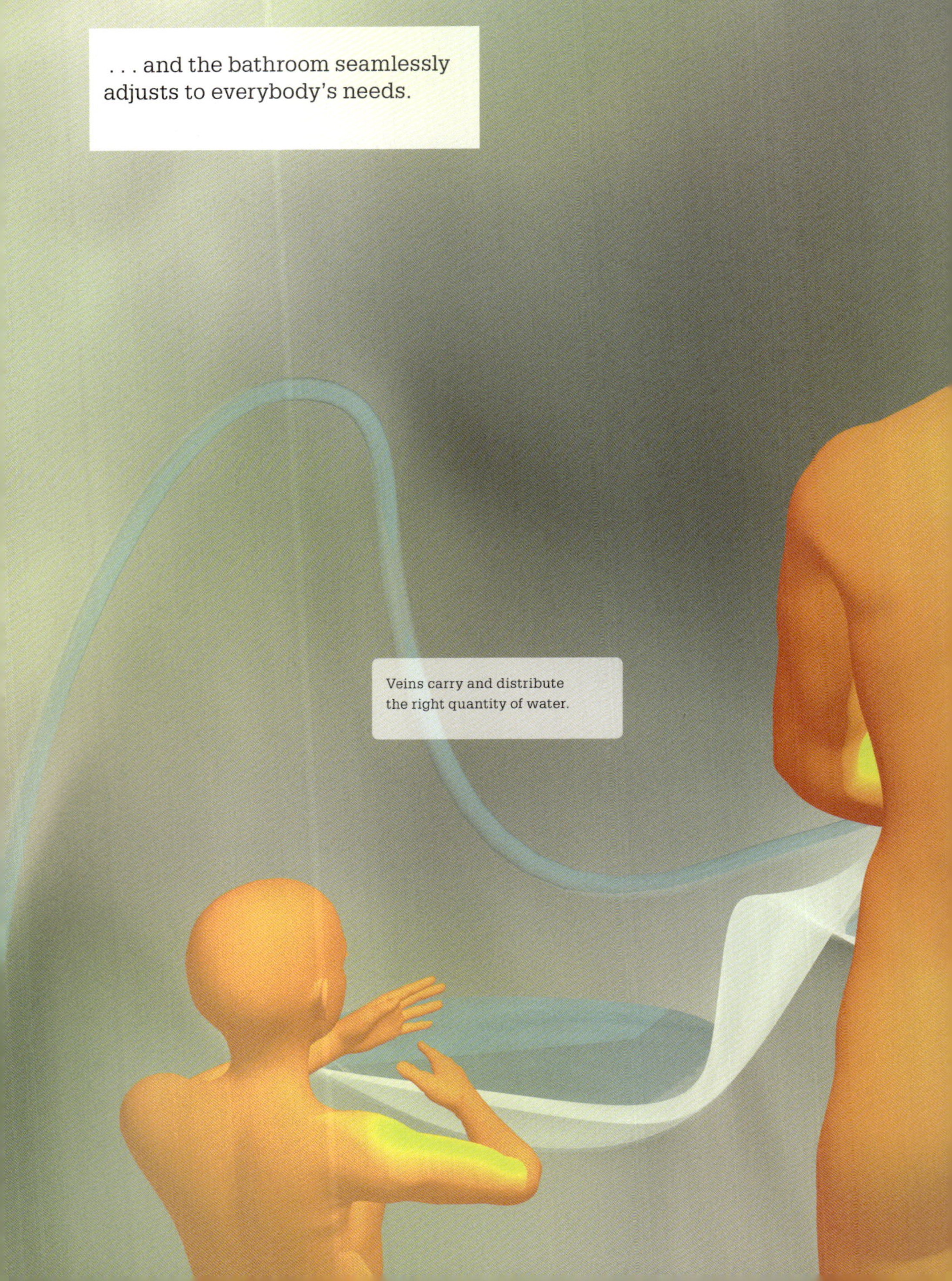
. . . and the bathroom seamlessly adjusts to everybody's needs.
Veins carry and distribute the right quantity of water.

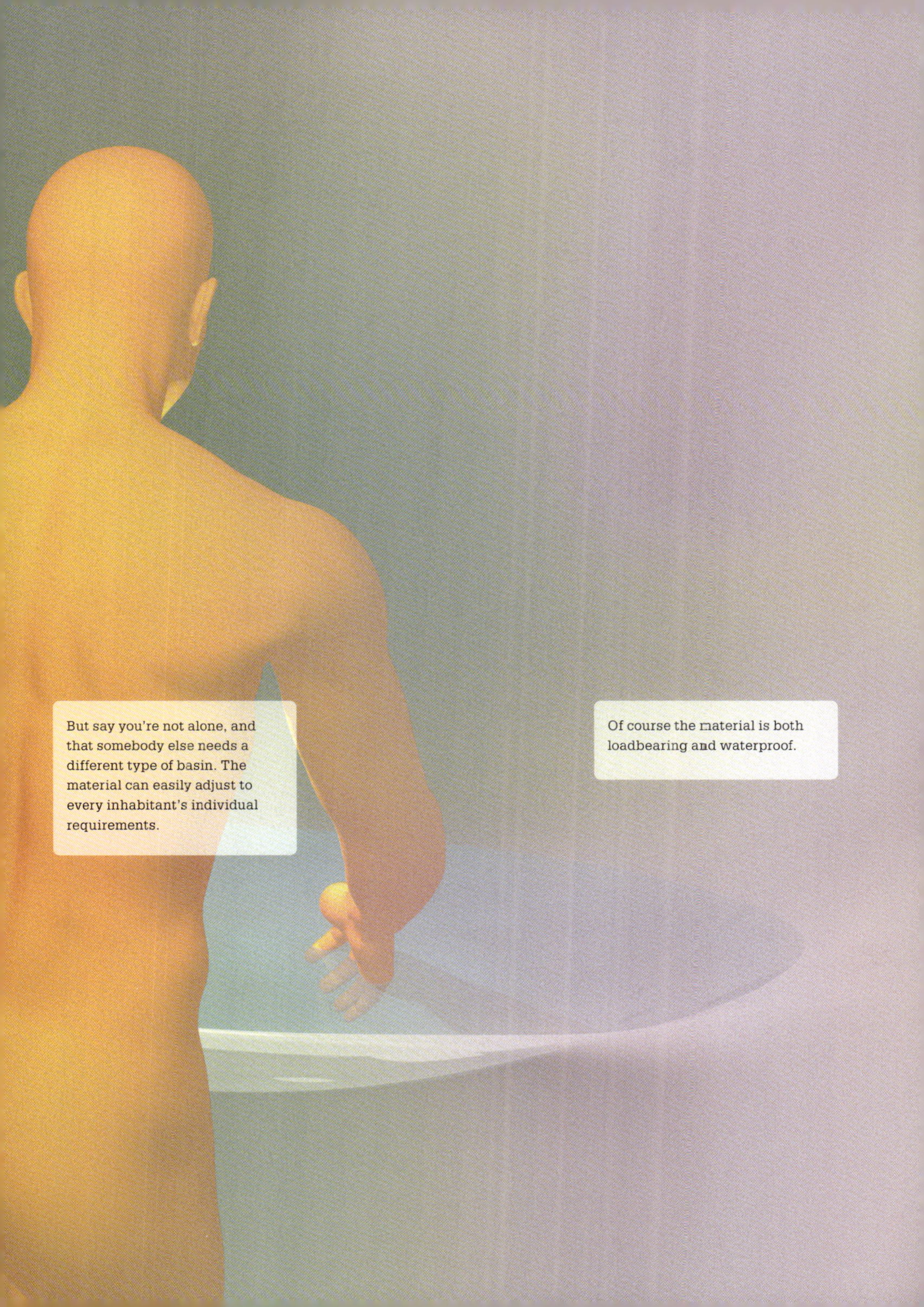

But say you're not alone, and that somebody else needs a different type of basin. The material can easily adjust to every inhabitant's individual requirements.

Of course the material is both loadbearing and waterproof.

And why not have
a morning swim?

With one smooth gesture, your washbasin becomes a pool.

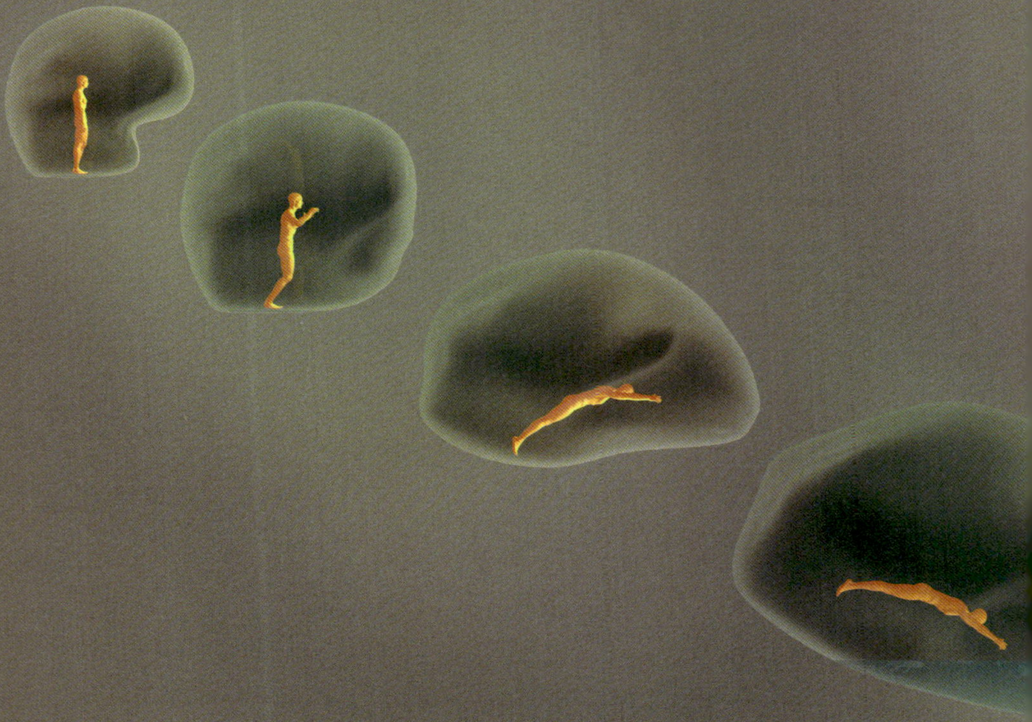

Your environment reads your movement and anticipates the next step.

You can blindly trust the environment to react to your desires.

A pool filled with water is there when you need it.

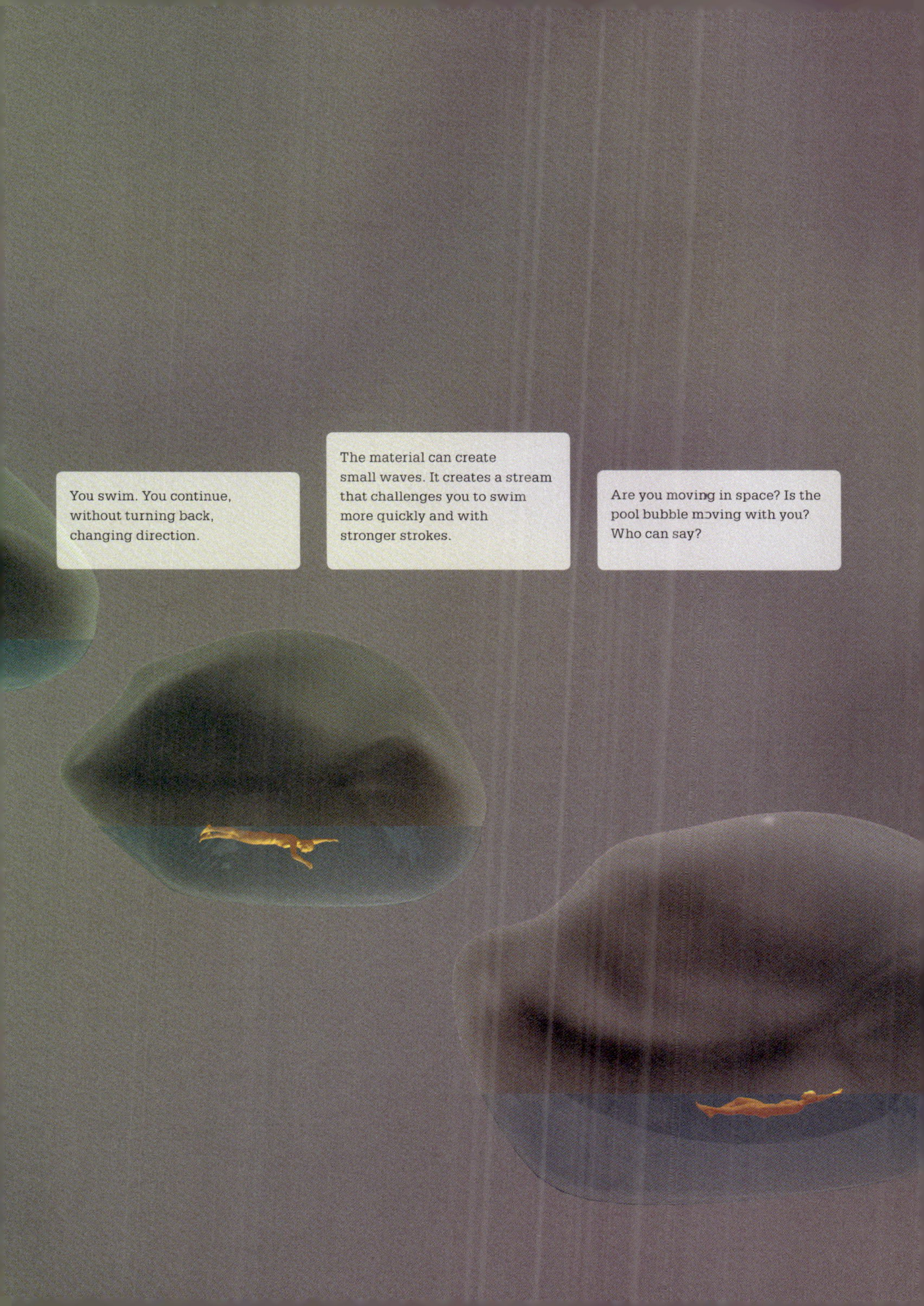
You swim. You continue, without turning back, changing direction.
The material can create small waves. It creates a stream that challenges you to swim more quickly and with stronger strokes.
Are you moving in space? Is the pool bubble moving with you? Who can say?

Your pool becomes a cave and
sunlight shines in from above.

Another swift movement turns your pool into a bathtub, and then into a shower.

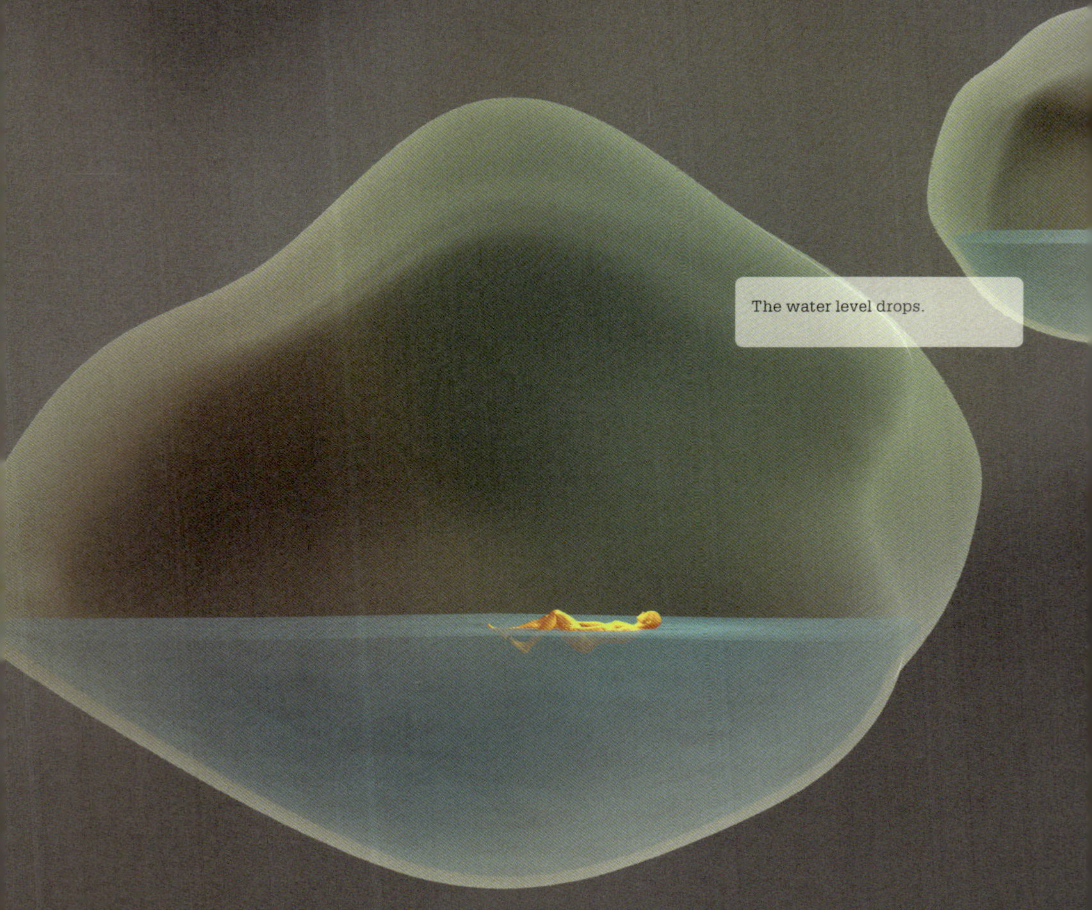

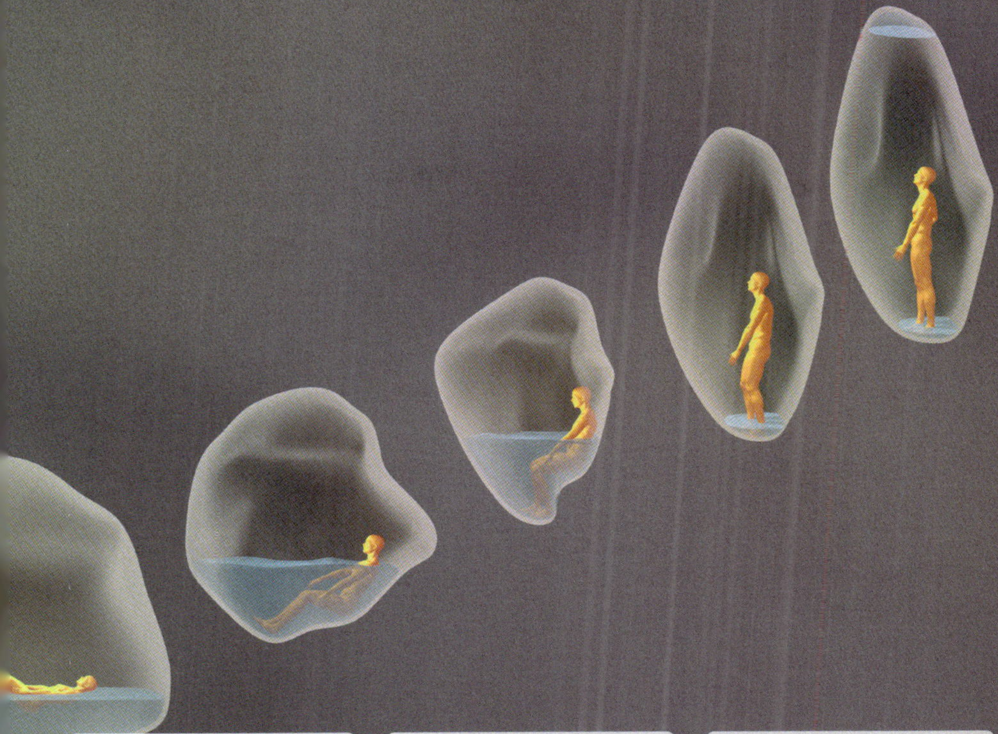

The material pushes you up.

Water gathers on top of you.

And like a small waterfall, water pours down from above, creating a refreshing shower.

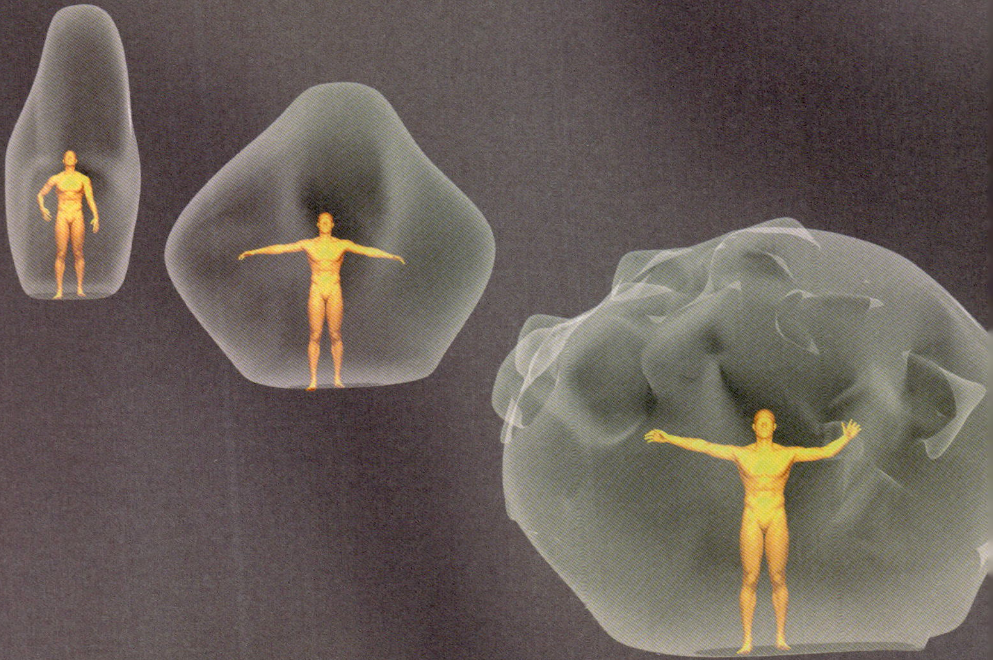

The water is pumped away.

Quick vibrations allow the material to heat up air. Pumping movements create a flow of warm air that dries your body.

You dry off and prepare to dress.

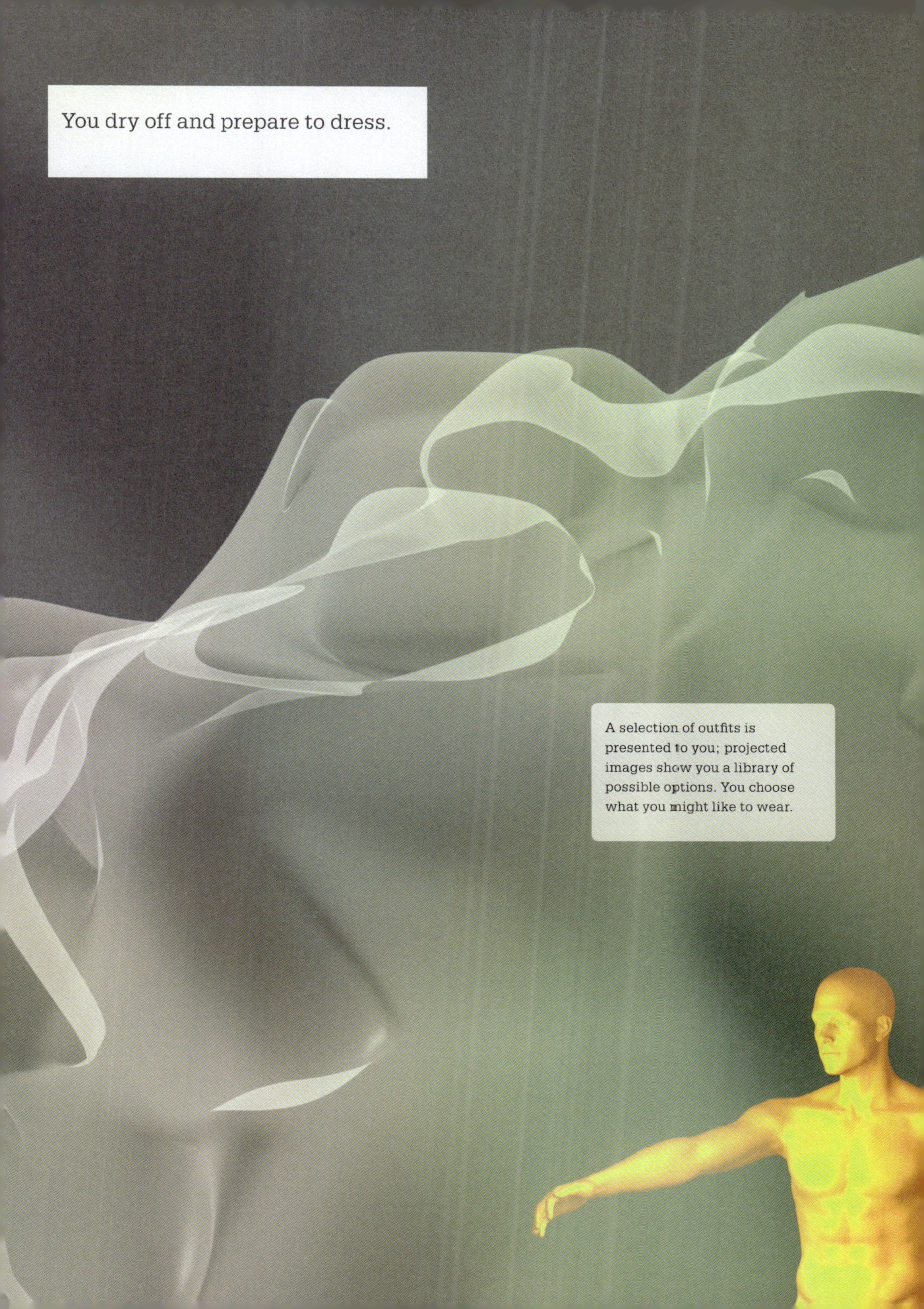

A selection of outfits is presented to you; projected images show you a library of possible options. You choose what you might like to wear.

The clothes you've chosen are moulded around your body.

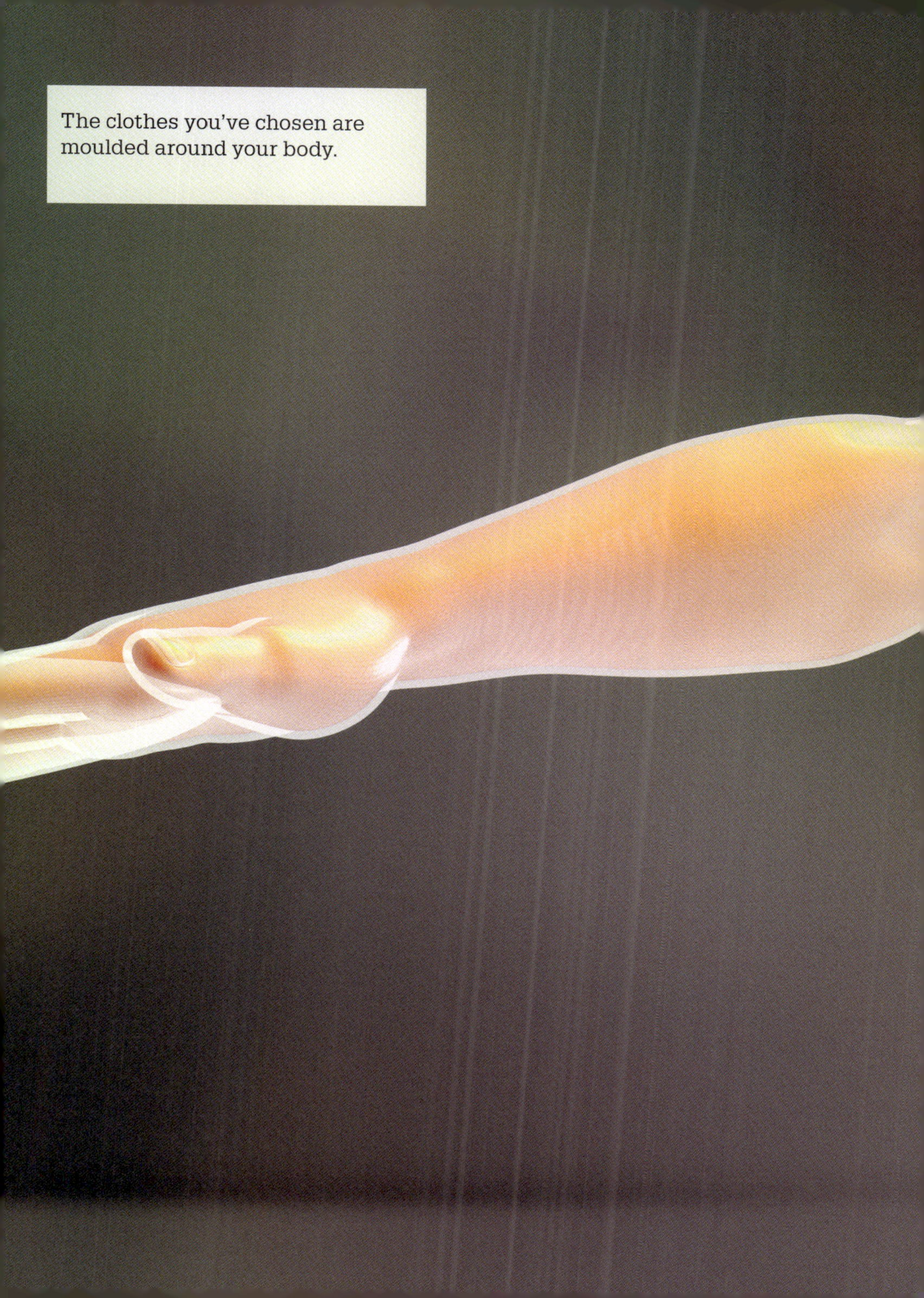

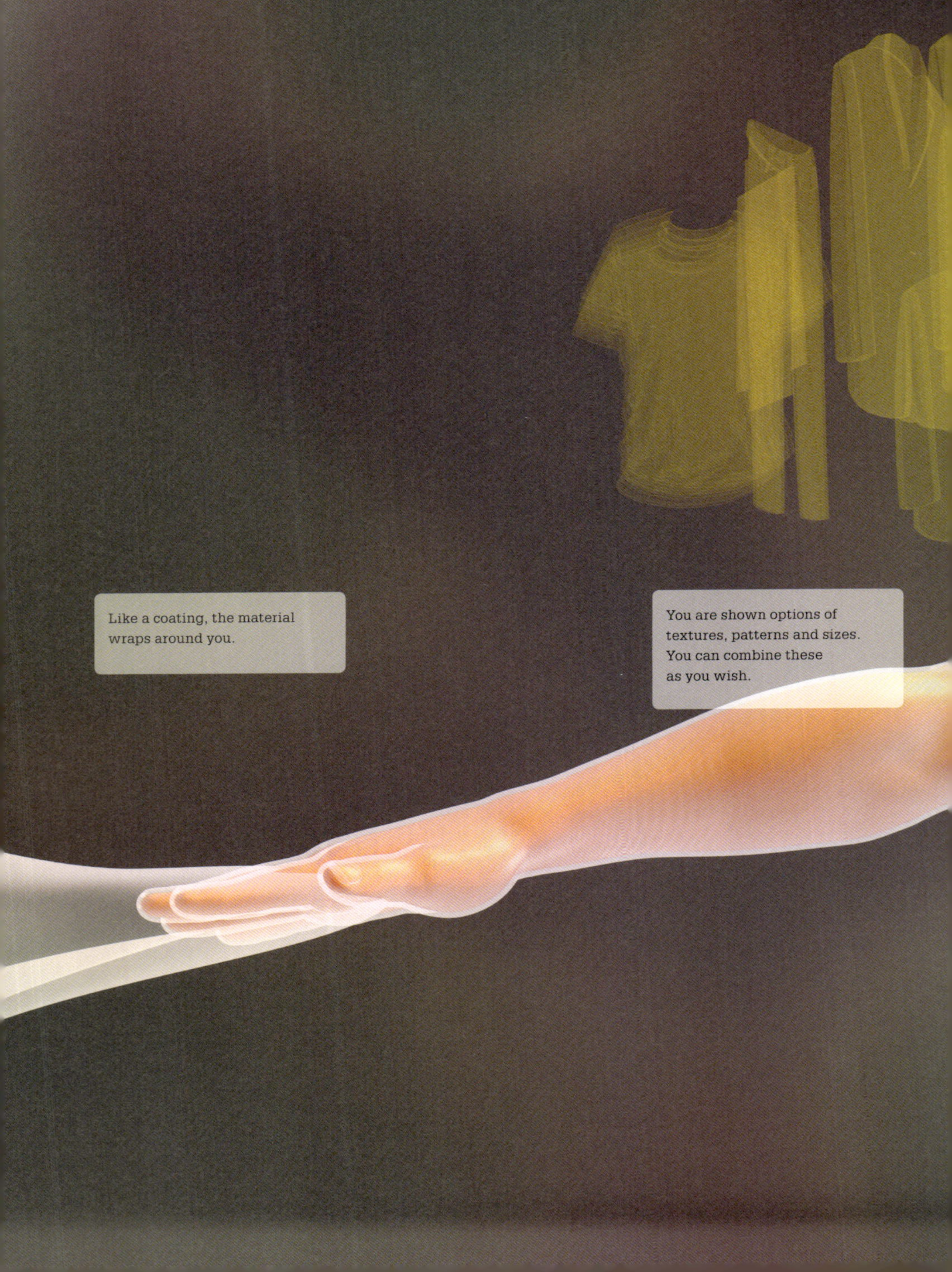
Like a coating, the material wraps around you.
You are shown options of textures, patterns and sizes. You can combine these as you wish.

The functional, protective skin can become a garment suited exactly to your taste.

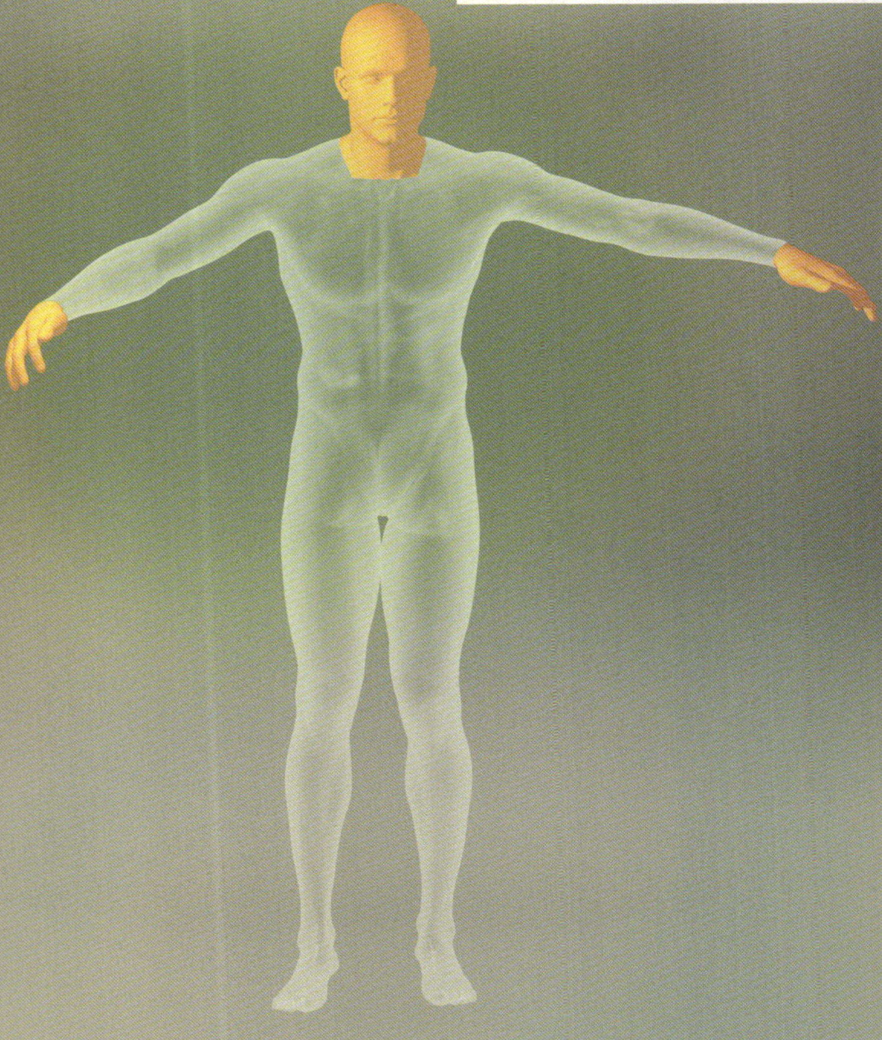

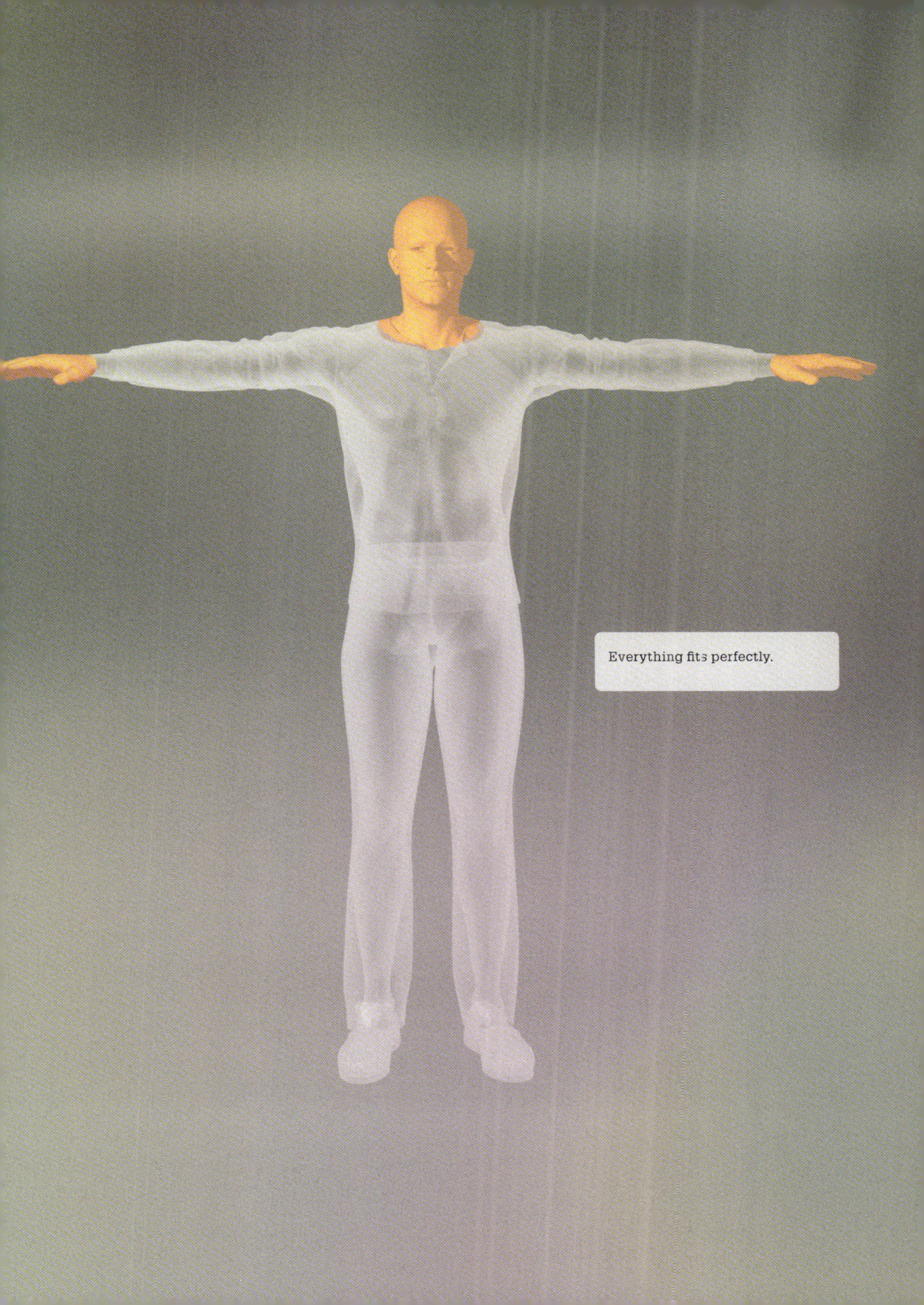
Everything fits perfectly.

Let's have breakfast now!

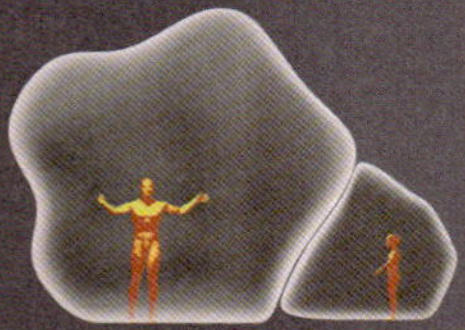

Being together means sharing spaces.

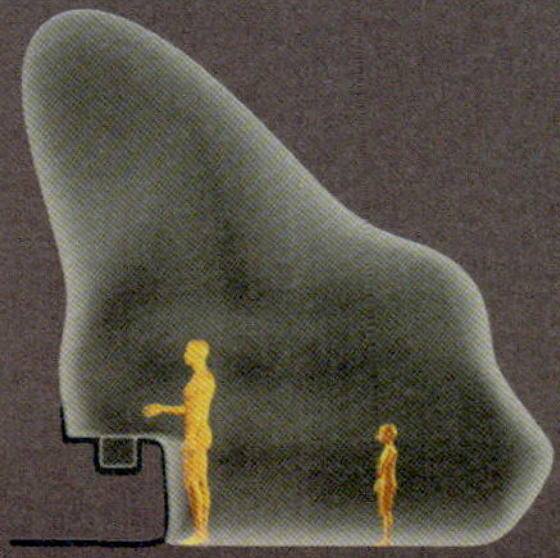

Sun enters through an opening.

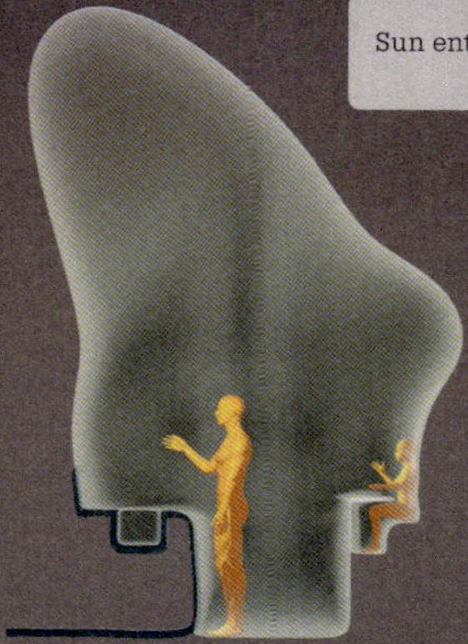

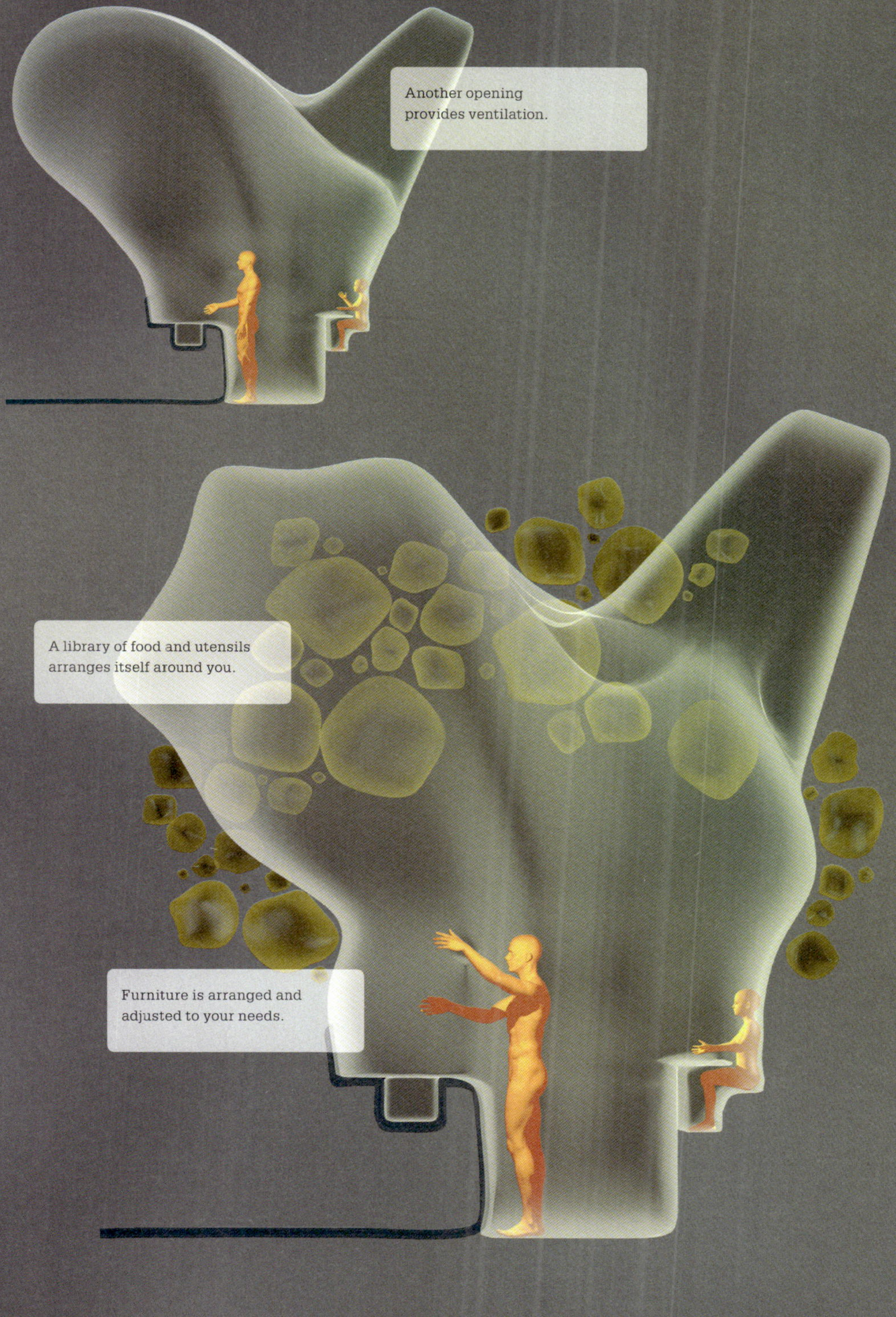
Another opening
provides ventilation.
A library of food and utensils
arranges itself around you.
Furniture is arranged and
adjusted to your needs.

You select what you need
and start to prepare your meal.
Bacon
Oregano
Chocolate
Banana
Veins provide you with water,
transporting it through
peristaltic movement.

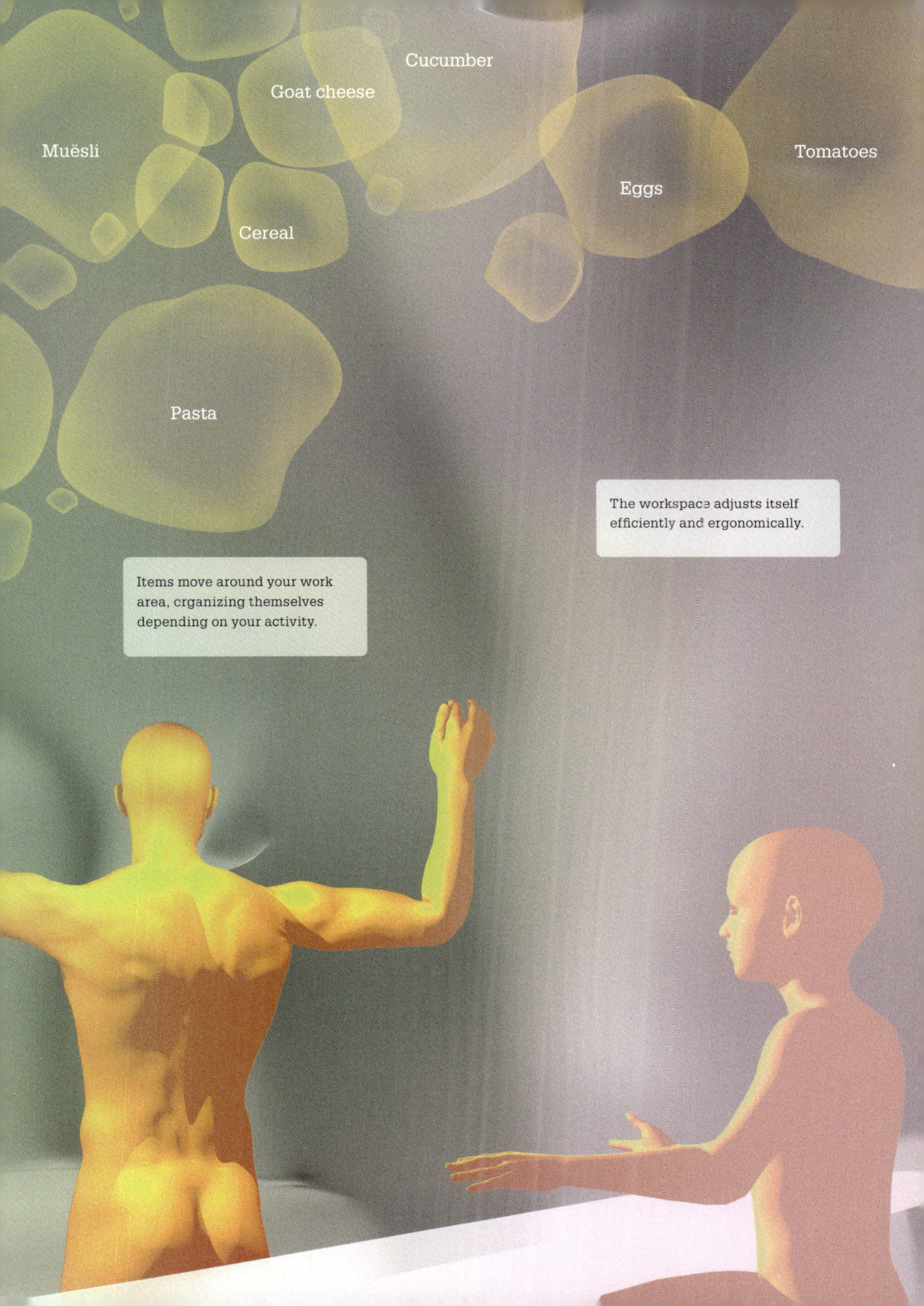
Cucumber
Goat cheese
Muësli
Tomatoes
Eggs
Cereal
Pasta
The workspace adjusts itself efficiently and ergonomically.
Items move around your work area, organizing themselves depending on your activity.

The material heats the water to boil.
Fine pleats in the material move rapidly against each other, creating heat.

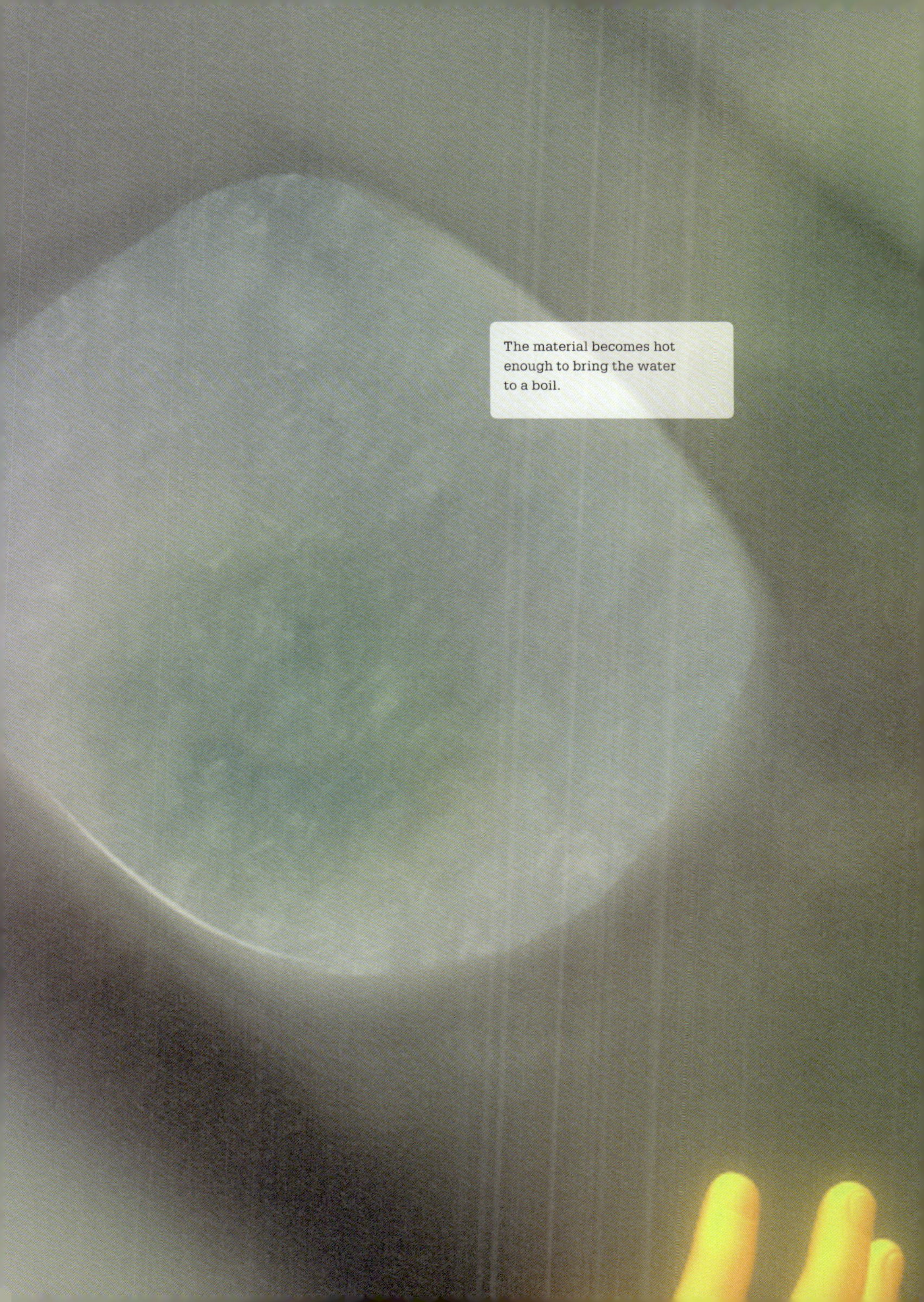

The material becomes hot enough to bring the water to a boil.

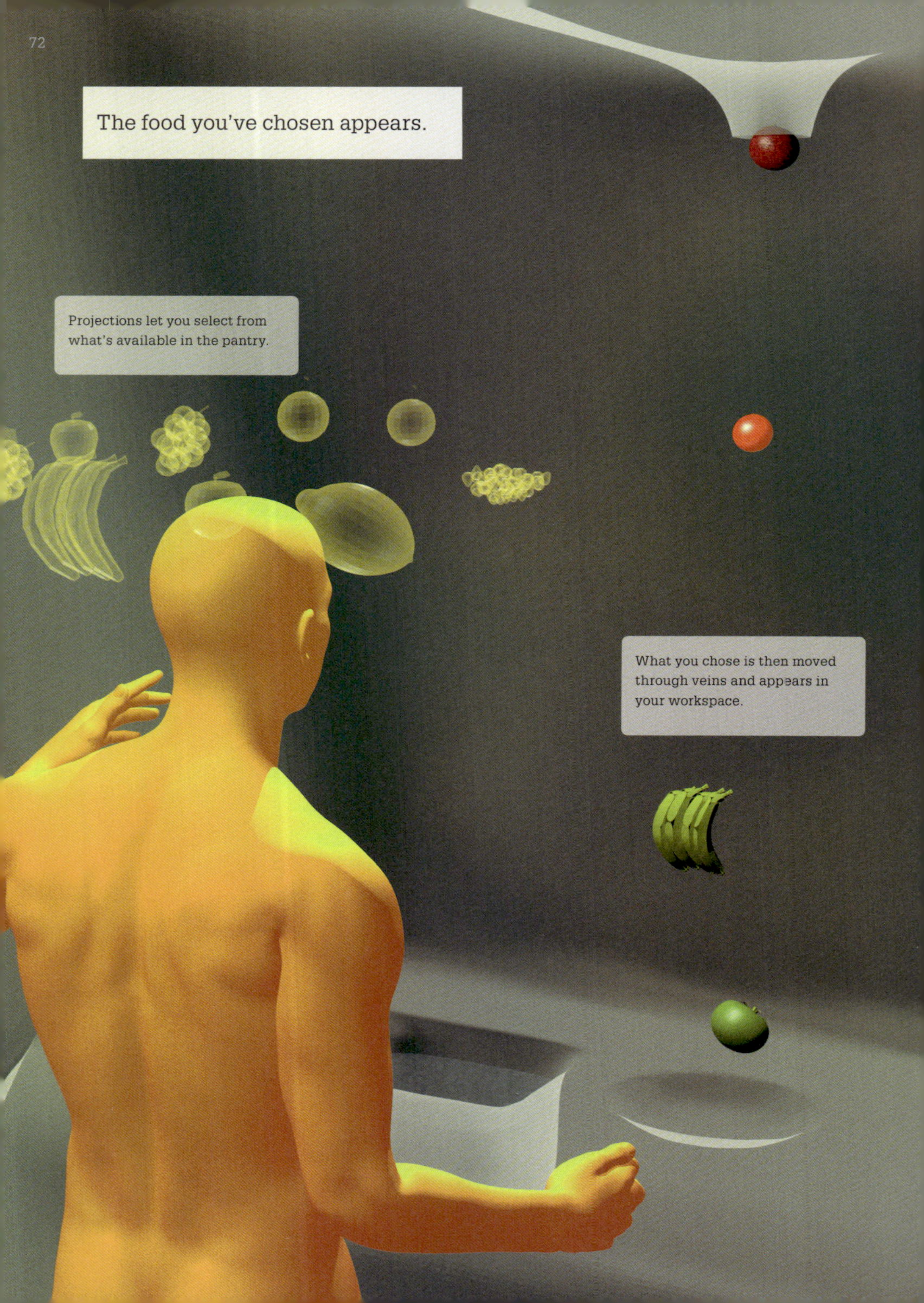
The food you've chosen appears.
Projections let you select from what's available in the pantry.
What you chose is then moved through veins and appears in your workspace.

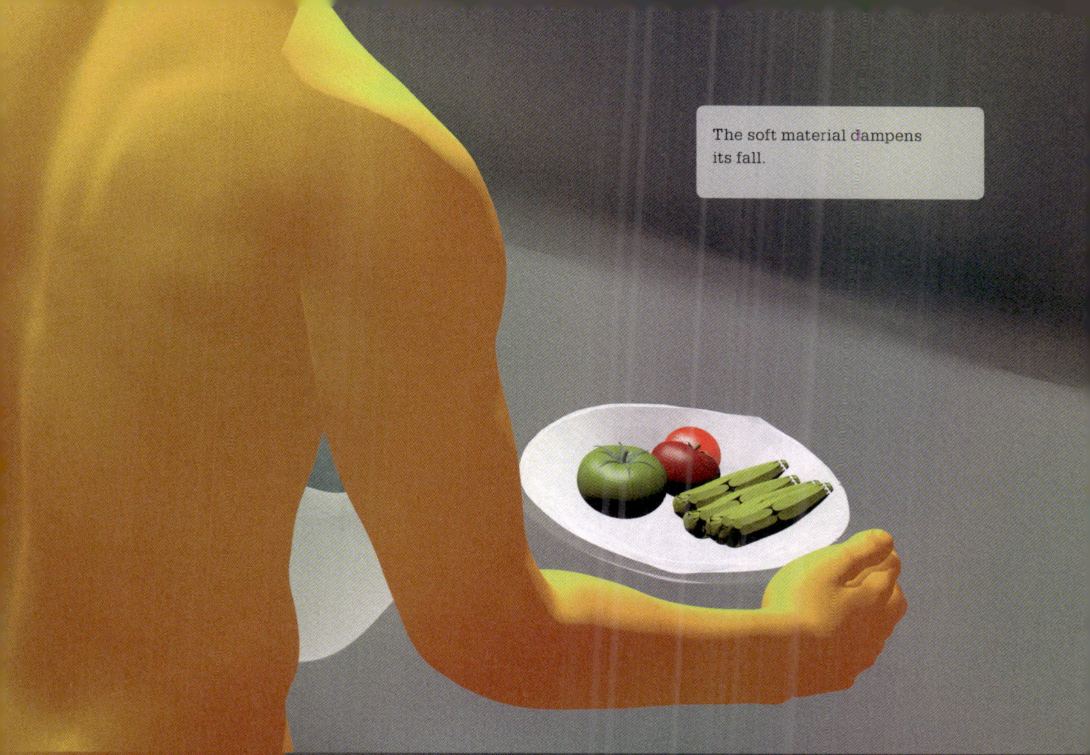
The soft material dampens its fall.

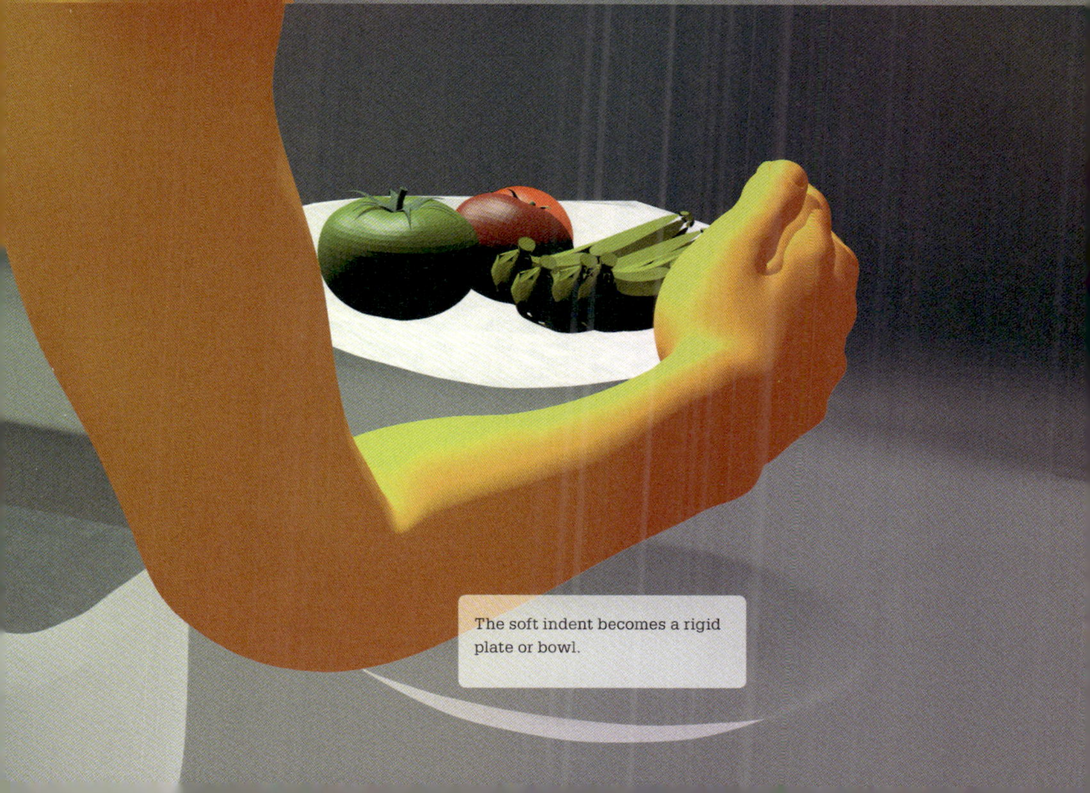
The soft indent becomes a rigid plate or bowl.

The space can be more than ergonomic and efficient.

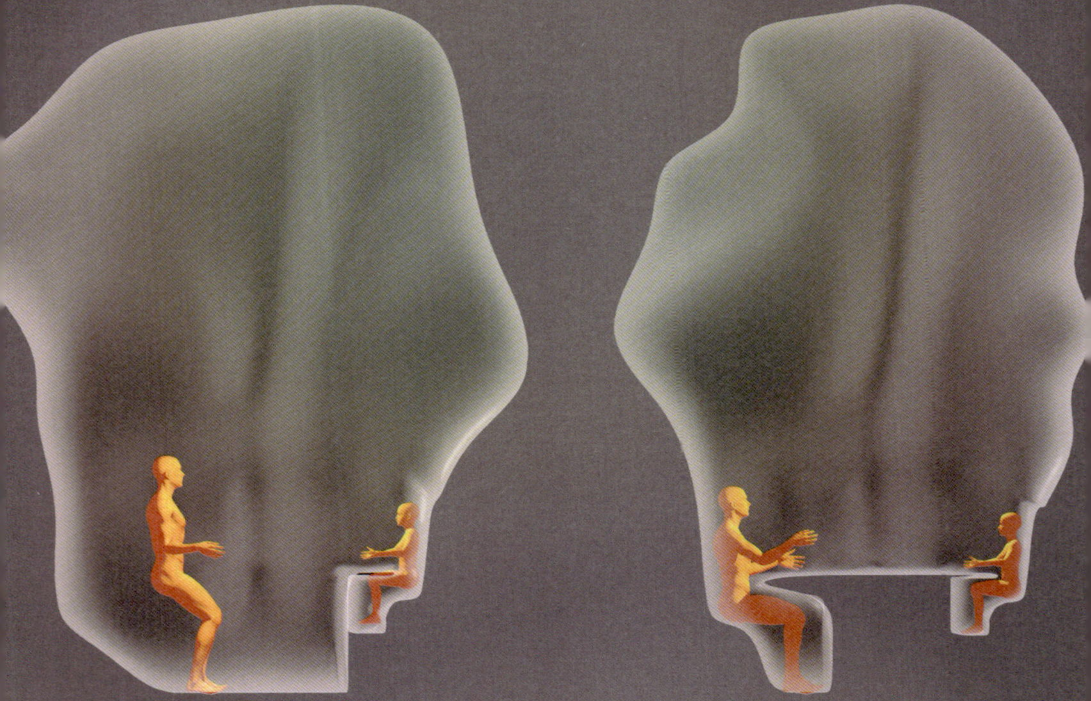

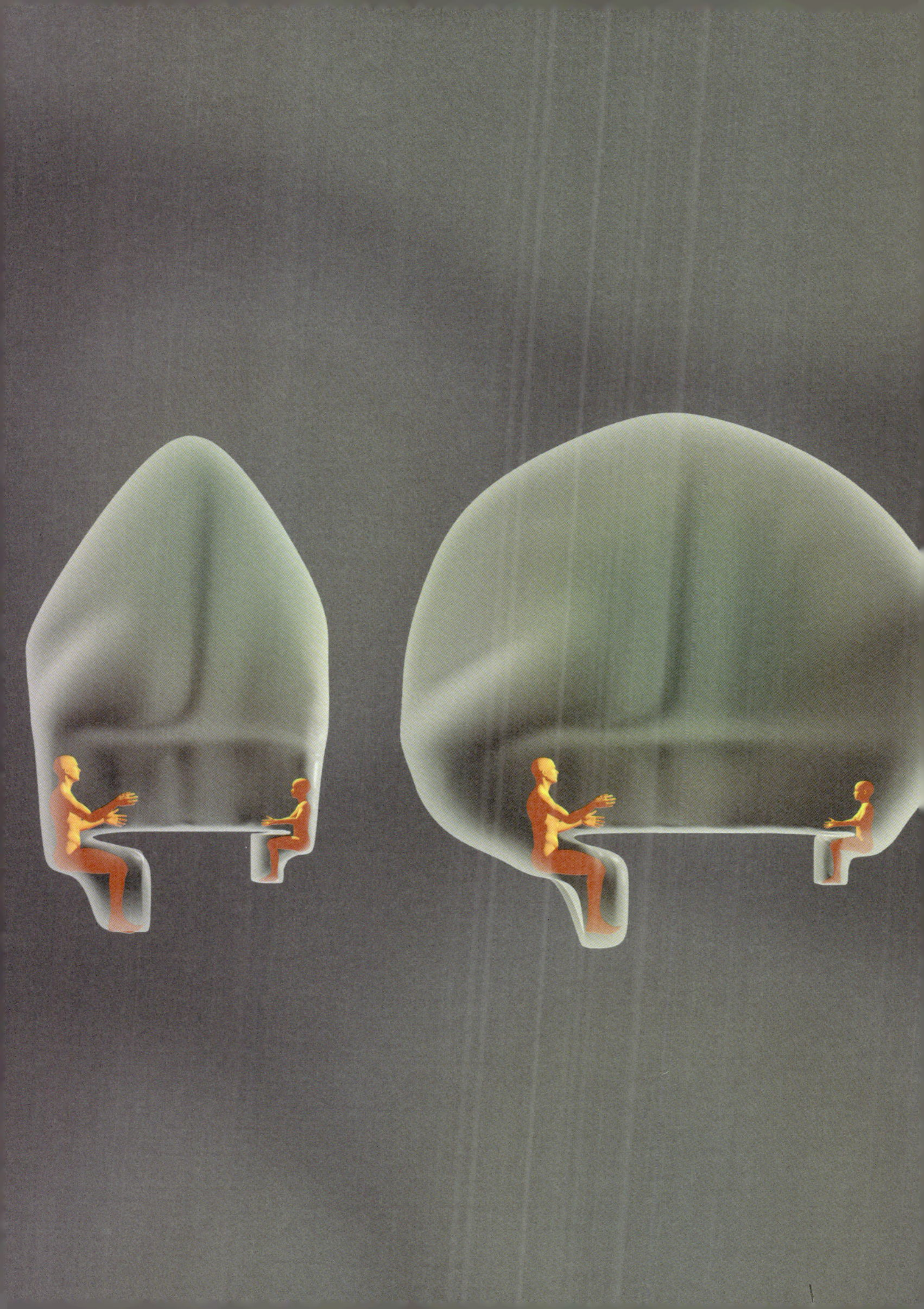

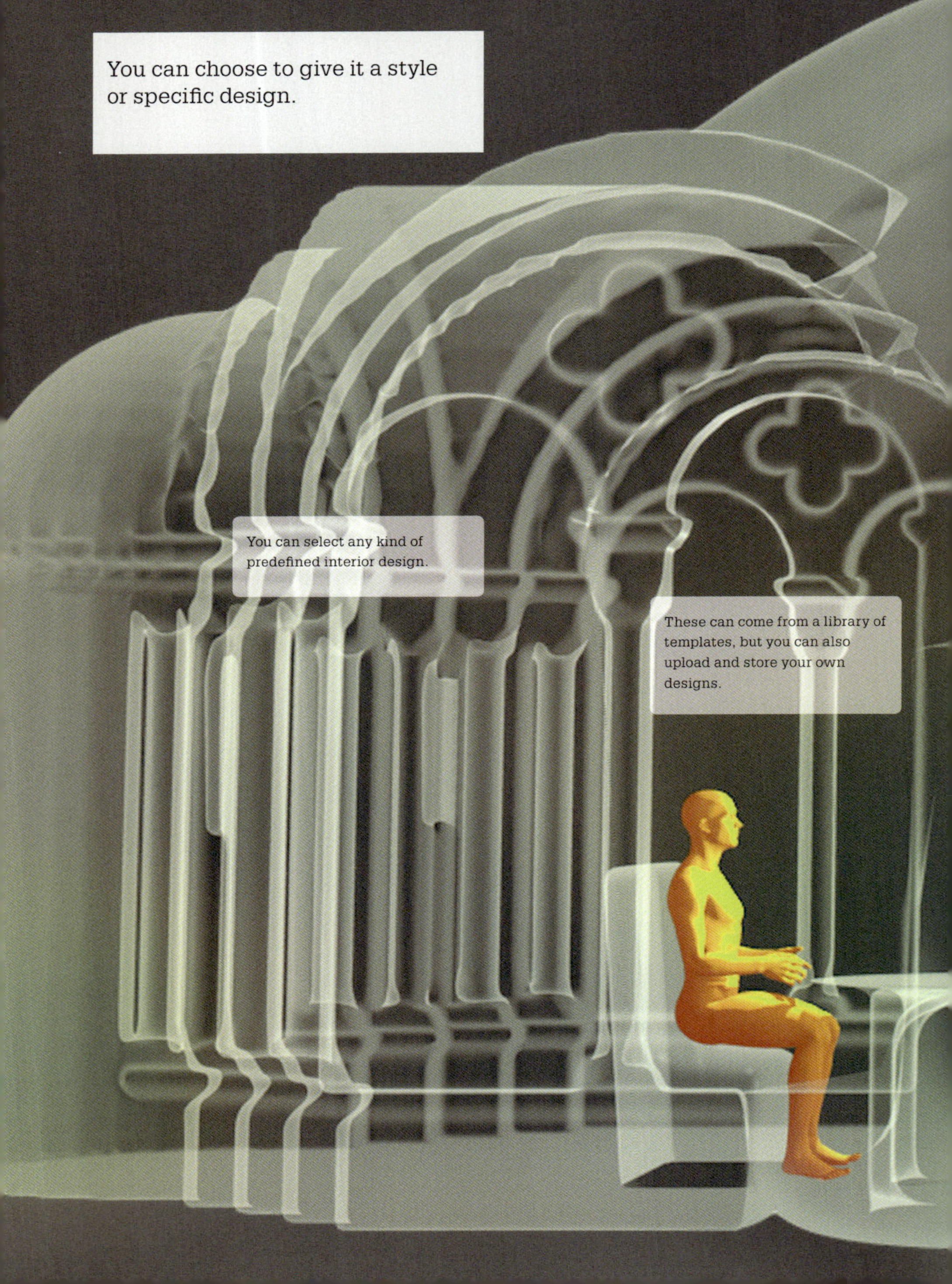
You can choose to give it a style or specific design.
You can select any kind of predefined interior design.
These can come from a library of templates, but you can also upload and store your own designs.

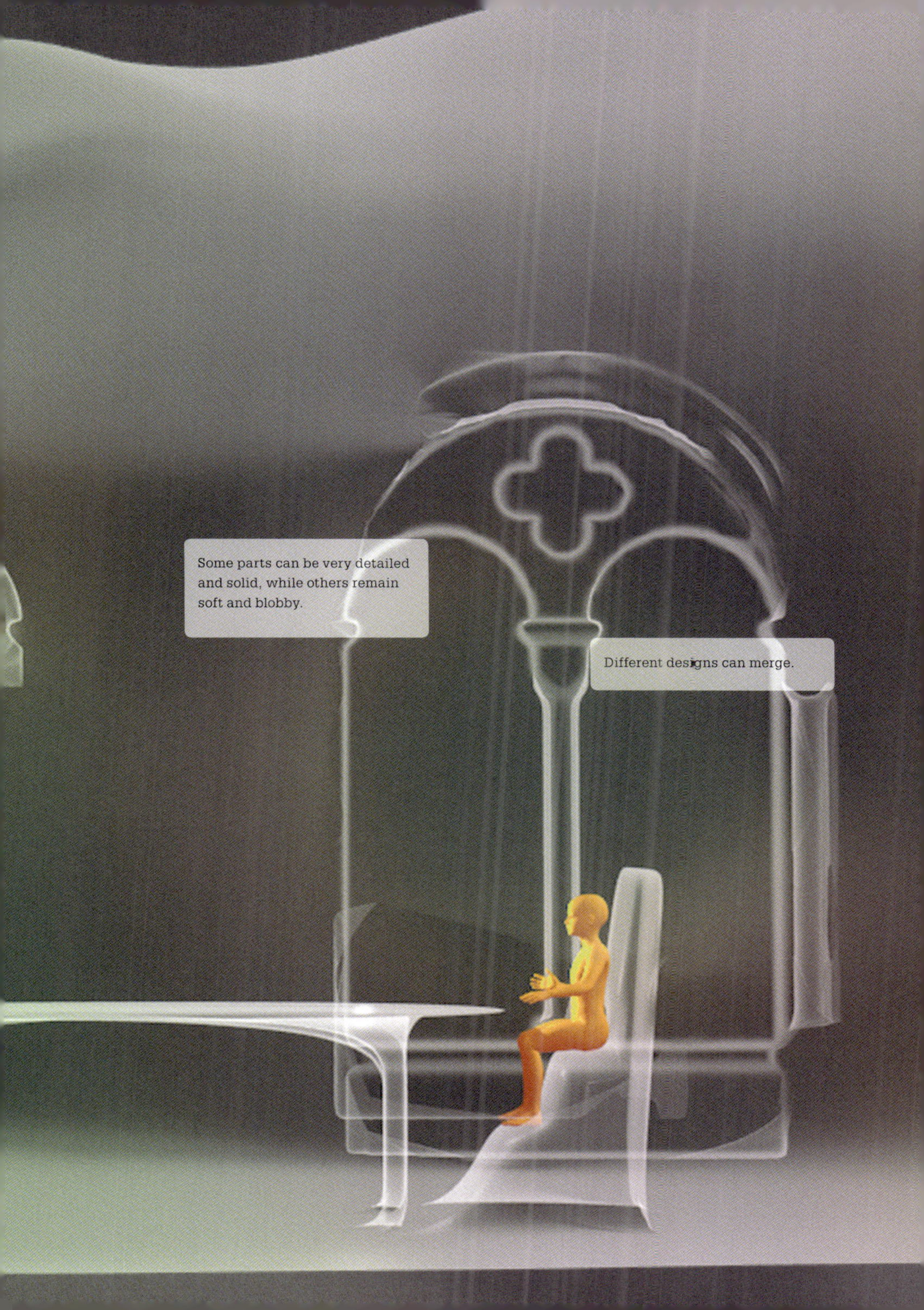
Some parts can be very detailed and solid, while others remain soft and blobby.
Different designs can merge.

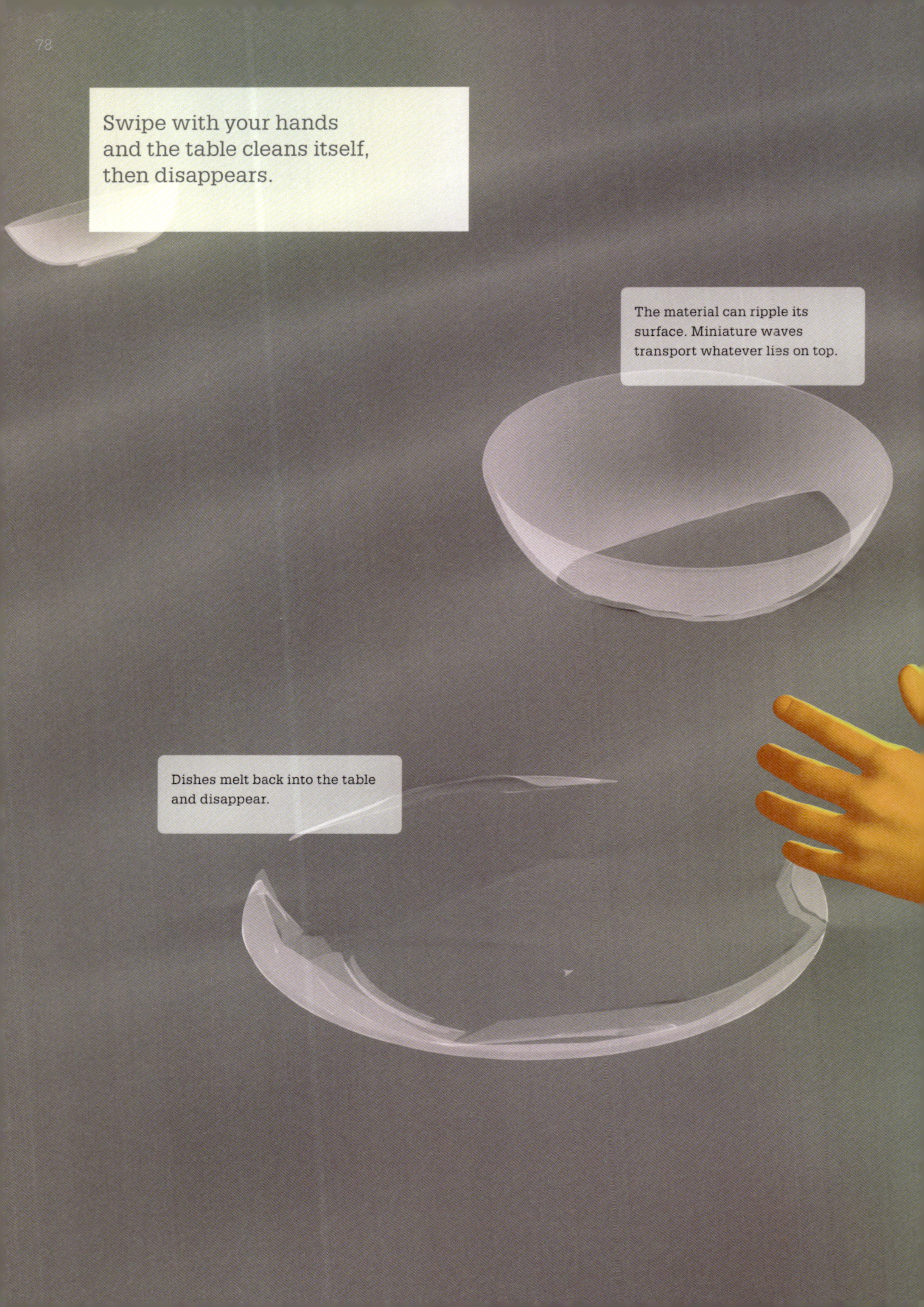

Swipe with your hands and the table cleans itself, then disappears.

The material can ripple its surface. Miniature waves transport whatever lies on top.

Dishes melt back into the table and disappear.

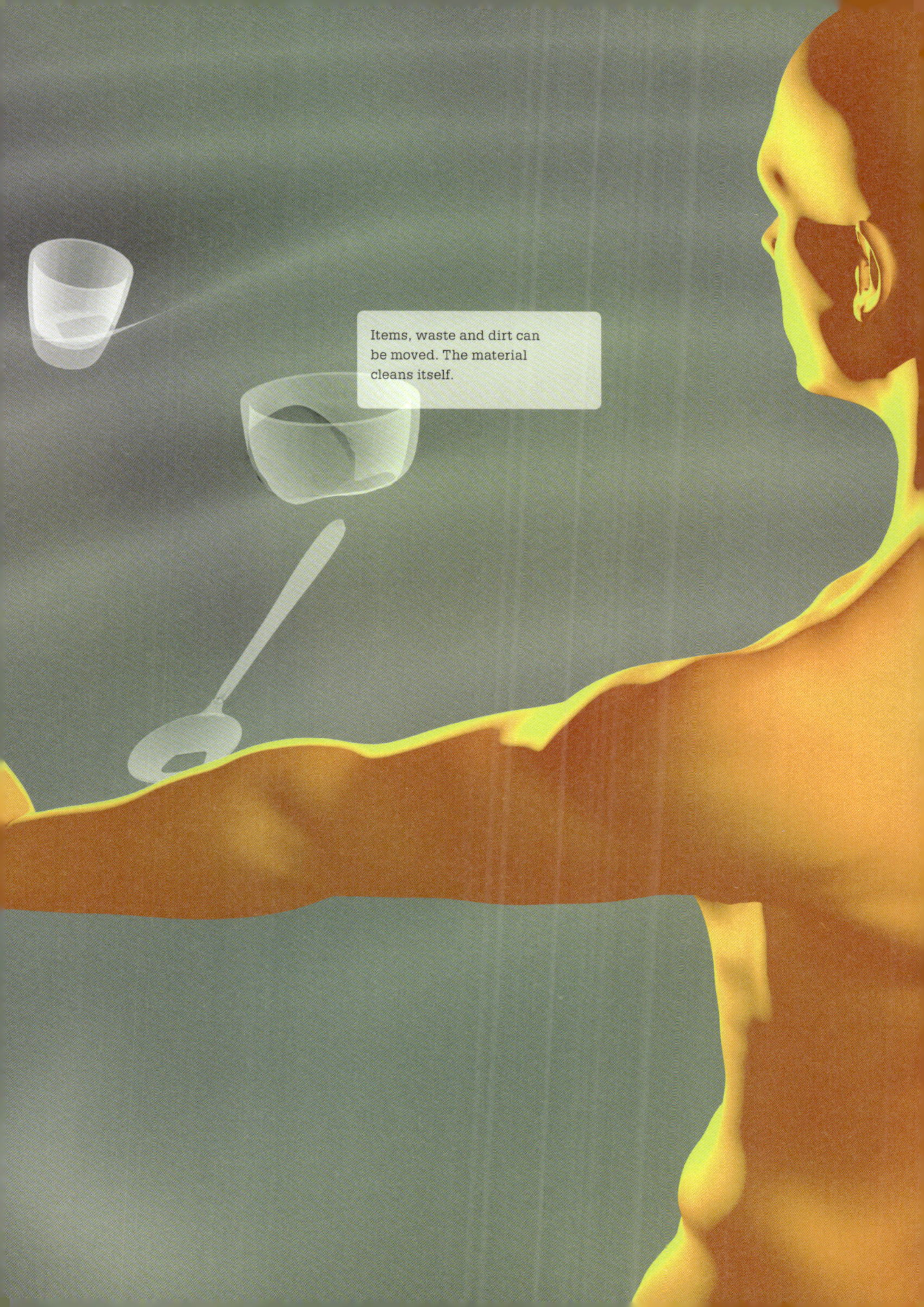
Items, waste and dirt can be moved. The material cleans itself.

Cleaning up has
never been so easy.

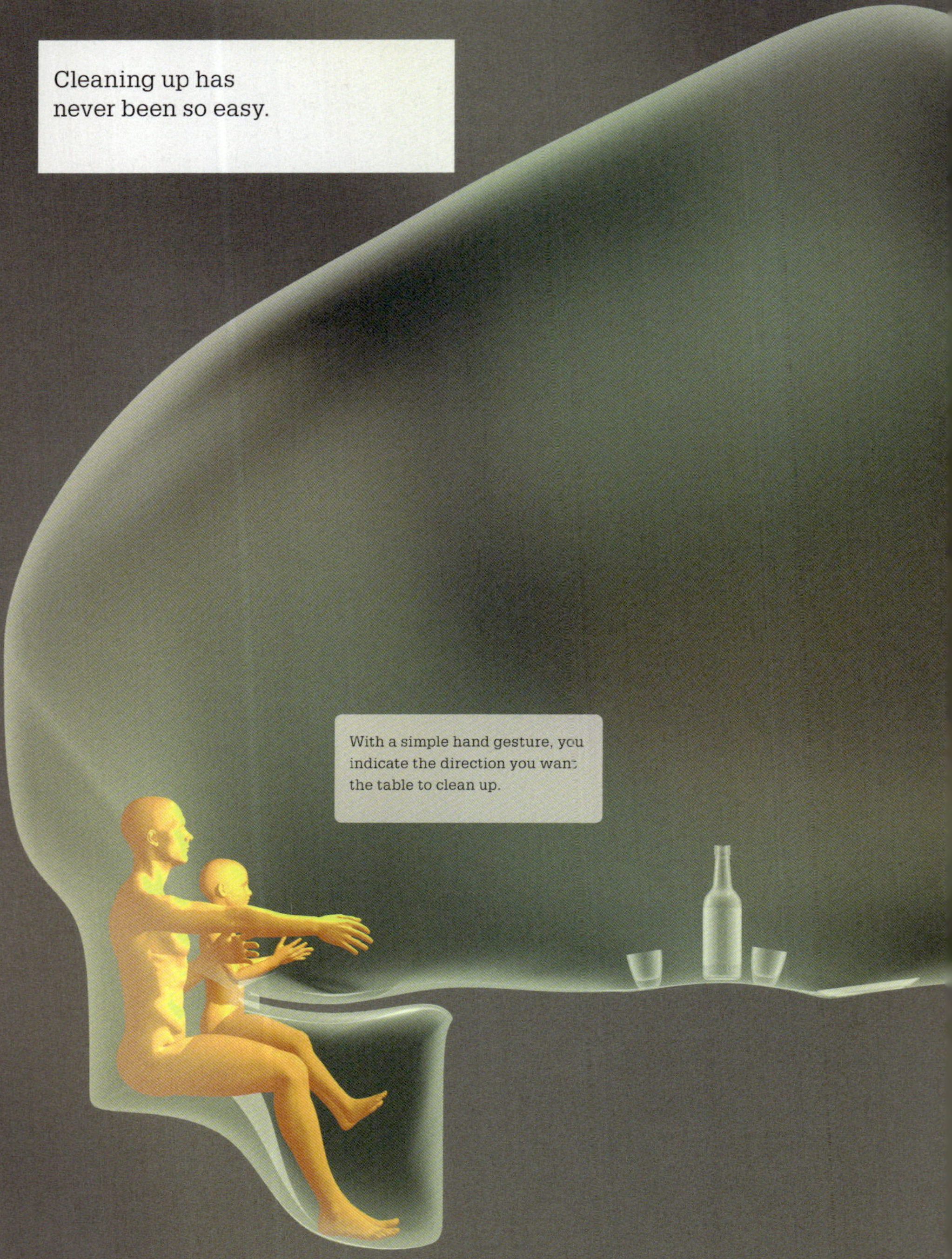

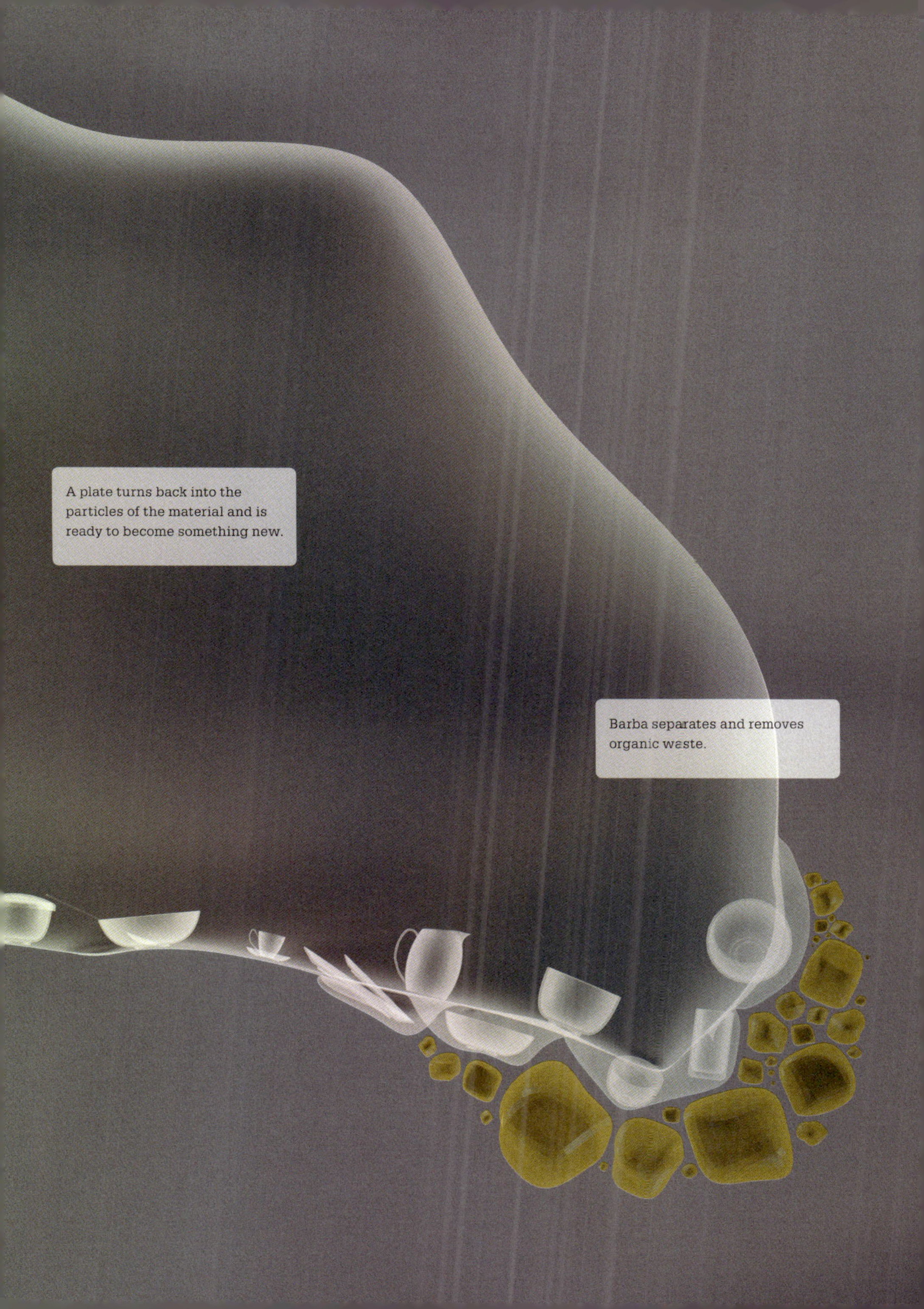
A plate turns back into the particles of the material and is ready to become something new.
Barba separates and removes organic waste.

Different people's spaces can merge and then separate again.

You can all go separate ways.

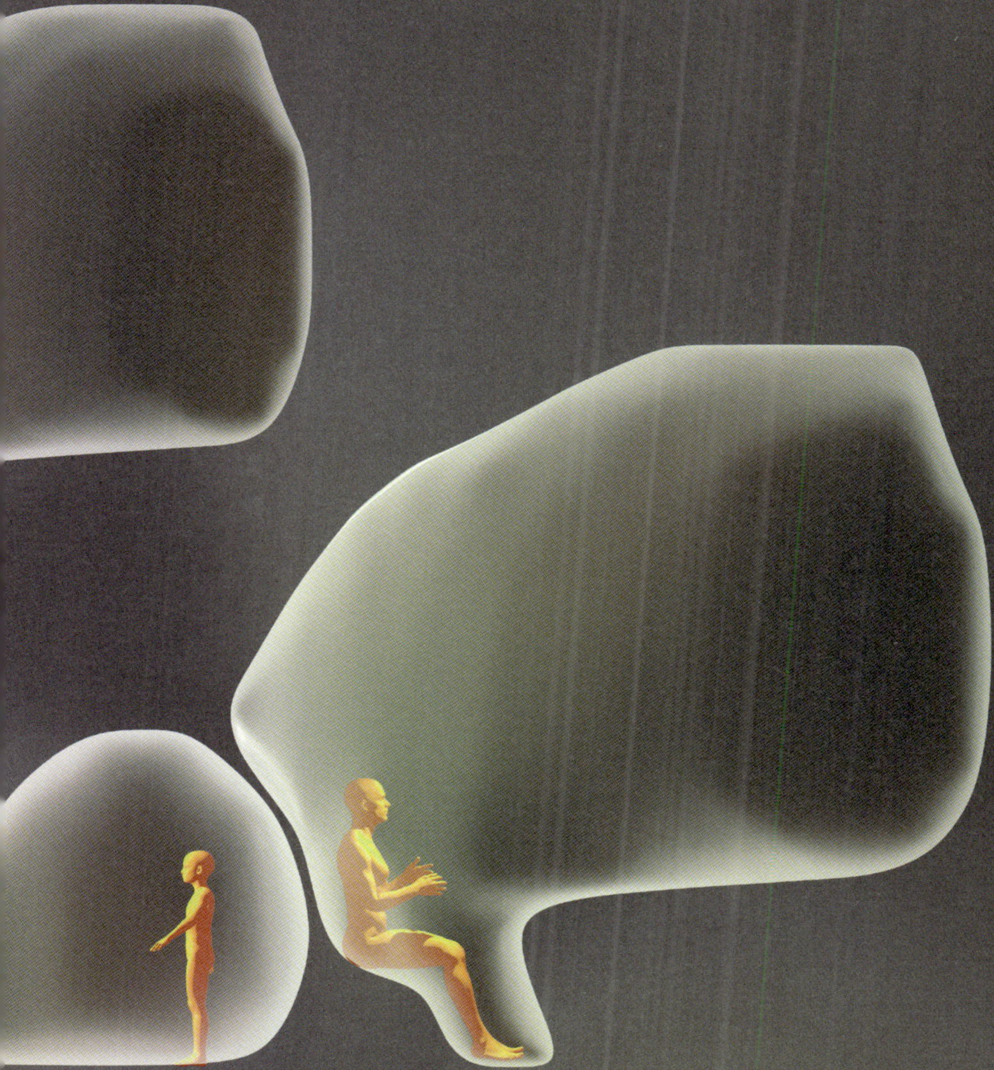

But would we be alone in this space? Or would others be close? Would spaces touch?

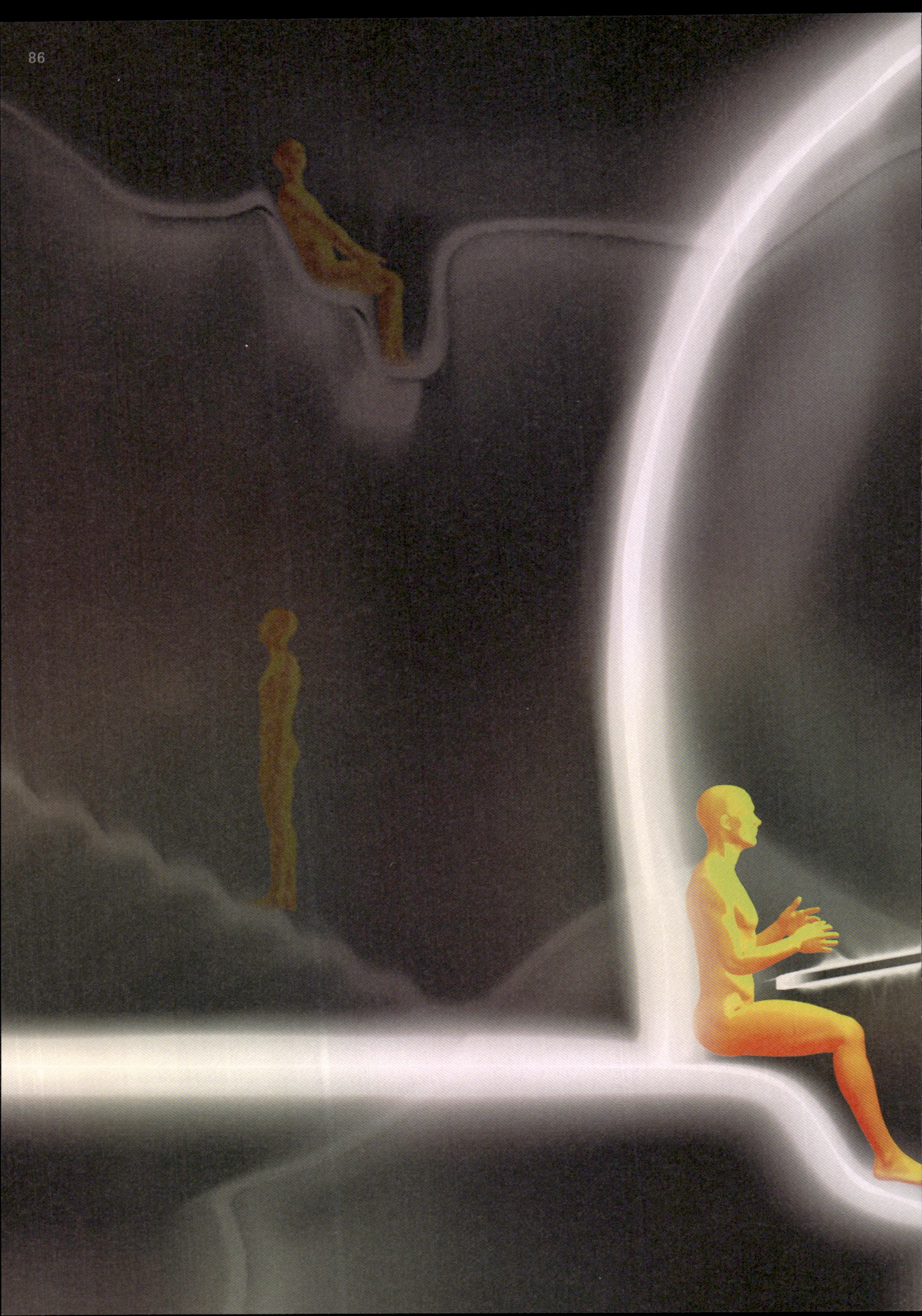

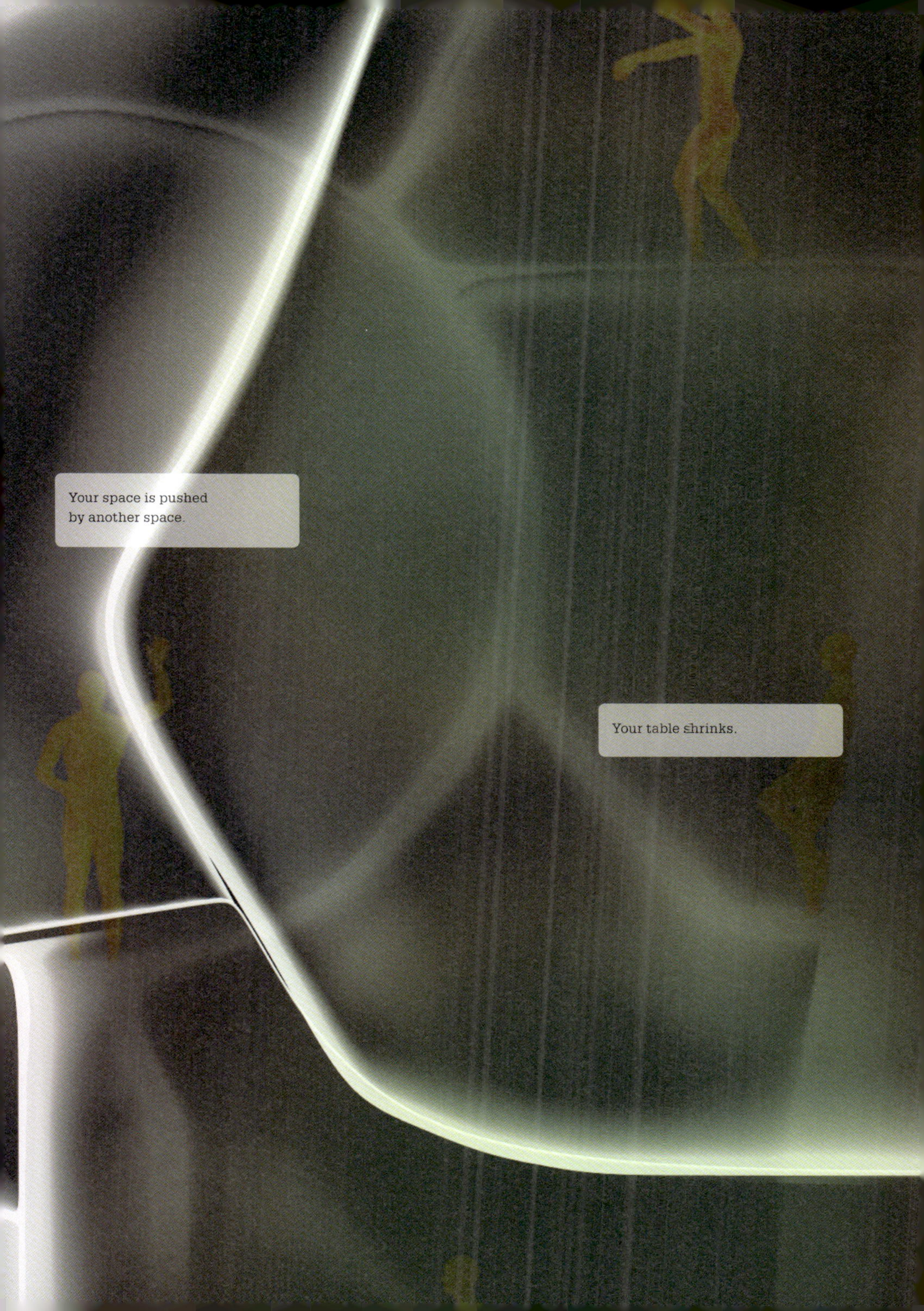
Your space is pushed by another space.
Your table shrinks.

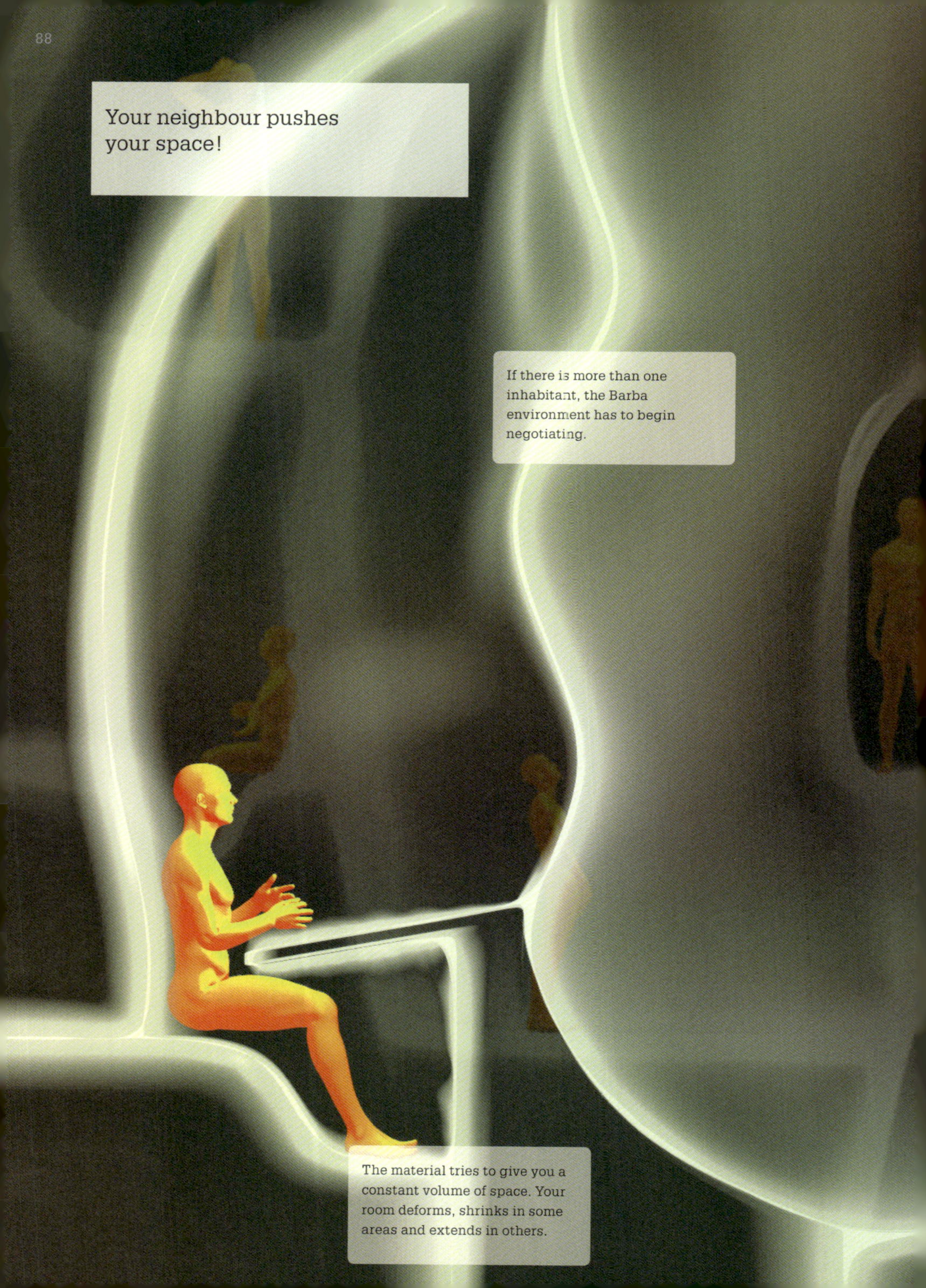
Your neighbour pushes your space!
If there is more than one inhabitant, the Barba environment has to begin negotiating.
The material tries to give you a constant volume of space. Your room deforms, shrinks in some areas and extends in others.

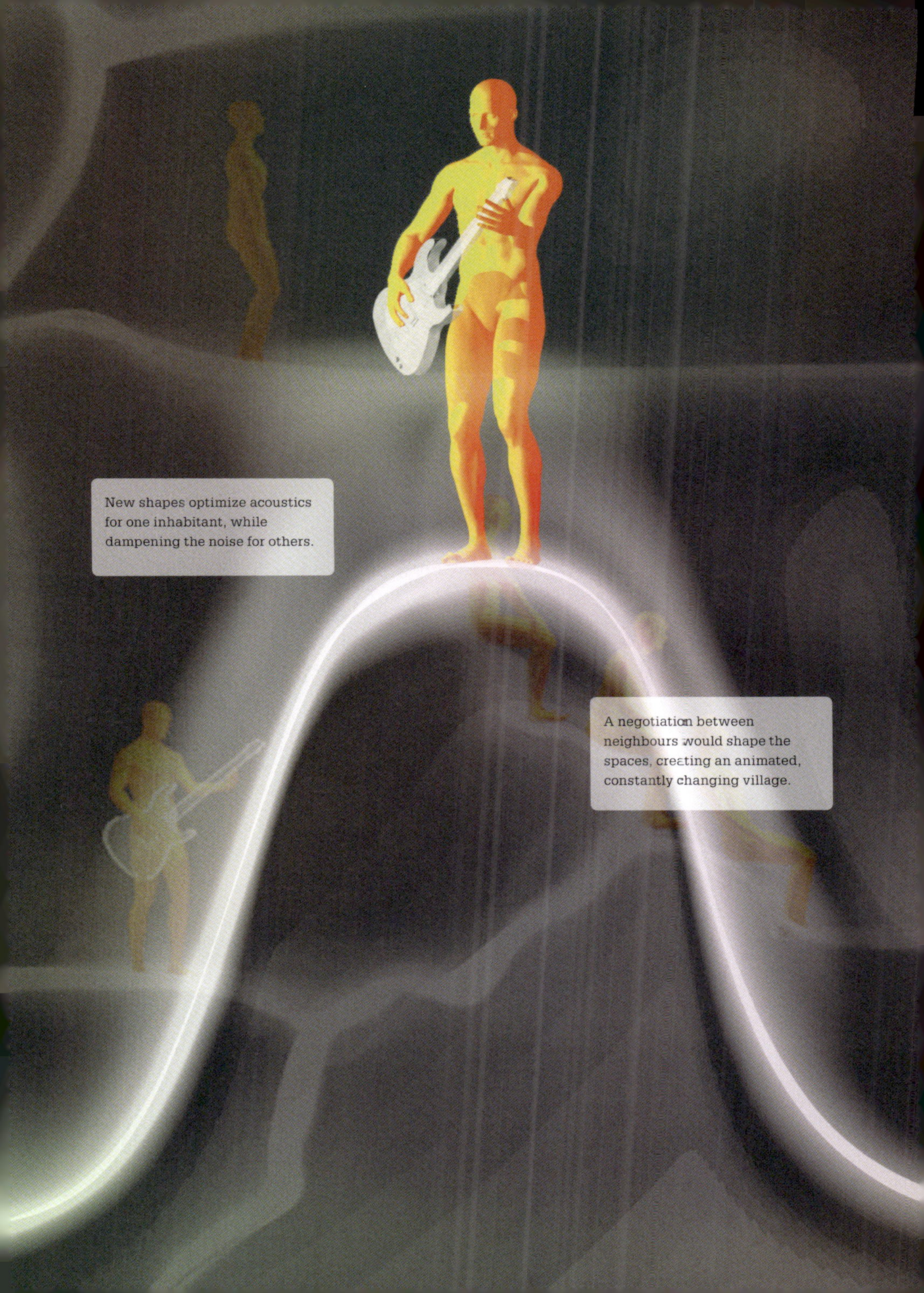
New shapes optimize acoustics for one inhabitant, while dampening the noise for others.
A negotiation between neighbours would shape the spaces, creating an animated, constantly changing village.

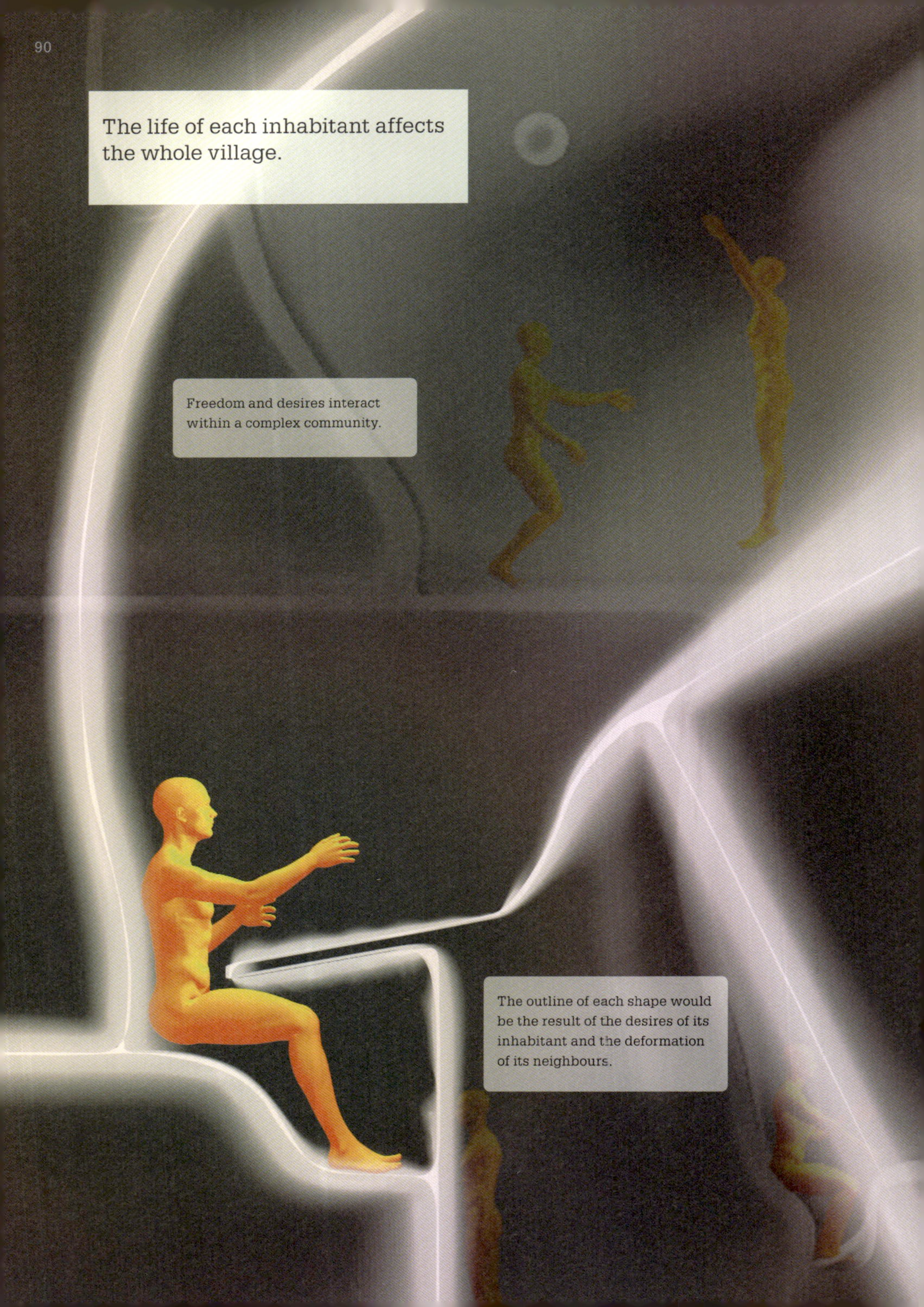
The life of each inhabitant affects the whole village.
Freedom and desires interact within a complex community.
The outline of each shape would be the result of the desires of its inhabitant and the deformation of its neighbours.

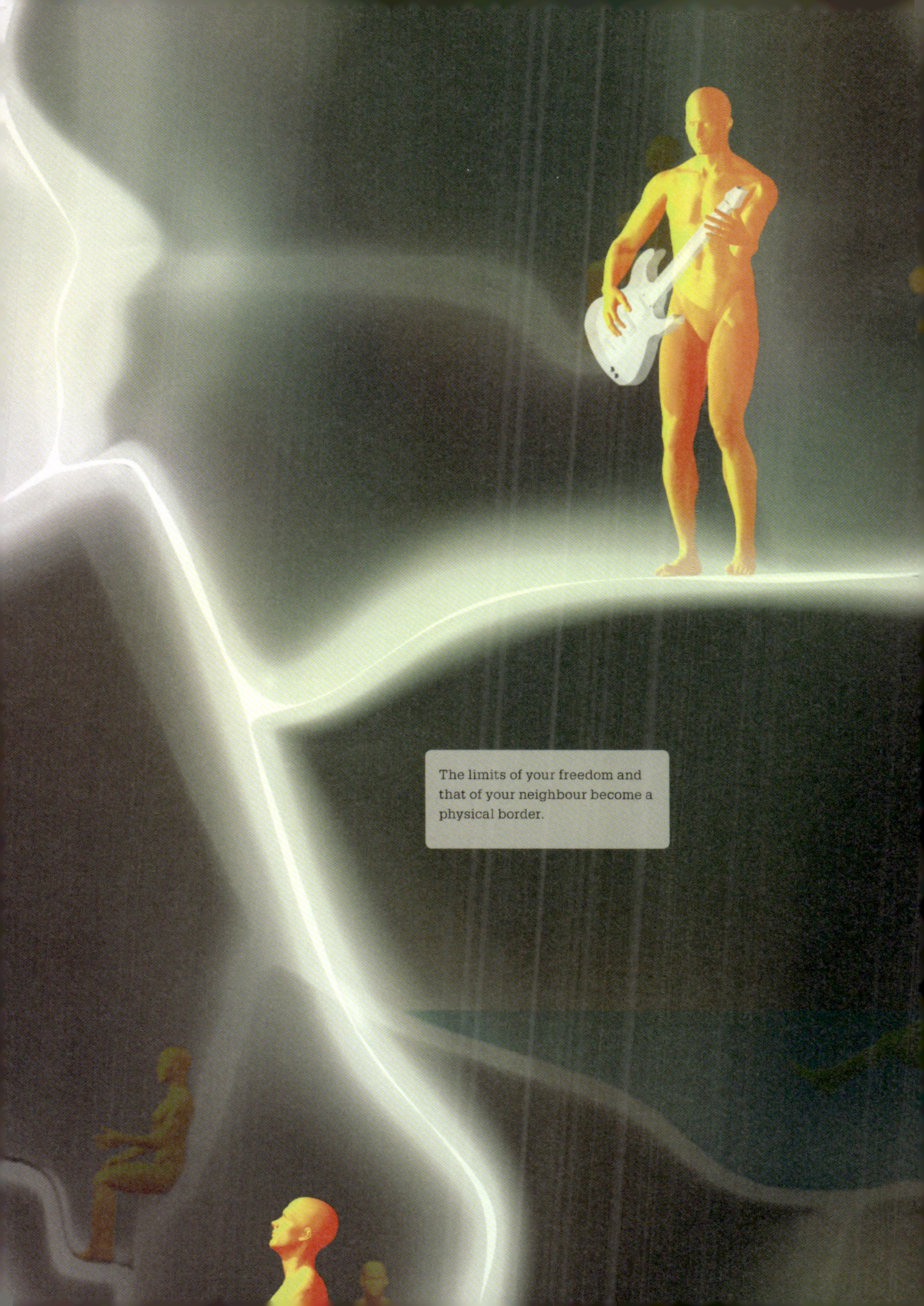
The limits of your freedom and
that of your neighbour become a
physical border.

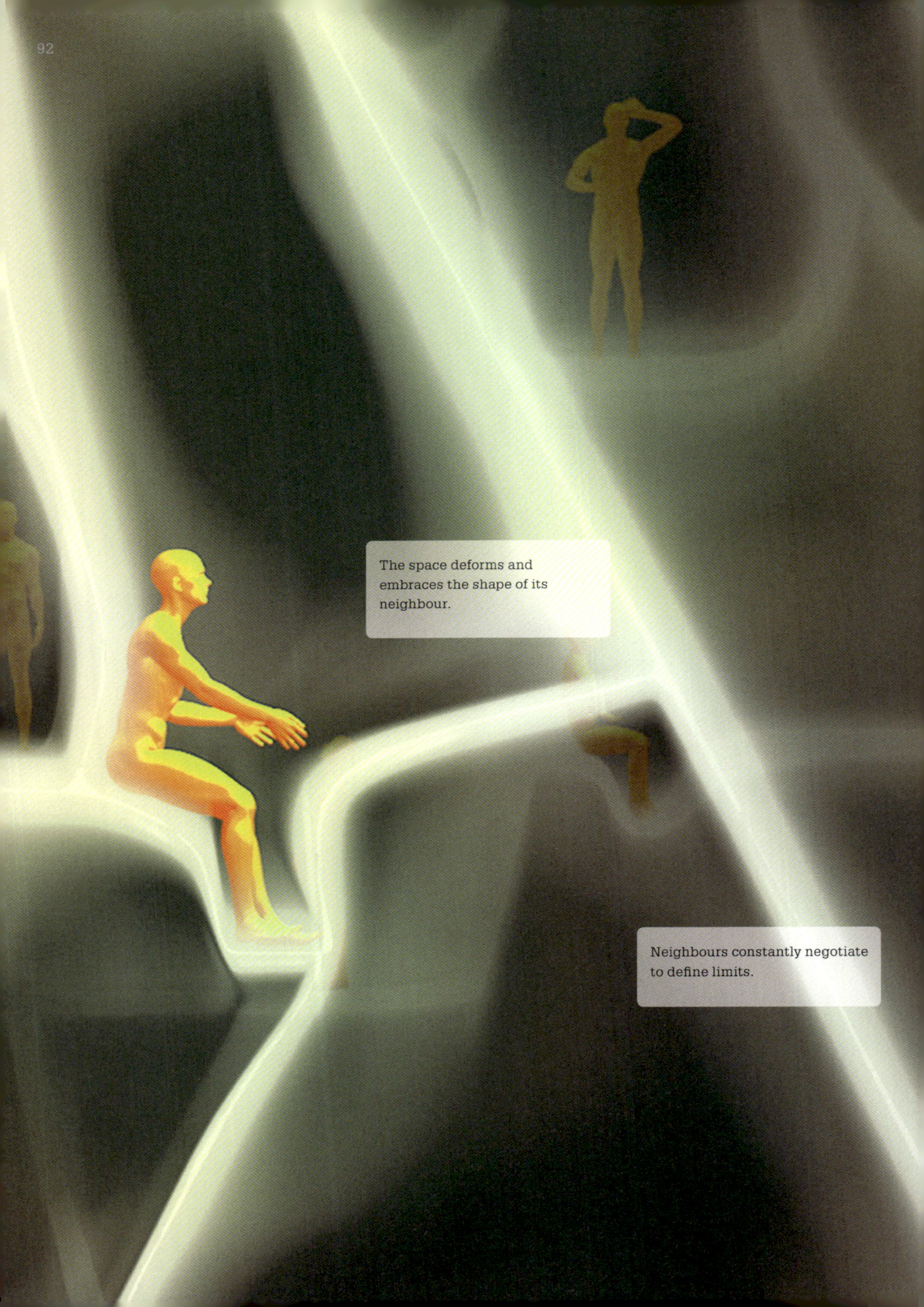
The space deforms and embraces the shape of its neighbour.
Neighbours constantly negotiate to define limits.

By gazing in one direction, one creates a view cone that pushes the ceiling to reach sunlight.

While sleeping, the material tries to shrink down to use minimum space.
Individual actions move like waves through the environment. A simple gesture deforms and affects the whole village.

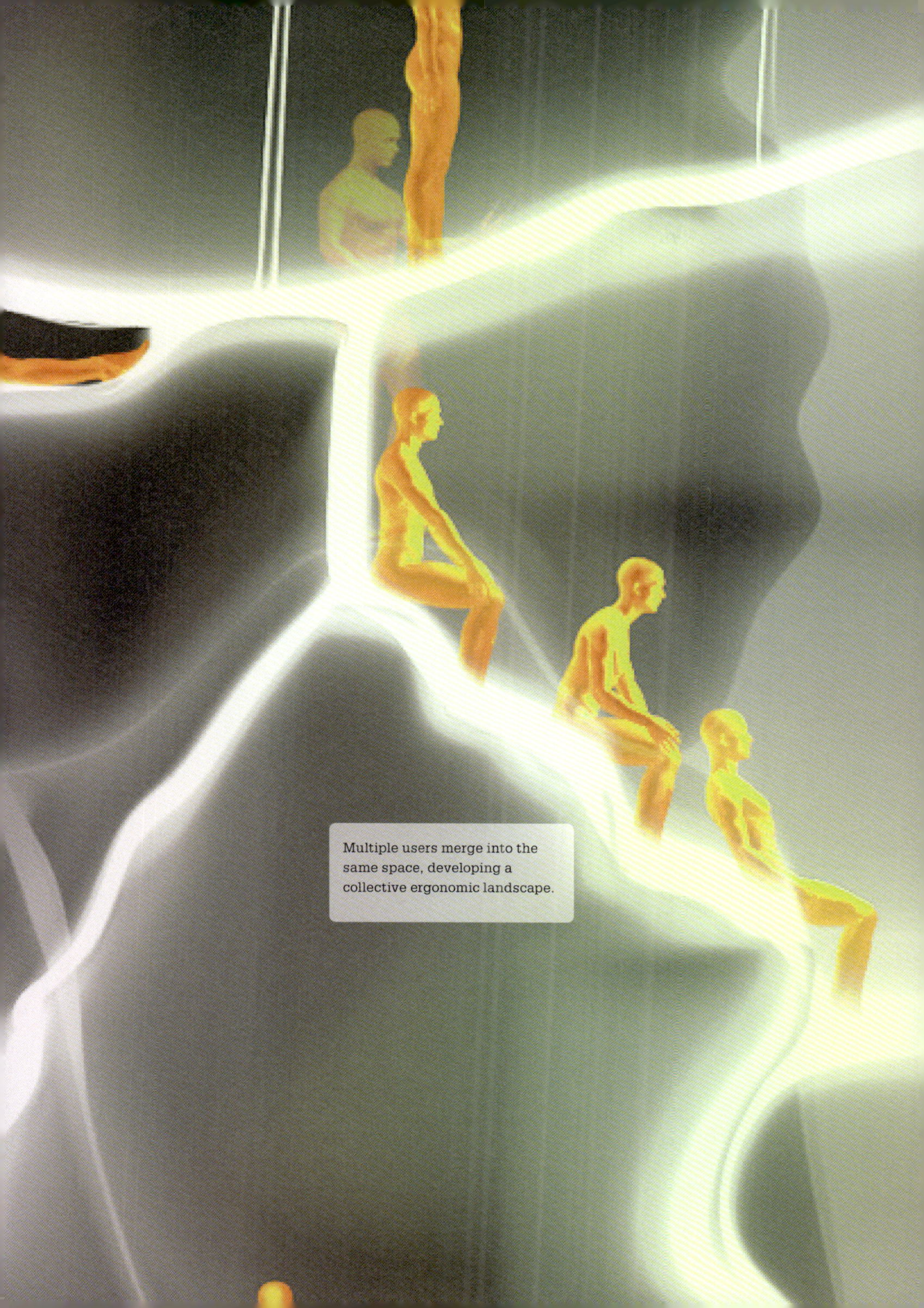

Multiple users merge into the same space, developing a collective ergonomic landscape.

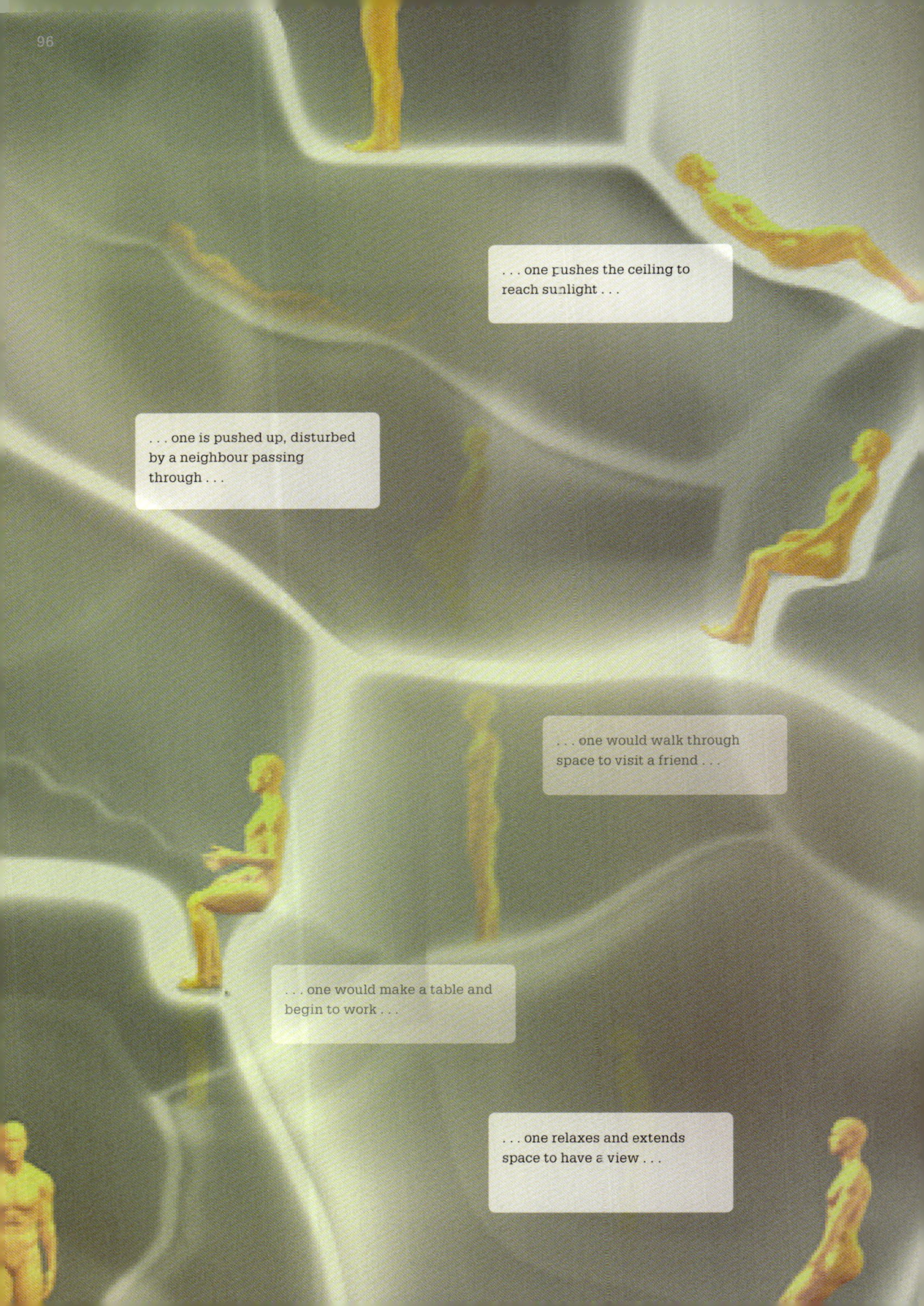
. . . one pushes the ceiling to reach sunlight . . .
. . . one is pushed up, disturbed by a neighbour passing through . . .
. . . one would walk through space to visit a friend . . .
. . . one would make a table and begin to work . . .
. . . one relaxes and extends space to have a view . . .

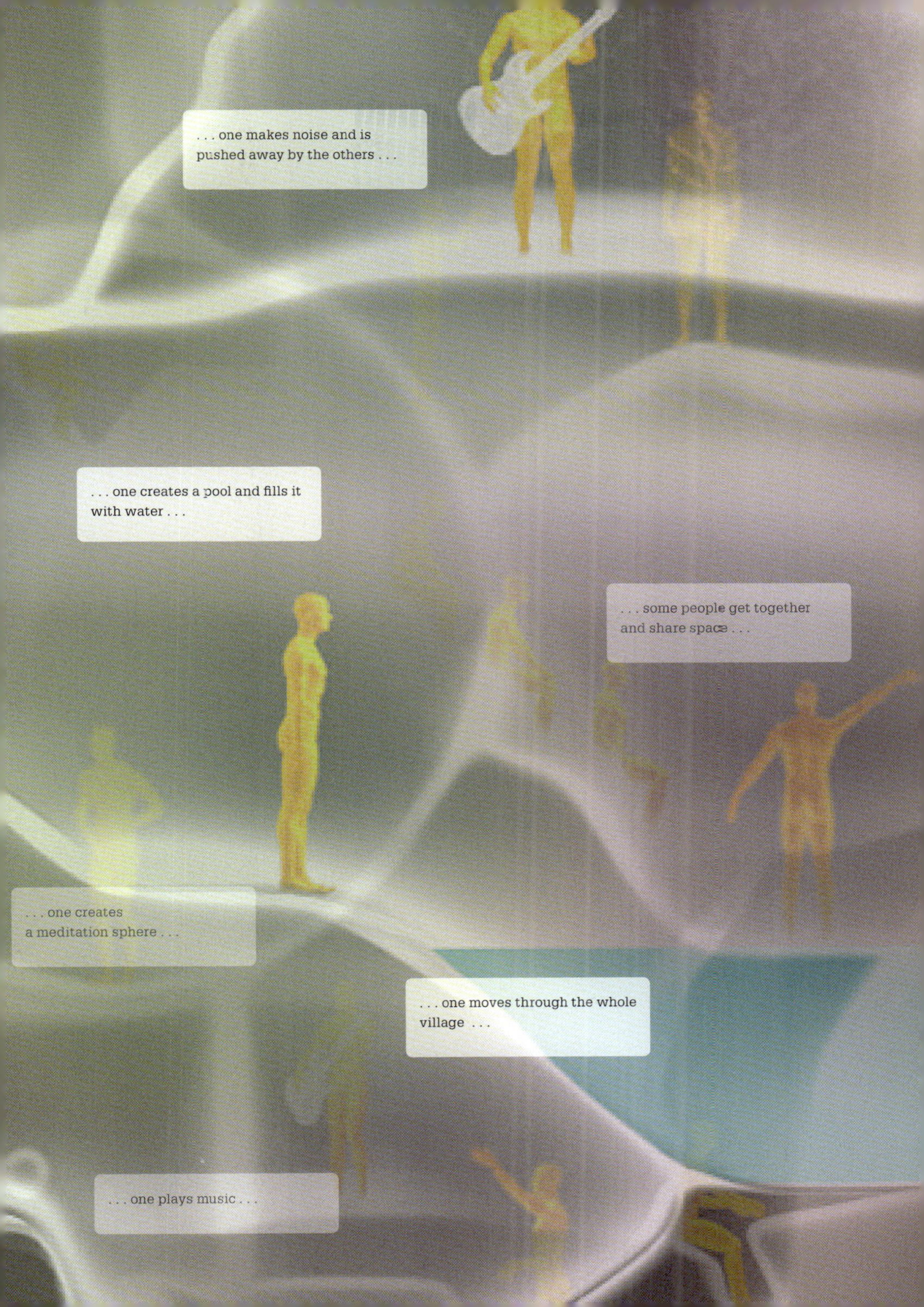
. . . one makes noise and is
pushed away by the others . . .
. . . one creates a pool and fills it
with water . . .
. . . some people get together
and share space . . .
. . . one creates
a meditation sphere . . .
. . . one moves through the whole
village . . .
. . . one plays music . . .

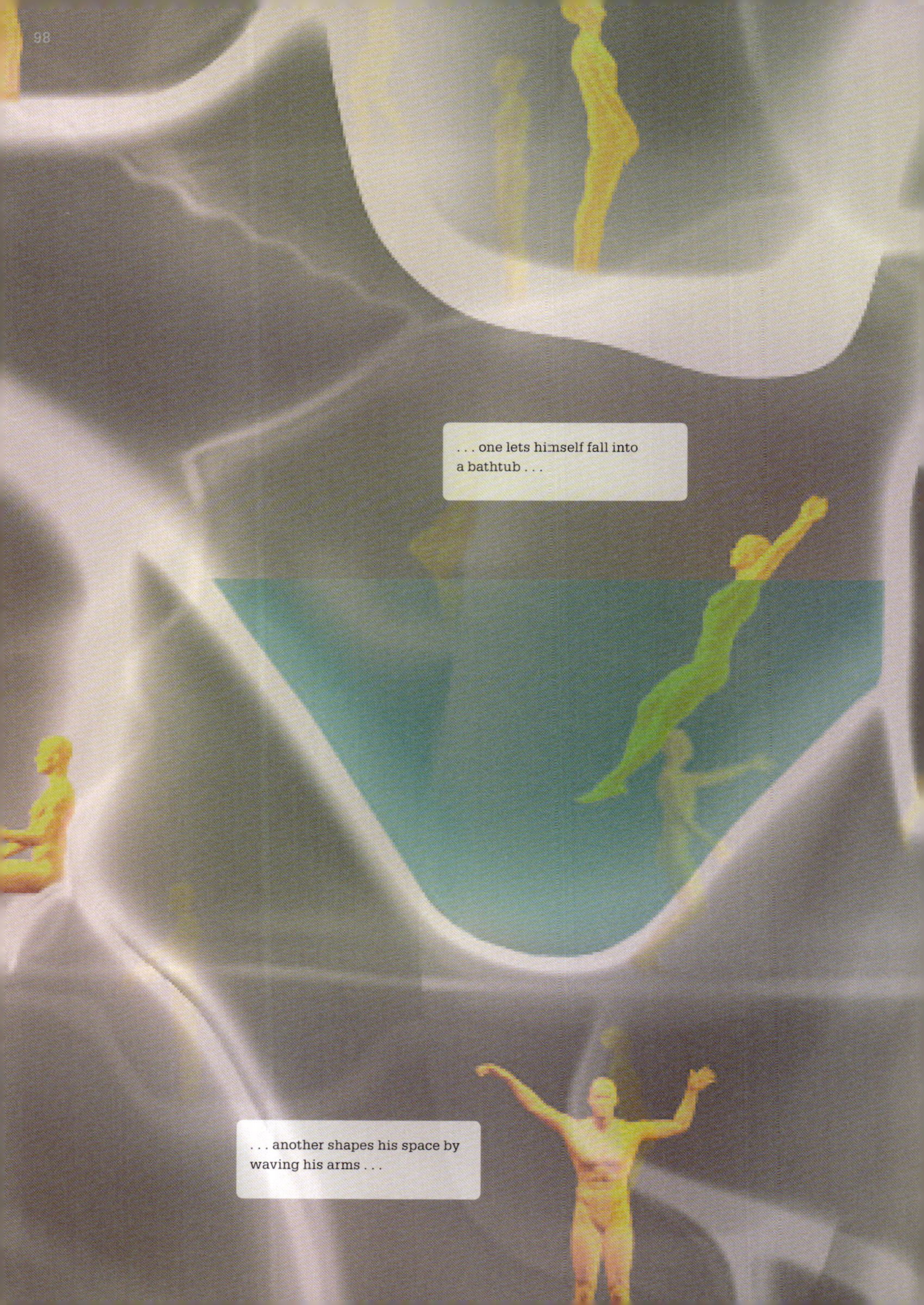
. . . one lets himself fall into a bathtub . . .
. . . another shapes his space by waving his arms . . .

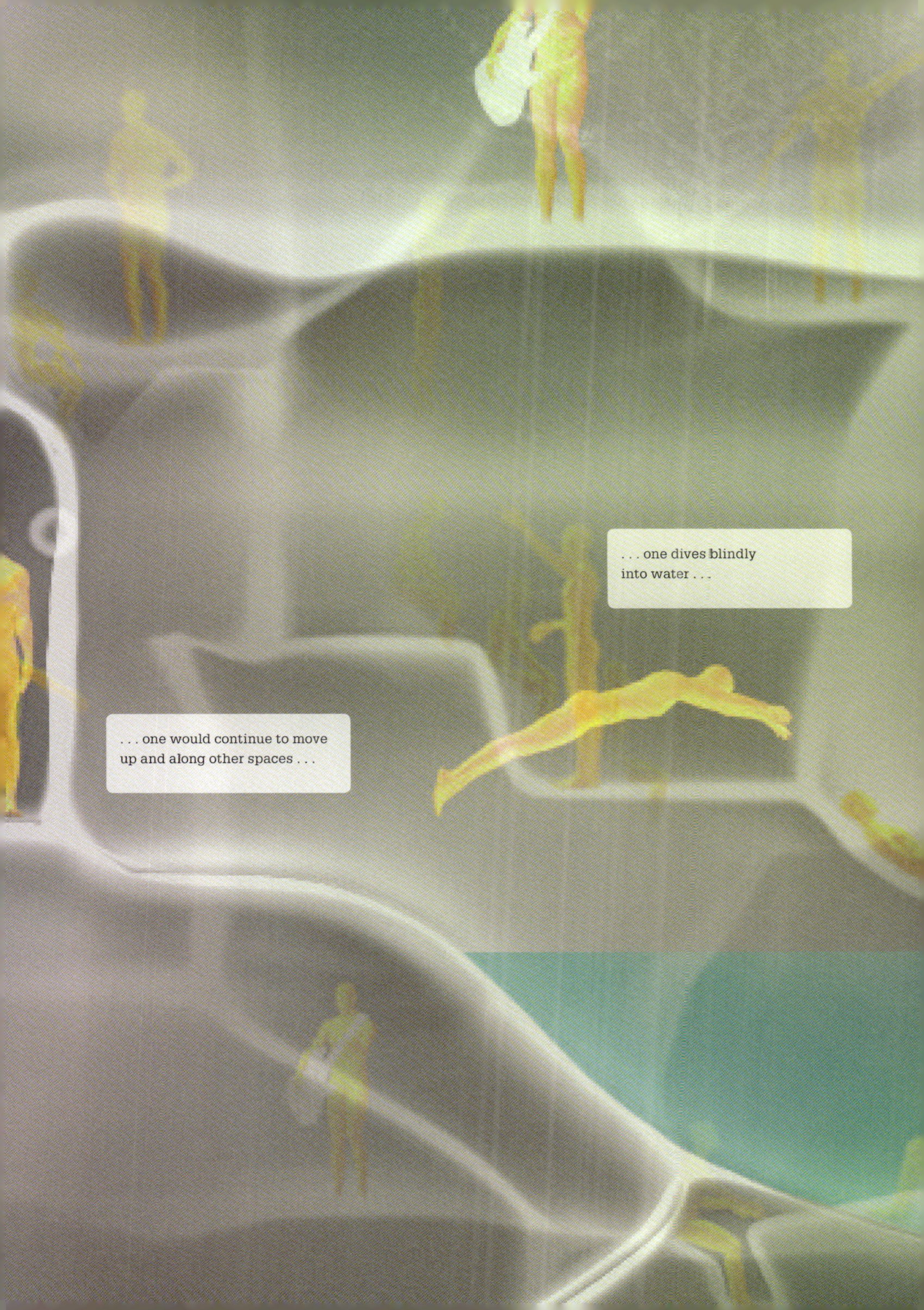
. . . one dives blindly into water . . .
. . . one would continue to move up and along other spaces . . .

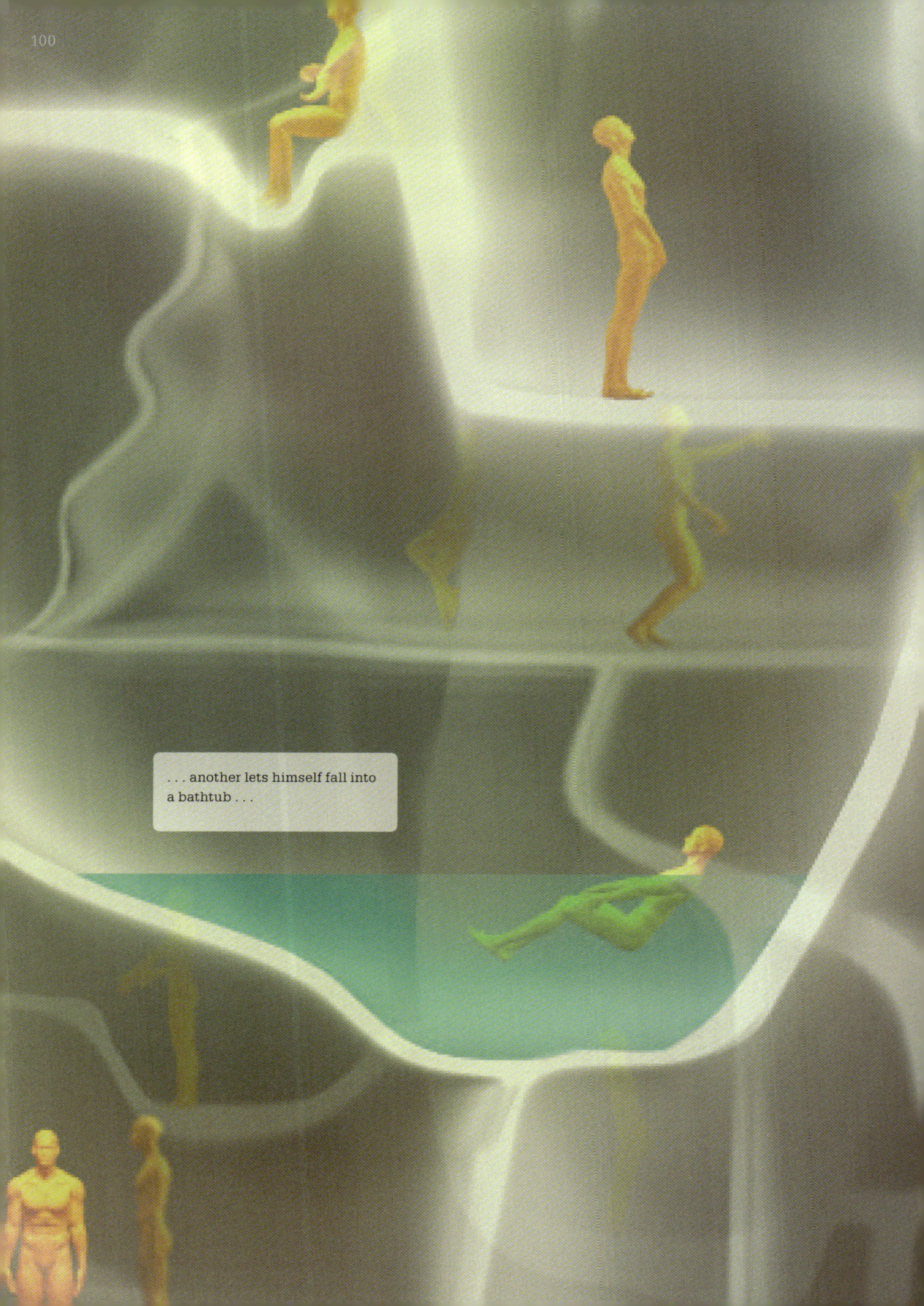
. . . another lets himself fall into
a bathtub . . .

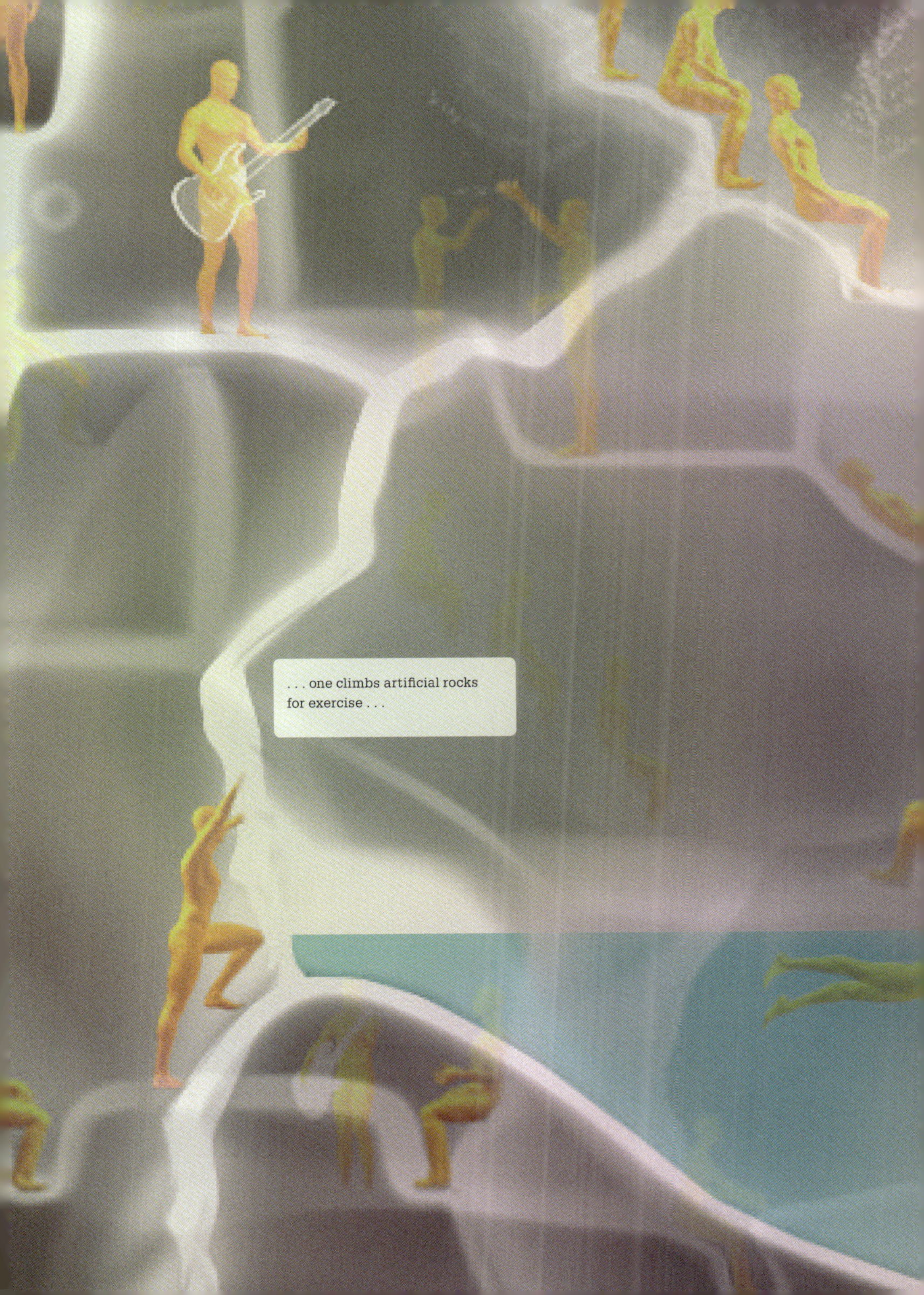
. . . one climbs artificial rocks
for exercise . . .

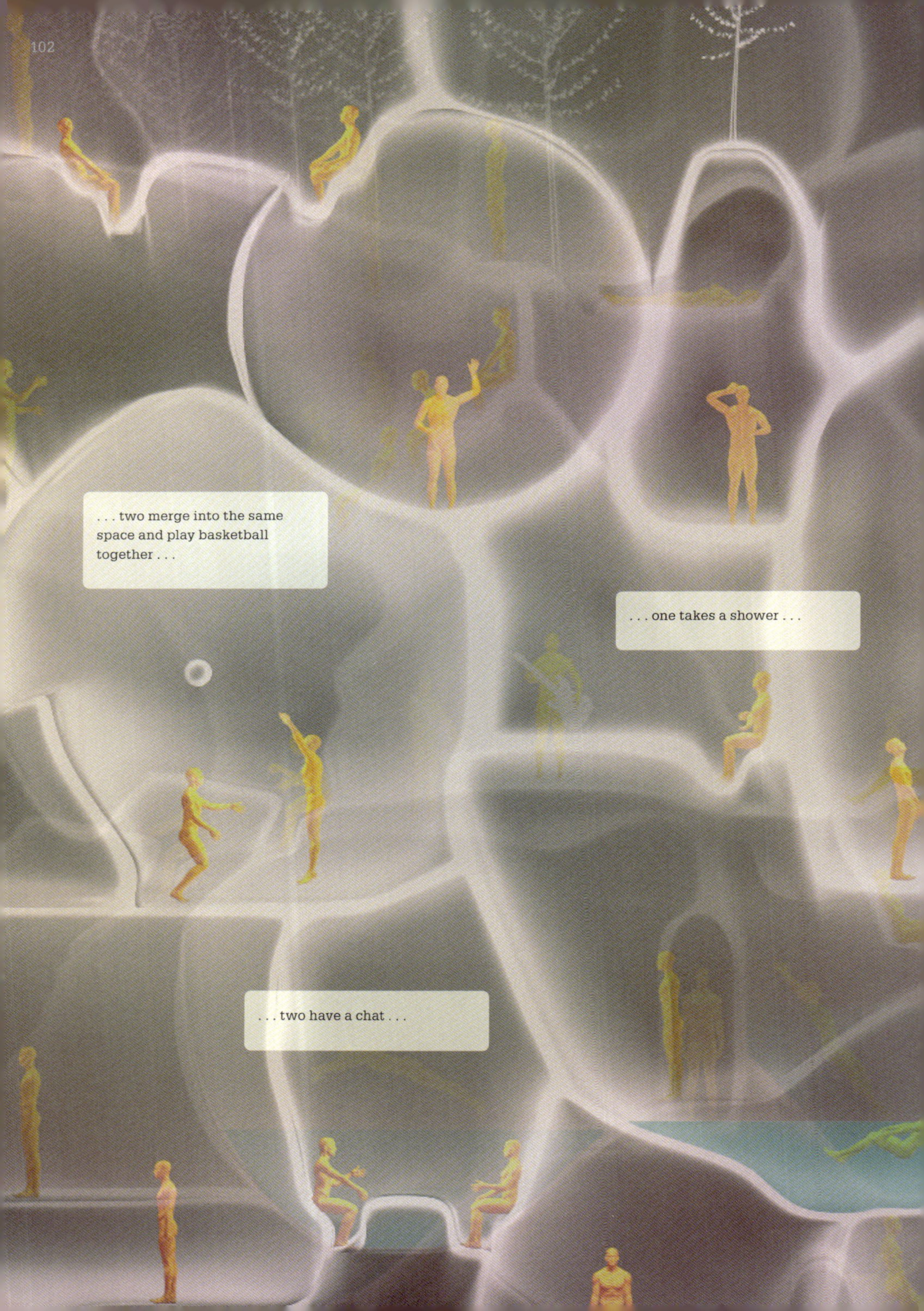
. . . two merge into the same space and play basketball together . . .
. . . one takes a shower . . .
. . . two have a chat . . .

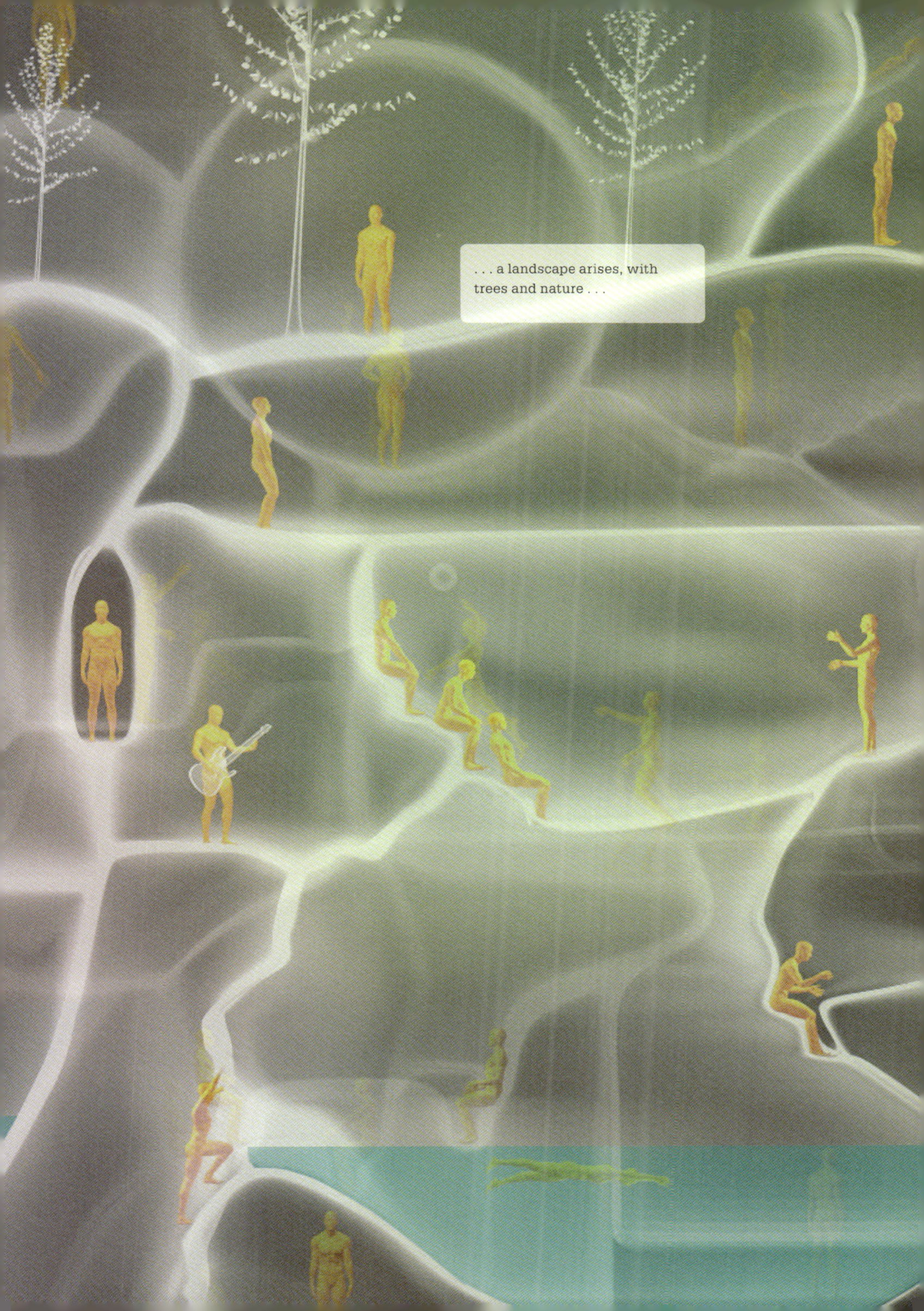
. . . a landscape arises, with
trees and nature . . .

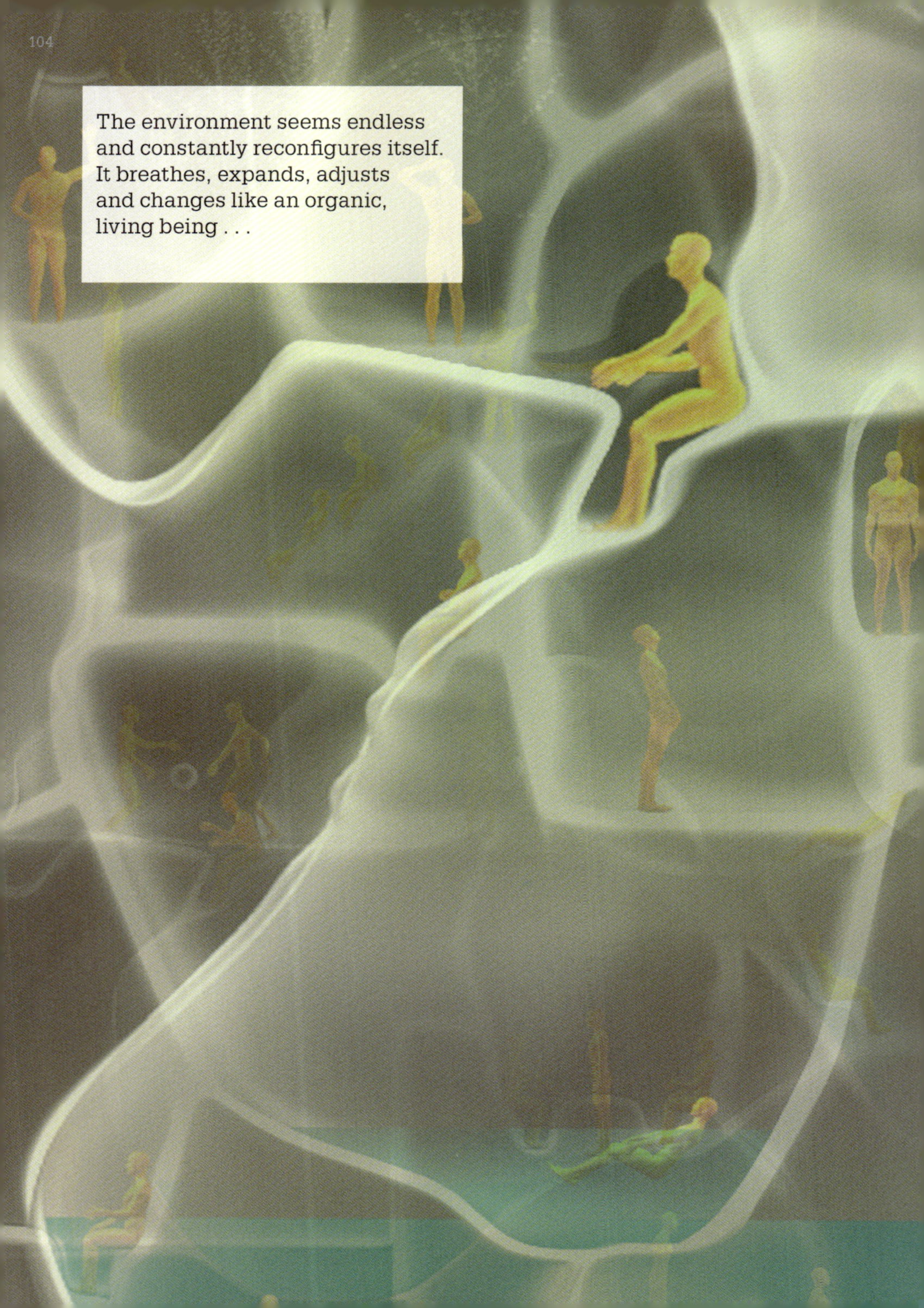
The environment seems endless and constantly reconfigures itself. It breathes, expands, adjusts and changes like an organic, living being . . .

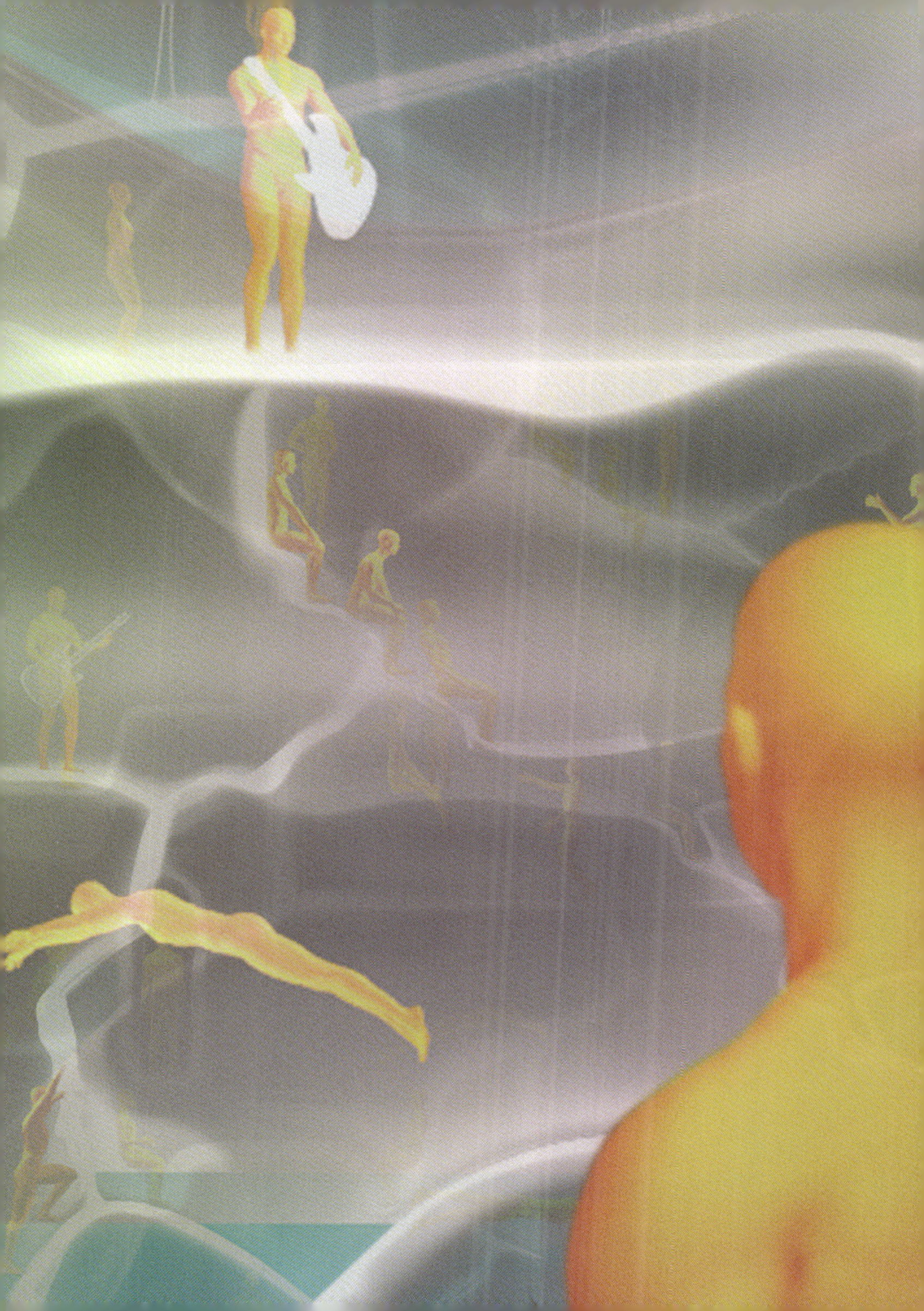

How can we realize our vision of Barba space and the fully adaptable environment? We start with the individual body and visualize the space an individual moving person needs and creates. We can use this to start imagining interaction between individuals. We then add specific tools to create furniture or openings. Eventually, tools for structure, services and light are developed to support Barba houses and villages.

While in this chapter the fully adaptable environment is a virtual space, the tools employed within it are real. Connected to motion-tracking devices, they allow us to see and imagine how it might feel to live in a fully adaptable environment.

The tools presented here can be imagined in use in many different scenarios. They are general tools, things that can be used to analyse, create, plan or design in varied relation to the human body and its flexible spaces.

Space Makers

The Why Factory and ETH Zürich: Adrien Ravon, The Why Factory research team and the Transformer at ETH Zürich studio

The Why Factory research team:
Adrien Ravon with Ulf Hackauf, Sander Mulders, Stefan Wülser, Kaegan Walsh, Arslane Benamar, Lex te Loo

Transformer at ETH Zürich studio:
Paolo Birrer, Tania Carl, Riccardo Caruso, Melina Cerfeda, Elena Chestnova, Nicholas Ganz, Marcel Hodel, Léone Kündig, Marisa Muscionico, Sebastian Oswald, Kai Peter, Pascal Ruckstuhl, Enrique Ruiz, Paula Schilliger, Vera Schmidt, Henry Stehli, Ying Yi Tan, Yangzom Wujohktsang
Teachers: Winy Maas, Ulf Hackauf, Adrien Ravon, Stefan Wülser

Muybridge Revisited

Tool of Measurement

Arslane Benamar, Paolo Birrer, Tania Carl, Melina Cerfeda, Stavros Gargaretas, Ulf Hackauf, Lex te Loo, Sander Mulders, Sebastian Oswald, Kai Peter, Adrien Ravon, Henry Stehli, Kaegan Walsh, Stefan Wülser

The flexible, programmable, transformable Barba material we've speculated about potentially allows for a new and very precise way of defining architectural spaces, interiors and objects. It would require a new, dynamic ergonometry. Today, we are accustomed to sitting on a chair with an average height, walking through corridors of average width and sleeping in beds of average length. With Barba, spaces and objects fit perfectly to the ergonomic needs and the spatial desires of the user. Reading the human body is thus essential.

In 'Muybridge Revisited', we therefore take the body as the starting point for defining space. In this interactive installation, we use a motion-tracking device to record and visualize human bodies and movement.

In the Barba scenario, 'Muybridge Revisited' would be the first step in developing interaction between the user and the fully adaptable environment.

Today, it can be used for explorations of the relation between body, space and time. It can be part of artistic choreography as well as useful in ergonomic studies that start from the body.

Watch the movie:
www.thewhyfactory.com/barba

The human body forms the starting point for the interaction between the user and his transformable environment. We developed an installation that allows the user to experience her own movements and the space she occupies or creates through immediate feedback. In the late nineteenth century, American photographer Eadweard Muybridge worked on a series of stop-motion photographs of animals and humans in motion. With these photographs, he revealed details that the human eye had never before been able to see. His work influenced not only scientists, but also artists and figures in popular culture. Like Muybridge's photographs, our study provides an accurate, real-time analysis of the body's positions and movements. It helps us to understand how we move and how we occupy space.

For this first study, we work with a simplified representation of the body, a virtual skeleton. We visualize the skeleton and play with it. The trace of the body is shown. Movements in space become visible. The data is recorded. 'Muybridge Revisited' is an interactive installation that can track, record, save and overlay the movements of the body. It records the image of the skeleton for a few seconds before it fades away, depicting a brief history of the user's movements and enacting a phenomenon we've thus referred to as 'inertia'.

The inertia introduced within this installation could be an important element in developing the reactivity of the fully adaptable environment. By analysing past movements, Barba could predict those that followed, extending space in accordance with what it anticipated. The quicker the movement is, the larger the space of anticipation would be. The fully adaptable environment would require more complex algorithms than we've used so far in order to really allow for useful anticipation. It would also require distributed and smaller sensors for complex motion tracking.

As an installation capable of use today, 'Muybridge Revisited' reveals patterns, analyses movement and, moreover, notably provides a visual experience of a body's movements and the space they create.

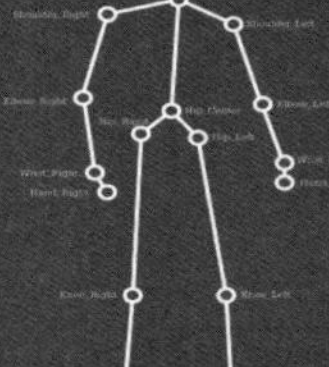

In order to track, measure and record bodily movements in space, we need sensors. For our research, we choose the Kinect motion-tracking device.[1] This tool uses infrared sensors to scan environments, detect depth in space, recognize the human body and its movements, and translate this information into a reduced number of nodes that together form a simplified skeleton.

This virtual skeleton can be used not only in games but also in other software. We chose the three-dimensional-modelling software Rhinoceros with several plug-ins for visual programming, most importantly Grasshopper and Firefly.[2] The complex data generated by a human body moving in space is reduced and made workable for our experiments. Every ten milliseconds, the position of each of the skeleton's single nodes is recorded, stored and visualized. An established limit of data collection determines when these traces expire and when information fades away.

The installation uses one Kinect for the skeleton tracking, Rhinoceros for three-dimensional modelling, Grasshopper for visual scripting and Firefly to connect the Kinect with the scripts in Grasshopper.

1 Kinect® is a motion-tracking device originally developed by Mircosoft® as an input for video games. It can effectively serve other applications as well. See <www.microsoft.com>.

2 Rhinoceros® is a three-dimensional-modelling software. See <www.rhino3d.com>. Grasshopper® is a graphical algorithm editor designed as a plug-in for Rhinoceros. See <www.grasshopper3d.com>. Firefly® is a set of comprehensive software tools developed to connect Grasshopper with different input devices. See <www.fireflyexperiments.com>.

The Halo Maker

Experience it!

Arslane Benamar, Paolo Birrer, Tania Carl, Melina Cerfeda, Stavros Gargaretas, Ulf Hackauf, Lex te Loo, Sebastian Oswald, Kai Peter, Adrien Ravon, Henry Stehli, Kaegan Walsh, Stefan Wülser

The 'Halo' visualized in this installation could be described as a virtual space that surrounds and follows the body in movement, defining its outline and describing its minimum space. And this 'Halo' can be extended to represent the space of movements, activities and desires. In Barba, the Halo is an important tool for defining and shaping space and for making visible the interaction between user and the fully adaptable environment.

The 'Halo Maker' is an interactive installation that measures personal space in real time. It builds upon 'Muybridge Revisited' by extending the skeleton beyond the human body such that it appears outside, like an aura or halo. This Halo reveals a space around the body, something akin to what American anthropologist Edward T. Hall described as 'personal space'.[1] In our Barba scenario, the 'Halo Maker' would work as a tool to create space and as a means to interact with the material. Today, it can be used for general studies on spaces related to the human body. It can work as a tool to design spaces that start from the body.

Watch the movie:
www.thewhyfactory.com/barba

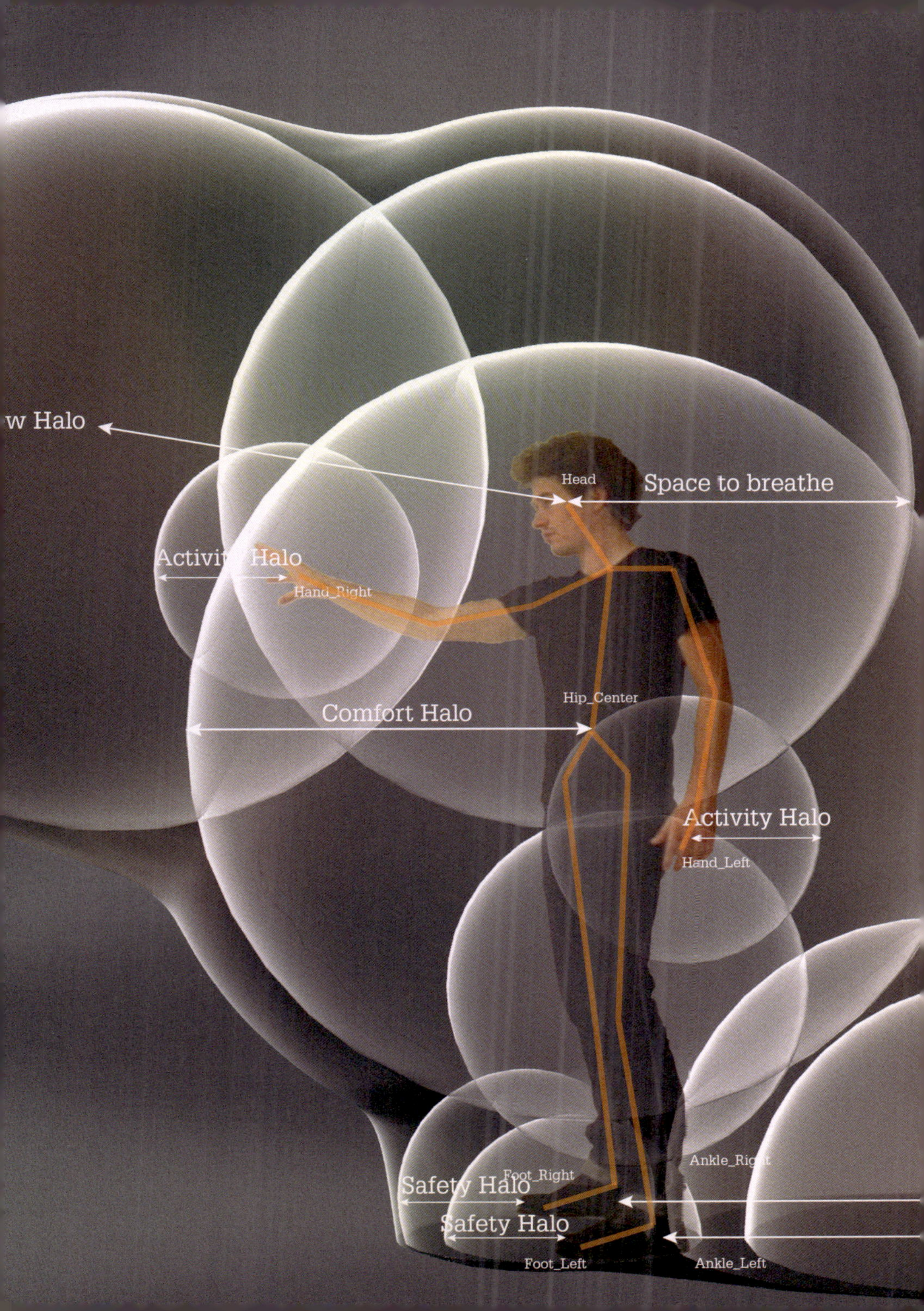
w Halo
Head
Space to breathe
Activity Halo
Hand_Right
Hip_Center
Comfort Halo
Activity Halo
Hand_Left
Ankle_Right
Foot_Right
Safety Halo
Safety Halo
Foot_Left
Ankle_Left

The Halo is, in its simplest form, an offset of the body that describes the immediately surrounding space. When it extends, it can carry other meanings and visualize other requirements. We can imagine comfort halos, halos of inertia and anticipation, visible and audible halos, and so on.

The 'Halo Maker' builds upon the virtual skeleton from the previous chapter and attaches 'metaballs' to the skeleton's nodes. [2] These metaballs are connected with a mesh, resulting in a three-dimensional envelope that encompasses the human body. A threshold factor controls how precisely the envelope follows the metaballs. The virtual skeleton with the attached metaballs moves with the user in real time, creating a constantly changing space. For the installation, we project the skeleton onto a screen. The user sees how the skeleton repeats his movements and how the Halo around it changes shape.

Different gestures are programmed to modify the Halo. For example: holding your hand above waist height increases the 'head metaball'. Holding your hand at one height increases all metaballs and thus the whole Halo. Turning your head moves the 'View Halo'.

The installation uses Kinect for the skeleton tracking, and Rhinoceros, Grasshopper, Firefly and Weaverbird. [3]

1 Edward T. Hall, *The Hidden Dimension* (New York: Anchor Books, 1966).

2 Metaballs are a computer graphics technology invented by Jim Blinn in the early 1980s.

3 Weaverbird® is a plug in for mesh transformation in Grasshopper and Rhino. See <http://www.grasshopper3d.com/group/weaverbird>.

The 'Comfort Halo' would translate basic spatial needs and desires. It would be a buffer against claustrophobia, a mean to ensure that the responsive material remains at a comfortable distance from the body. It could extend based on movement, pulse or breathing. The 'View Halo' would be created through the gaze. By staring at one point, the material would take the form of a cone extending along the view line. What we've called the 'Inertia Factor' would help to create a more stable surrounding environment. By adding time and inertia to the Halo, the fully adaptable environment would react in less hectic ways; the space it created would be smoother. These different halos would serve as the main means of interaction between the user and his environment. The particles of the material could wrap around the Halo, never intruding upon it but using it as a kind of mould. Physical space thus becomes a direct manifestation of the virtual, personal, comfortable and desired space around a person.

For the development of 'Muybridge Revisited', we produced a catalogue of movements. By overlapping these movements to create a smooth transition from image to image, we explored the transition between different activities.

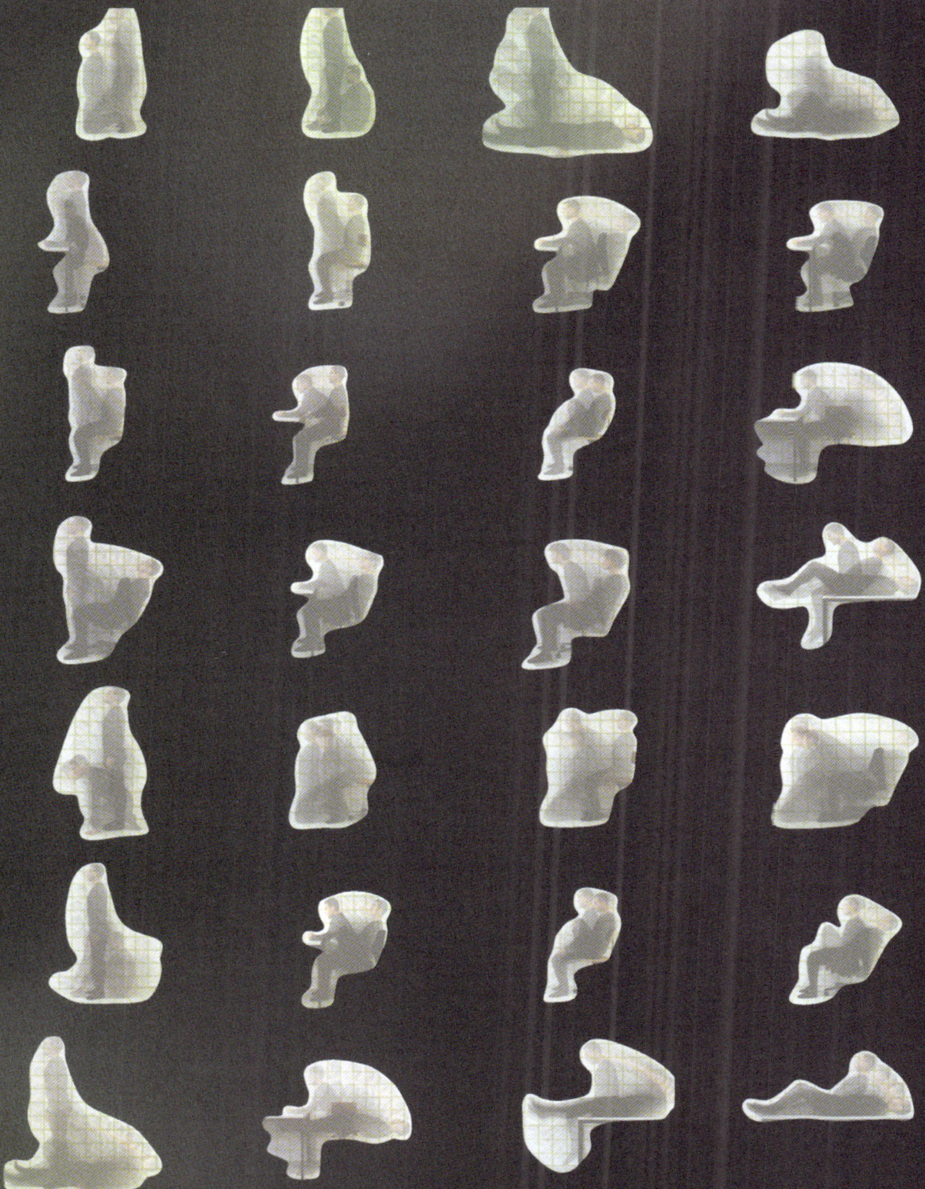

By translating these images into contours, we illustrated minimum ergonomic shapes based on different postures or activities.

Negotiator

Fight for Space

Arslane Benamar, Riccardo Caruso, Stavros Gargaretas, Ulf Hackauf, Lex te Loo, Marcel Hodel, Marisa Muscionico, Adrien Ravon, Pascal Ruckstuhl, Ying Yi Tan, Kaegan Walsh, Yangzom Wujohktsang, Stefan Wülser

The 'Halo Maker' can visualize how halos interact. It can show how individual personal spaces can touch and overlap, leading to conflict or cohabitation. The 'Negotiator' can modify two conflicting halos in different ways. One can push the other away. One can shrink and the other extend. Both can form a border where they touch. Or they can merge, creating one collective Halo. The 'Negotiator' is an interactive installation in which two users can watch their halos collide and engage in a fight for space. The installation reveals the process of spatial negotiation, conflict and compromise between multiple users.
In the Barba scenario, the 'Negotiator' would be an essential tool for cohabitation. It could allow multiple inhabitants to change the size and shape of their personal spaces and thus allow personal spatial freedom while maximizing density. Today, the 'Negotiator' can be used to visualize and experience one's personal space in relation to that of others. It can be applied in installations and could also be used for design in dense architectural spaces.

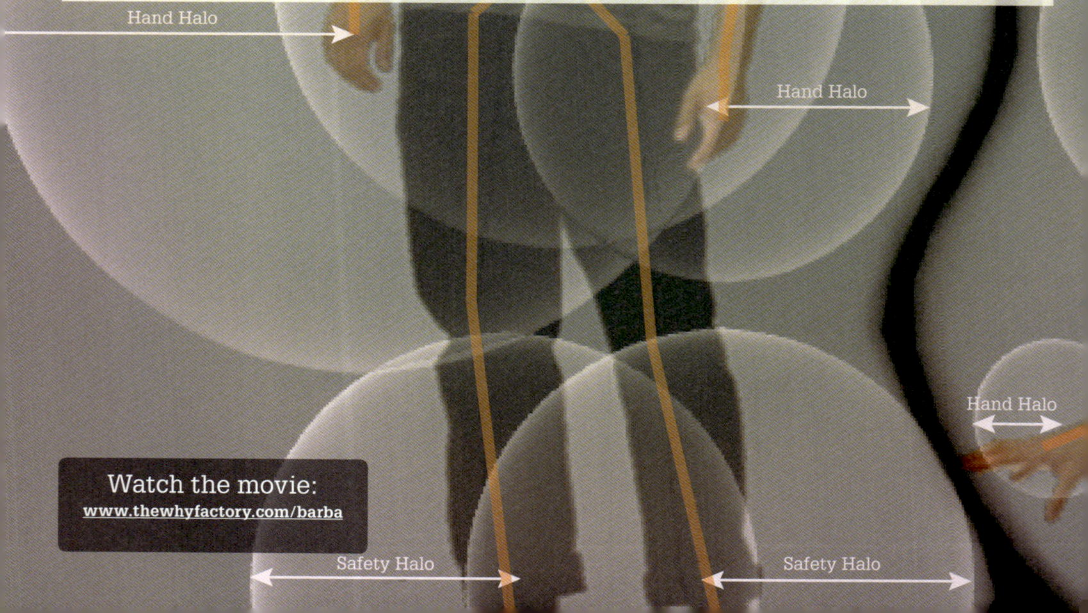

Watch the movie:
www.thewhyfactory.com/barba

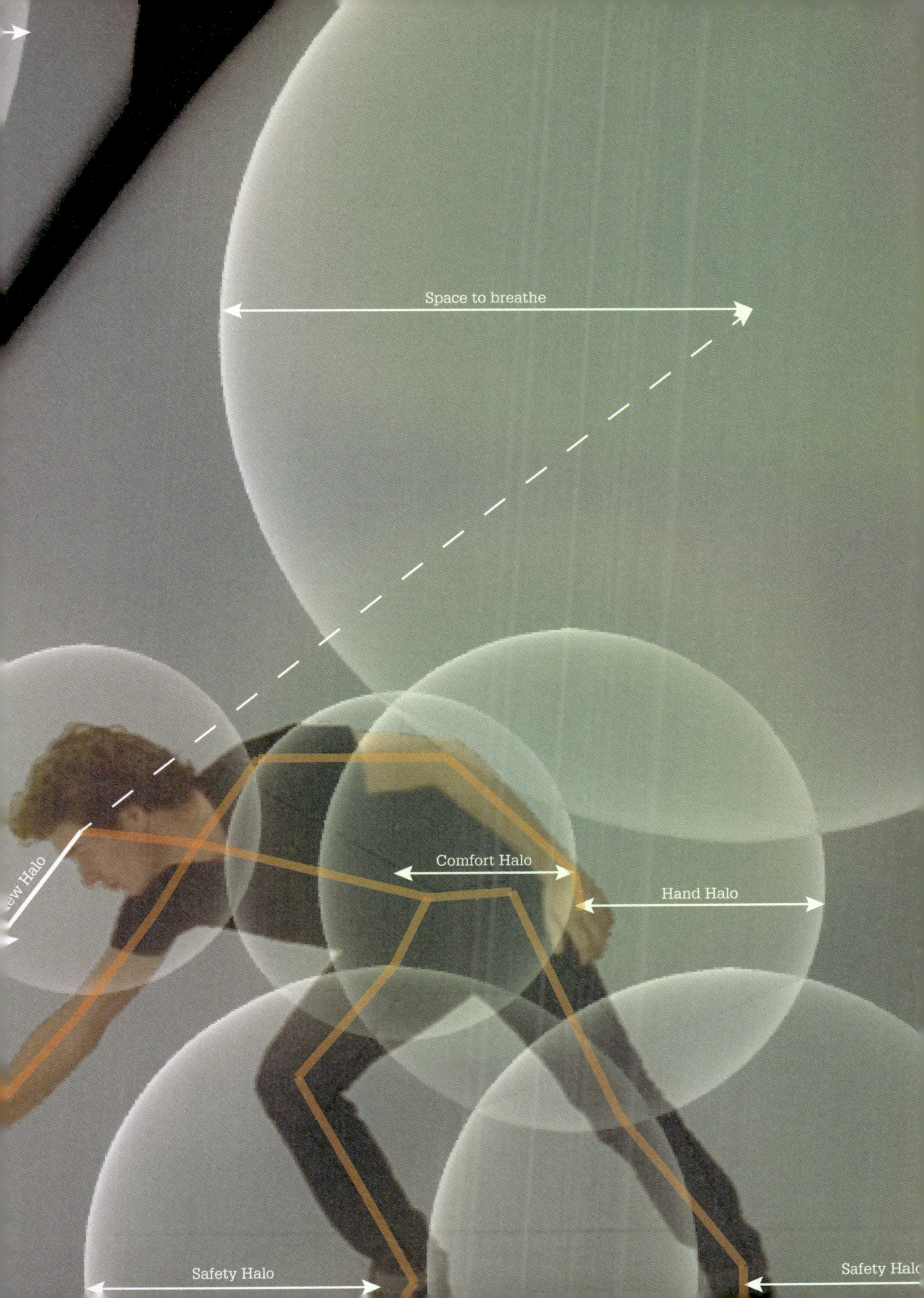

Space to breathe
Comfort Halo
Hand Halo
Safety Halo

Conflicts between multiple users can be visualized as overlapping halos. These can be spatial halos – I want my room to extend into yours, for example – but could also, for example, be acoustic halos – as in, I want to keep your noise out of my space. As the halos change with the users' movement and activity, the overlaps and conflicts between users would constantly change.
The 'Negotiator' can react to these conflicts in different ways:
The user who moves more could dominate. His Halo would push the other's Halo away. Each person would need to constantly fight to maintain the integrity of her personal space. If a neighbour were to become too intrusive, for example, a person could push him back. His Halo could in turn extend to the other side, solving the conflict by moving away.
Or we could watch personal space shrink in reaction to a neighbour's activity. The quieter a person is, the smaller her Halo would be.
Another configuration takes the form of a win-win or lose-lose situation. If both individuals are fighting with the same strength and intensity, then a straight line can form where the two halos overlap, something

like the shape created when two soap bubbles touch. This shape of compromise could be reorganized in real time but could also react much more slowly. As a result, a steady line of separation between the users would appear, a more standardized organization with average volumes. If Barba learns from the conflicts between its inhabitants, it can offer optimized compromises between the different users' desires. New spatial qualities and architectural typologies could emerge from such a scenario.

The 'Negotiator' is an interactive installation based on the principle of metaballs. It uses three-dimensional Voronoi diagrams – which separate a plane or a volume populated by a collection of points into cells or regions – to define the limit between two spaces. These regions are defined by the proximity of the different points. The regions represent the zones of influence of each point in relation to the ensemble of all points.
As in the 'Halo Maker', spheres are attached to each of the skeleton's nodes. The user's movements control the radius of each sphere, and the limit separating personal space is defined by the sum of the zones of influence of each point in relation to the ensemble.

The installation use Kinect, Rhinoceros, Grasshopper, Firefly and Weaverbird.

Space Moulder

Sculpting in Time

Arslane Benamar, Paolo Birrer, Tania Carl, Melina Cerfeda, Stavros Gargaretas, Ulf Hackauf, Lex te Loo, Sander Mulders, Sebastian Oswald, Kai Peter, Adrien Ravon, Henry Stehli, Kaegan Walsh, Stefan Wülser

With the previous tools, the Halo always forms an offset of the user. You can increase or shrink your Halo, you can use it to push another user's space to the side or you yourself can be pushed, but as yet you can't touch it. The 'Space Moulder' therefore introduces the possibility of overriding the Halo's reaction. The main override function allows you to freeze a part of the Halo, put its responsiveness on hold and keep it from retreating. This enables you to touch the Halo or even to push through it. You can also use this to consciously adjust the shape of your Halo and to mould your personal space.

The 'Space Moulder' is an immersive installation in which you can interact with, sculpt, freeze and deform your Halo. In this installation the user becomes a model maker and can sculpt her space in real time. She is free to create free forms and shapes with her hands, arms and her whole body.

In the Barba environment, this tool would allow the user to take control of the space around her, to change its appearance and turn the fluid body-based shapes into a sculpted, controlled space. A user could create objects from the material, and she could use the 'Space Moulder' to create furniture.

Today, the 'Space Moulder' can be used to shape virtual, three-dimensional shapes and spaces based on the individual human body.

Watch the movie:
www.thewhyfactory.com/barba

Head
Hand_Right
ulder_Right
Shoulder_Centre
Wrist_Right
Elbow_Right
Shoulder_Left
Elbow_Left
Hip_Centre
Hip_Right
Hip_Left
Wrist_Left
Hand_Left
Knee_Left
Knee_Right
Ankle_Right
Ankle_Left
oot_Right
Foot_Left

The installation allows you to override your Halo and to freeze or extend some parts by simply moving your hand. You can thereby define the intensity and precision of interaction with your own Halo. Several parameters, such as elasticity, response time and inertia define the Halo's behaviour. Based on gestures, movements and time, it allows its users to experience wholly different ways of interacting with the environment.
The base for this is the Halo developed and described in the previous experiments. The envelope, created from virtual skeleton and metaballs in the 'Halo Maker', is 'frozen'. It becomes a static, three-dimensional mesh that no longer moves with the individual body.
In this new mode, you could begin to consciously transform the virtual Halo. By moving your hands, you start sculpting your space.
Depending on the position of your hand and the distance from your body, this allows you to push or pull parts of the Halo. One hand becomes a sculpting tool, like a large virtual brush. The other hand can control the size of this brush, allowing you to make large changes quickly or to work more slowly or with more precision.
Time is added as a parameter that defines both the material's speed of reaction and the influence of your brush on it. The longer you hold the brush at one point, the larger its effect. Once again – and ever more – your body creates the space that surrounds you.

The installation is based on the principle of attraction and repulsion. The Halo created with the 'Halo Maker' is frozen or 'baked' in Grasshopper, creating an interconnected mesh composed by a series of points. The hand of the user represents a charge point from which a series of vectors depart until they touch a collection of points on the envelope.
The points closest to the hand are either attracted or repulsed. By changing the position of the other hand in height, the user can increase, decrease or invert the attraction or repulsion effect of his hand. Depending on the speed of her gestures, the user can interact more or less precisely with the envelope. A time count calculates how many points the user is selecting per second. The points selected increase their value along their vector each time they are selected.
The installation uses one Kinect for the skeleton tracking, Rhinoceros, Grasshopper, Firefly and Weaverbird.

Shape Library

Download and Save

Arslane Benamar, Paolo Birrer, Tania Carl, Melina Cerfeda, Stavros Gargaretas, Ulf Hackauf, Lex te Loo, Sander Mulders, Sebastian Oswald, Kai Peter, Adrien Ravon, Henry Stehli, Kaegan Walsh, Stefan Wülser

The 'Space Moulder' can be used to create objects and furniture. You can use it to push an indentation into a wall in Barba in order to then sit on it. You can use it to pull a hill from the floor, flatten it and lie on it. And, with more effort, you can carefully shape a whole table, complete with chairs. But to create all the objects you want in your room bit by bit is a lot of work. The 'Shape Library' therefore introduces the function 'store and upload shapes'. You would save a shape you created to a library and then would be able to upload shapes from your library so that they would instantly appear as part of your fully adaptable environment. These could be both shapes created before and shapes derived elsewhere or borrowed from someone else. Once activated, a shape can be modified by changing its scale, location or orientation. You can even stretch or bend shapes.

Say you want to sit and read a book. You could flip through the different types of armchairs included in the library. You could select one and move it to the place you wanted it. And then you could turn it, so that it would face the window. You might want to sit really comfortably, so you could recline more deeply, and the chair would recline with you. And then a friend might join you. You could pull at one side of the armchair and it would become a sofa for two. You could even change its surface. By altering the way it resists your body weight, the chair could gradually change from a rock-hard bench to a soft cushion into which you sink with your whole body. And you could program the behaviour of the furniture to sync with your activities and needs. Like: when you've nearly overslept, it could give you a little push and gently kick you out of bed.

The 'Furniture Maker' is an immersive installation with Kinect that allows for the placement, shaping, orientation, deformation and customization of virtual objects and furniture. In the Barba environment, it would be used to store, upload, exchange and modify complex shapes, furniture or complete interiors. Today, it can add functionality to the uses of the 'Space Moulder' and make it a tool for digital design and fabrication.

The 'Space Library' merges the freedom of user customization with the advantages of refined ergonomic design. The hidden intelligence and precise proportions of designed objects can be manipulated and adjusted to the specific needs of the user. Moreover, it allows fluent transitions from one object to another. Changing one parameter can smoothly deform a shower into a bathtub and then to a swimming pool. The shapes created are not merely translations of the body into a soft envelope, but rather possess precise parametric and algorithmic properties.
With the 'Space Library', you could interact with furniture in several ways.First, you can upload predefined furniture. You can select

To be able to select objects from the library, we added a display that will appear in the virtual space in front of the user. Programmed gestures allow the user to select different types of furniture from the objects projected on the display. In this installation, the user can activate a menu by raising her hand in front of her head. A menu then appears on a projected plane.
The orientation plane for the menu display is defined by the position of the user's head and shoulders. With her other hand, she can then flip through and select menu items. Once the item is selected, the user can place the item depending on her position in space and can specify its orientation by moving her hand around to define a directional vector between hand and body. The user could later specify its scale by increasing or reducing the diagonal distance between her two hands.

The installation uses Kinect for the skeleton tracking, Rhinoceros, Grasshopper, Firefly and Weaverbird.

from a large collection of predefined models. Libraries of three-dimensional models offer a large variety of (simplified) design objects: from Eames classic seating to recent homemade designs. You can upload such furniture and then position and orient it in a virtual space with defined gestures of your arms. Once the furniture is in position, you can adjust its shape. With gestures, you can scale it and stretch or squeeze it along different axes. Eventually, you use the functions of the 'Space Moulder' to modify the furniture. You can thereby mould and reshape the furniture with great precision. An infinite number of new designs can emerge, designs that are ever more ergonomic, more personalized, more efficient.

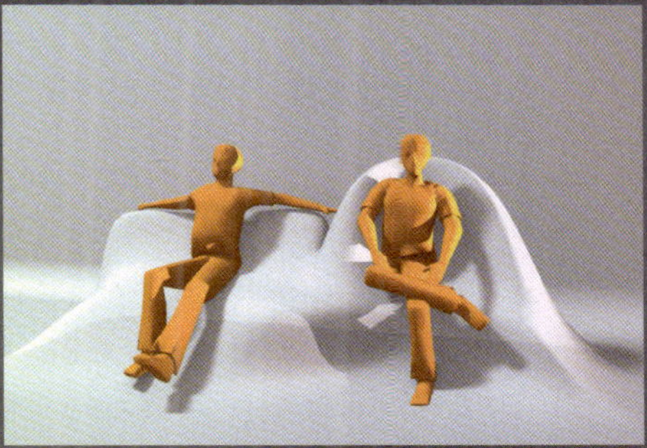

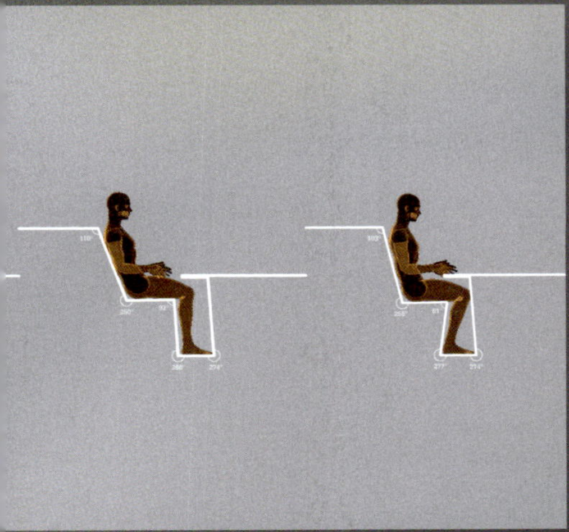

Window Maker

Open Up!

Arslane Benamar, Paolo Birrer, Tania Carl, Melina Cerfeda, Stavros Gargaretas, Ulf Hackauf, Léone Kündig, Lex te Loo, Sander Mulders, Sebastian Oswald, Kai Peter, Adrien Ravon, Enrique Ruiz, Henry Stehli, Kaegan Walsh, Stefan Wülser

We initially assumed that the user would be inside the fully adaptable environment, surrounded and sheltered by the Barba material. But a world would still exist outside one's personal space. We need openings: for light, to connect to other spaces and to step outside. How can we create an opening in Barba?

In the Barba environment, windows and doors are neither fixed nor standardized. You can create openings where and how you want them. You can draw openings, scale them and add patterns to them. Windows can be programmed to follow you and to change size according to the available daylight in order to provide you with just the right amount of light.

In the Barba environment, the 'Window Maker' would allow you to create different types of openings in the material. Today, the installation provides you with more tools for interaction in the virtual space.

Hand_Right
132 cm
Hand_Left

If you gaze in one direction, the View Halo would extend your space. But you might not want to extend the space, but rather open it up and look outside. There are several ways to do this. You can draw an area with your hands on the envelope surrounding you, and this drawn area would then become an opening. You can use your hands to define two opposite points of a rectangle, which would then become a window. Or you can have one hand define the centre of a circular opening and the other define the size of this round window. Once the opening is there, you can move it around by pointing at it.

The distance between your body and the surface on which the window appears could define the size of the opening. The further you move away from the window, the larger it becomes, bringing light deeper into your space. The window could also move around according to the position of the sun and the desired amount of sunlight.

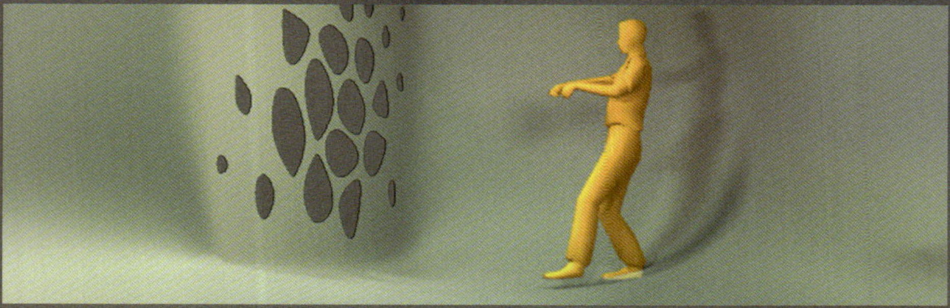

You could also upload a predefined pattern like a louvered opening, for example, which would modify the intensity of light. You could gradually turn a window with louvers into several small windows and vice versa. The overall transparency of the surface could thereby change according to the brightness and intensity of the sunlight. Windows could also appear automatically and be predefined based on your activity.

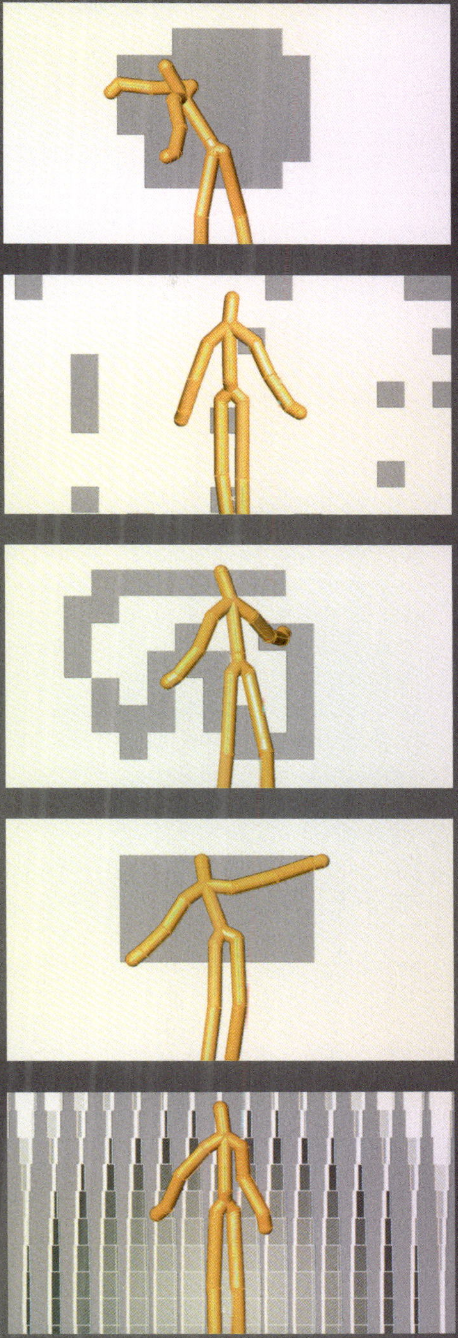

To be able to select objects from the library, we added a display that will appear in the virtual space in front of the user. Programmed gestures allow the user to select different types of furniture from the objects projected on the display. In this installation, the user can activate a menu by raising her hand in front of her head.
A menu then appears on a projected plane. The orientation plane for the menu display is defined by the position of the user's head and shoulders. With her other hand, she can then flip through and select menu items. Once the item is selected, the user can place the item depending on her position in space and can specify its orientation by moving her hand around to define a directional vector between hand and body. The user could later specify its scale by increasing or reducing the diagonal distance between her two hands.

The installation uses Kinect for the skeleton tracking, Rhinoceros, Grasshopper, Firefly and Weaverbird.

Packer

Densify!

Riccardo Caruso, Elena Chestnova, Nicholas Ganz, Ulf Hackauf, Marcel Hodel, Léone Kündig, Sander Mulders, Marisa Muscionico, Adrien Ravon, Pascal Ruckstuhl, Enrique Ruiz, Paula Schilliger, Vera Schmidt, Ying Yi Tan, Kaegan Walsh, Yangzom Wujohktsang, Stefan Wülser

As the first chapters of this book suggest, the Barba environment could both provide a large freedom in shaping domestic and personal space and, at the same time, allow for high-density living. We could live in bigger and more beautiful spaces, close to others, on a smaller footprint. To make this possible, Barba would need to constantly reorganize the spaces it creates to be as dense as possible. While individual spaces change shape, size and connections, the collective arrangement of spaces is reconfigured to provide maximum density. To simulate this, we developed the 'Packer'.

The 'Packer' is an interactive, virtual three-dimensional model able to simulate the process of self-organization of non-regular solids. A predefined set of volumes can be packed according to different parameters. The user can change these parameters and watch in real time as the volumes rearrange themselves to an optimized configuration.

In the Barba environment, the 'Packer' would rearrange different spaces in relation to each other, optimizing proximity, distance and density. Today, the tool can be used for the optimization of space use in different scales, or to visualize abstract relations, for example of social aspects or environmental effects.

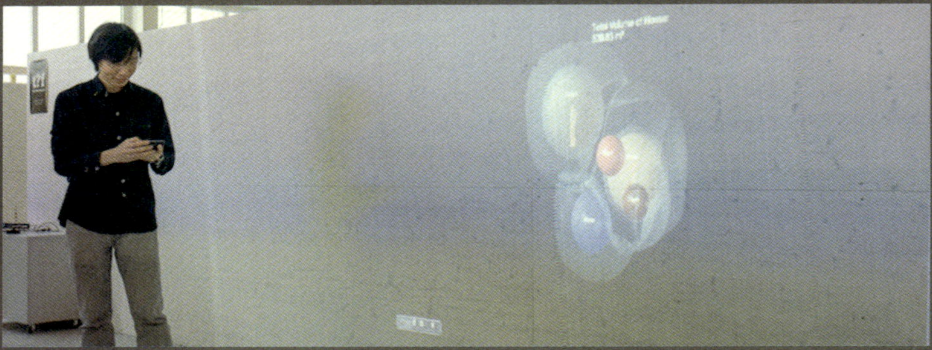

The 'Packer' is an interactive three-dimensional model. It represents inhabitants with different activities, requirements and desires as simplified volumes with parameters. The volumes are set in a virtual space. Unlike the previous tools, this virtual space is not directly linked to the user of the installation and his body. To engage with the model and manipulate it, we use smart phones as interfaces. The user surface of the smart phone allows the user to change parameters and packing strategies and to observe the result in real time on a projection. The user can thereby assign different connections between units. Users can also interact with and change the shape and volume of each unit. The 'Packer' tries to optimize the configuration of a collection of non-regular units. The model is based on particles' attraction and an irregular sphere-packing algorithm.
The installation has been developed using Grasshopper, Kangaroo plug-in and touch OSC. Users can test different modes of organization. The installation uses a smart phone as input device, touchOSc as the program running on the smart phone and Rhinoceros, Grasshopper, Firefly and Kangaroo for the virtual model.[1]

1 touchOSc is a modular control surface that runs on smart phones and tablets. <www.hexler.net/software/touchosc>. Kangaroo is a plug-in running on Grasshopper for interactive simulation, optimization and form-finding. <www.grasshopper3d.com/group/kangaroo>.

With this study, we no longer use one or two users and their bodies as a starting point. We now look at more users, groups of individuals and how their spaces can be accommodated in relation to one another.
The 'Packer' combines the spatial densification of a set of units with the optimization of the relations of these units towards each other. The relations can include, for example, social aspects (as in 'I want to live closer to you than to others'), hindering aspects ('if you want to play the drums, please move away from us') or environmental aspects ('during the day we all want to have a room with a view but at night we can be closer to each other'). These aspects are translated into spatial functions, such as attraction, repulsion, gravity or the formation of clusters. To optimize densification, a spatial limitation can be added into which all units have to fit at all times. We define this as a 'bounding box'. The different aspects of optimized relations are simulated in an interactive tool. Simplified

volumes hereby represent inhabitants of the fully adaptable environment and their spaces. Unlike in previous studies, users will not 'step into' this tool. Instead, they manipulate the requirements of the different inhabitants as well as the packing strategy, and are able to observe the results in real time. Different strategies are tested:
Shortest distance: The first option is to compact the units but give equal importance to the mutual connections between them. Everyone is connected to everyone. The distance between all units is minimized. All units tend to unite around a central point.
Shifting gravity: Another option is to explore the self-organization of the cells based on gravity. The cells stack one upon the other based on which is closest to the point of gravity. This point can be moved in space.
Arrangement by attraction and repulsion: Selected units can attract each other. This leads to arrangements that can optimize social or other functional relations. The opposite situation can also be introduced: units that repulse each other. This can, for example, represent noise or other hindrances. Both arrangements can be combined with the central arrangement or the shifted gravity, leading to a complex, optimized configuration.
Clustering: When the number of units becomes larger, stable clusters can appear in the dynamic environment. The formation of such clusters can be enforced to have less overall movement of individual units – to create an overall 'calmness' within the configuration. The clusters can appear according to activities (working or eating), environmental effects (noise versus quiet) or social groups (friends or families).

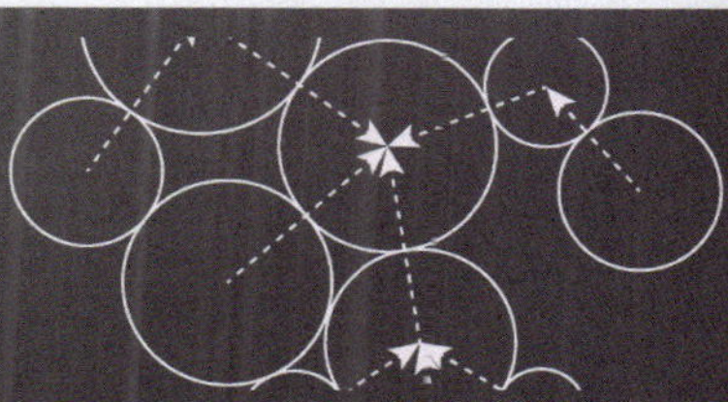

Social Network

Centred

Distributed

View Maker

Adapt to Environment

Elena Chestnova, Nicholas Ganz, Ulf Hackauf, Léone Kündig, Adrien Ravon, Enrique Ruiz, Paula Schilliger, Vera Schmidt, Kaegan Walsh, Stefan Wülser

The 'Packer' would allow individual units of the fully adaptable environment to arrange themselves as a dense yet functional Barba neighbourhood or village. But how can we make sure that every inhabitant gets the natural light he or she desires? And how could every inhabitant have a view? And how can sunlight and view follow the inhabitants in a neighbourhood capable of dynamic reconfiguration? The 'View Maker' adds the parameters of view and light to the previous study. The model is a solver, able to reconfigure the units of the village to determine their optimum organization, while taking density and other relations from the 'Packer' into account.
In the Barba environment, the 'View Maker' would provide light and views and thus contribute to the environment's liveability.
Today, it is a tool that can be used for light analysis and directional conflict resolution at a number of different scales, and regarding view or building orientations.

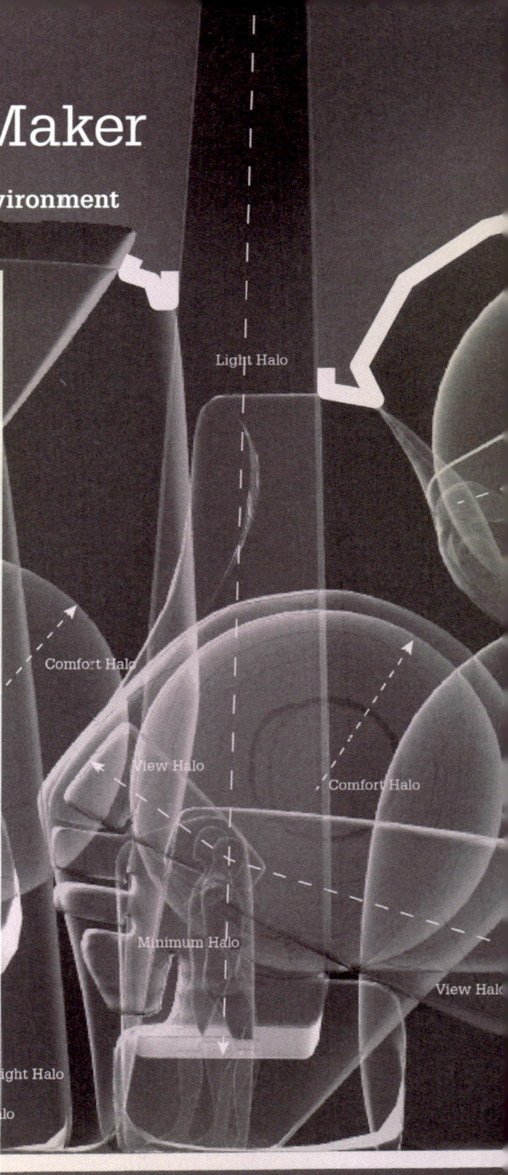

ort Halo
View Halo
alo
Light Halo
Minimum Halo
Comfort Halo
View Halo

The 'View Maker' model is based on the principles of problem-solving theory. In this model, each view cone can be specified by a vector direction. As long as there is no intersection with another view cone or another cell, no conflict appears.
In order to determine each optimum situation, the model uses Galapagos to calculate and test different possibilities of agency with the goal of avoiding any intersection. The model is also based on isovist, exposure analysis to calculate views and direct sunlight. [1]
The installation uses Rhinoceros, Grasshopper, Kangaroo, Galapagos and Weaverbird. [2]

1 'An isovist is the set of all points visible from a given vantage point in space and with respect to an environment'. See M.L. Benedikt, 'To Take Hold of Space: Isovists and Isovist Fields', Environment and Planning B 6(1) 1979, 47–65.
2 Galapagos is an evolutionary solver plug-in for Grasshopper. <http://www.grasshopper3d.com/group/galapagos>.

In the 'View Maker', units are moved and reorganized to determine an optimum configuration in which each inhabitant receives as much light as possible during the day. In this model, we calculate the optimized position regarding the maximum sun exposure for each unit per hour. Units are moved in space to find a place with sufficient sun exposure. The units can also rotate to turn dedicated surfaces towards the sun. The 'sun' is thereby represented as a magnetic force that attracts dedicated surfaces. As the exposure angle changes, the units reshuffle to receive optimal light and still remain as dense as possible. Views are added to the models as

cones. A view can be qualified as a cone that can extend along all other units of the configuration towards an imaginary 'outside' or landscape. A view can be defined by its length, such that you look at other units but keep enough space between yourself and the next obstruction. We might develop more advanced requirements regarding what qualifies as a 'nice view'.
The optimization of sunlight distribution and views may contradict the actions of the 'Packer', as some units may have to disengage from the village to receive sunlight. The two makers must constantly negotiate to find a compromise.
The 'View Maker' can compare several strategies for solving view and sunlight requirements. The units could be pushed away and constantly reshuffle their position in space. They could shrink and deform in order to let sunlight pass through them. The whole cluster of units could orient themselves towards the sun. Each strategy has its effect on how calm and stable or dynamic and hectic the neighbourhood appears to behave.

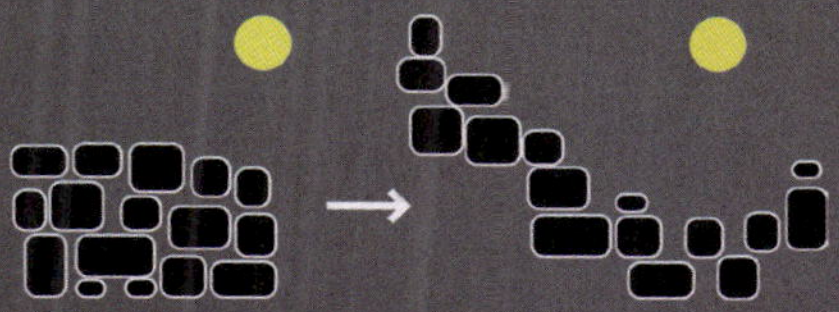

Reshuffle

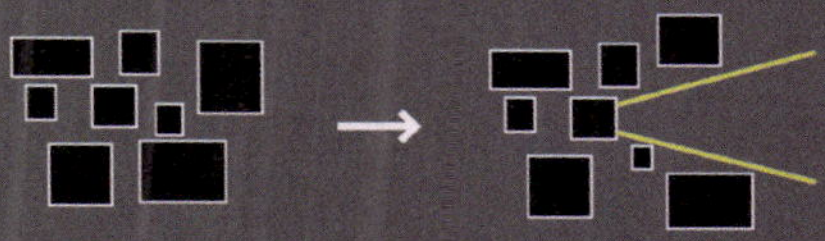

Push

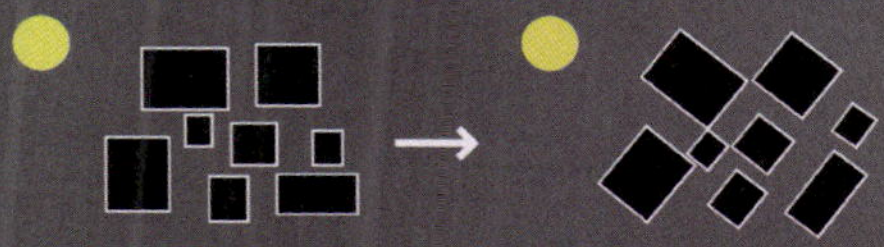

Orient

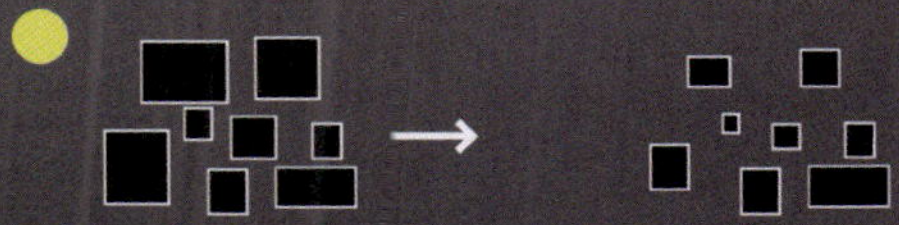

Shrink

Stabilizer

Make It Stand!

Elena Chestnova, Nicholas Ganz, Ulf Hackauf, Léone Kündig, Adrien Ravon, Enrique Ruiz, Paula Schilliger, Vera Schmidt, Kaegan Walsh, Stefan Wülser

So far, we allowed units to freely arrange themselves in space to optimize their spatial relations, density, views and sunlight. But if the Barba environment were to become real, the units could not fly in space. They would need to be supported. To maintain dynamic flexibility would require an equally flexible load-bearing structure. The 'Stabilizer' allows us to test different models for this.

In this chapter, we developed an interactive three-dimensional model to test different ways of stabilizing the fully adaptable cluster of units and explore the implication on its organization.
In the Barba environment, the 'Stabilizer' would be necessary to support the different units without compromising the dynamics of the structure. Today, the tool can be used to inspire load-bearing structures for configurations on different scales, from buildings to furniture or smaller objects.

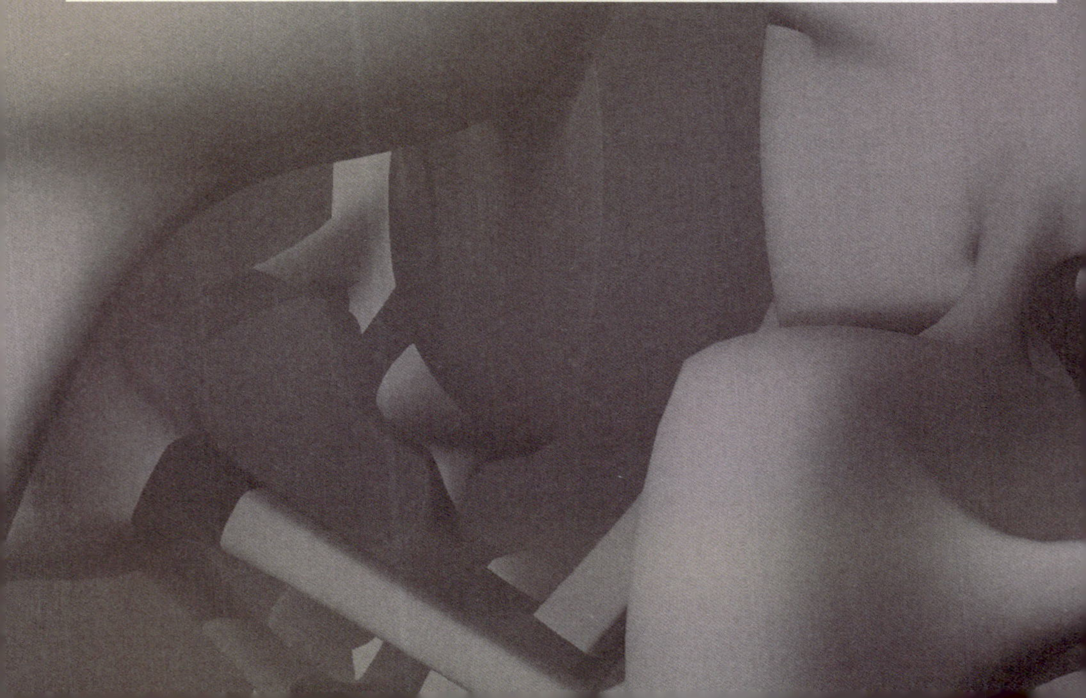

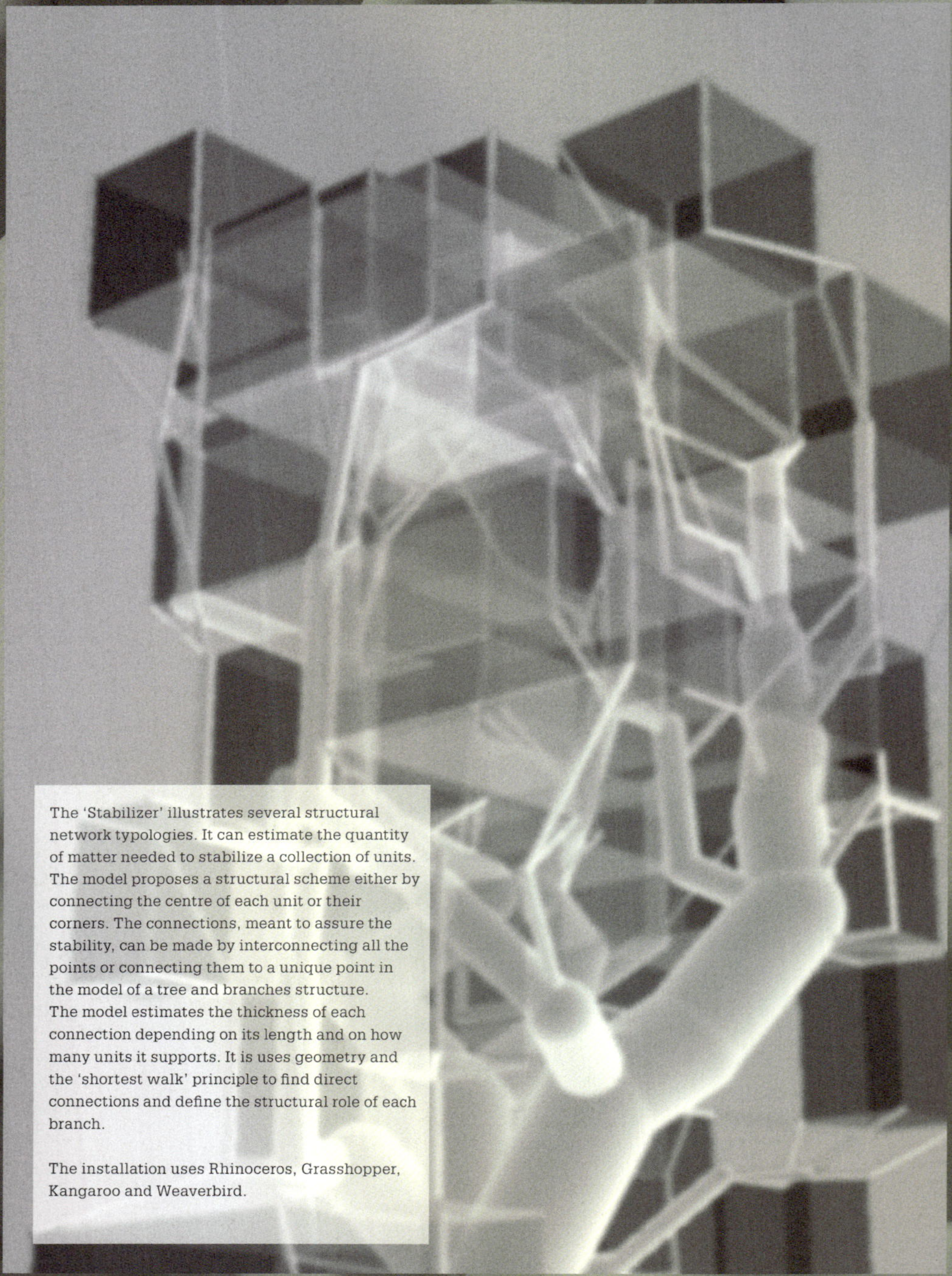

The 'Stabilizer' illustrates several structural network typologies. It can estimate the quantity of matter needed to stabilize a collection of units. The model proposes a structural scheme either by connecting the centre of each unit or their corners. The connections, meant to assure the stability, can be made by interconnecting all the points or connecting them to a unique point in the model of a tree and branches structure. The model estimates the thickness of each connection depending on its length and on how many units it supports. It is uses geometry and the 'shortest walk' principle to find direct connections and define the structural role of each branch.

The installation uses Rhinoceros, Grasshopper, Kangaroo and Weaverbird.

The 'Packer' tests different strategies for adding a supportive, potentially load-bearing structure to a configuration of units. It can thereby build upon the results from the 'Packer' and the 'Illuminator' and illustrate how these could become physical with structural integrity.
The first strategy is 'stacking'. The structure would be incorporated in the units themselves. Units would stand on each other and would not need extra supports. By changing thickness, the material would assume the function of walls and floors to hold the units together.
'Supporting' would mean that columns are added incidentally and as needed. These would change position and size to negotiate between structural requirements and views or sunlight.
'The Tree' would be a model in which structural reinforcement would appear as veins on the surface of the units, connecting to a larger, supporting trunk if necessary.
'The Truss' would be based on a system in which the vertices of the units would all be connected to a structure network. This would mean that every single unit would be separately connected to the whole structure system, which would generate the connection to the ground and the stability of the village as a conglomerate of different elements.

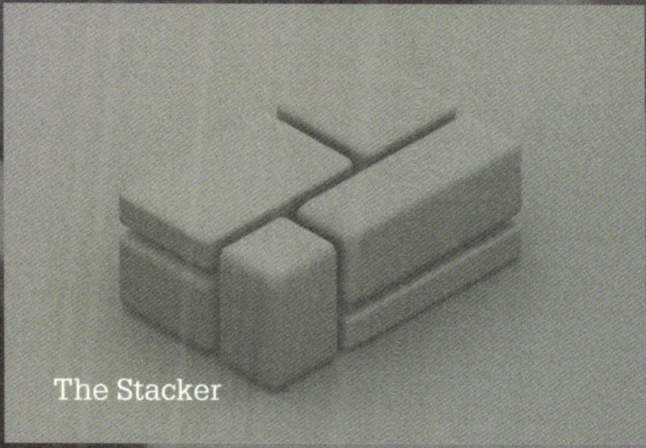

The Stacker

The Supporter

The Tree

The Truss

Vein Maker

Infrastructure On Demand

Elena Chestnova, Nicholas Ganz, Ulf Hackauf, Léone Kündig, Adrien Ravon, Enrique Ruiz, Paula Schilliger, Vera Schmidt, Kaegan Walsh, Stefan Wülser

To make Barba a functional living environment, we need an infrastructure that might bring water, power – and possibly goods – to the individual units. It goes without saying that static pipes and canals would not be sufficient in a dynamic environment. We would need an equally flexible system to replace pipes, wires and even stairs or elevators. We imagine this as a system of veins of different sizes. Small veins could transport water. Larger veins could transport waste and sewage. Even larger veins could be corridors, staircases or elevators. Veins filled with water could transport electricity. This network of flexible veins could change position and diameter to adapt to the different activities of the inhabitants of the fully adaptable environment. The 'Vein Maker' is a tool meant to simulate this system of adaptable veins. It explores which strategies can make the transportation of water, waste, goods and people efficient. In the Barba environment, the 'Vein Maker' would take care of the described networks. Today, it is a tool for simulating distribution and logistics on different scales.

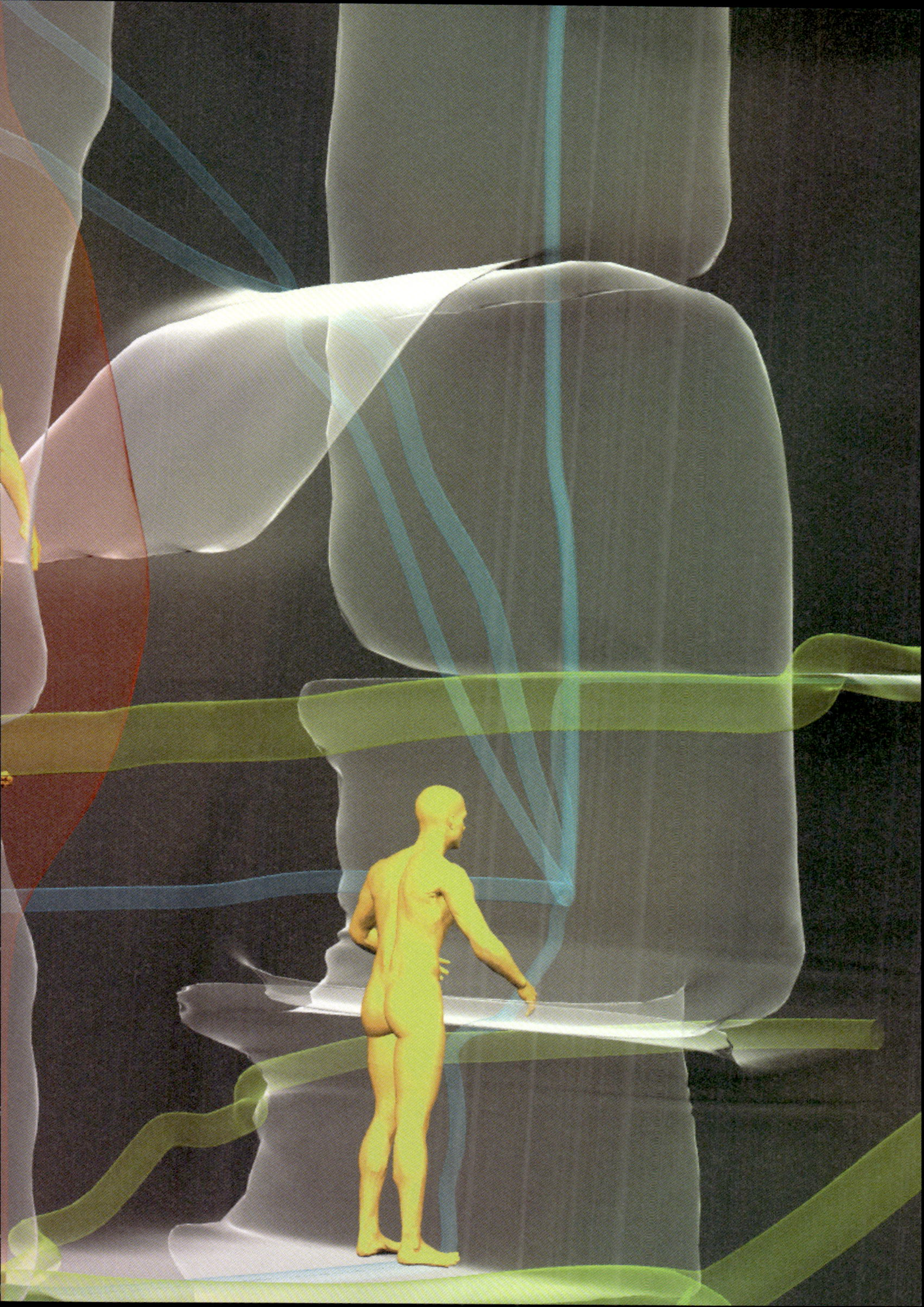

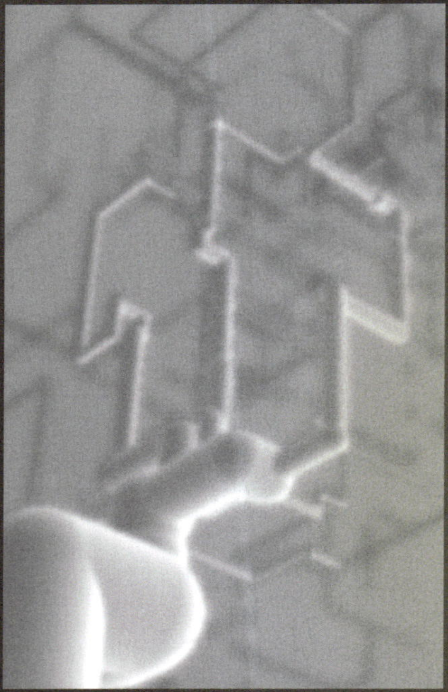

The 'Vein Maker' would be an adaptable circulatory system, transporting air, water, waste, goods and people. The veins would be flexible in thickness and length. They would constantly optimize the organization of flows, reconfiguring themselves in order to find the shortest path from one point to the other without creating any conflict. Veins could appear in the material. The material could create cavities, which would become pipes of different sizes to transport water, people and goods. Veins would constantly move, appear and disappear as needed. They could extend until reaching their destination point. They could change route, or take some detour if something were blocking them.

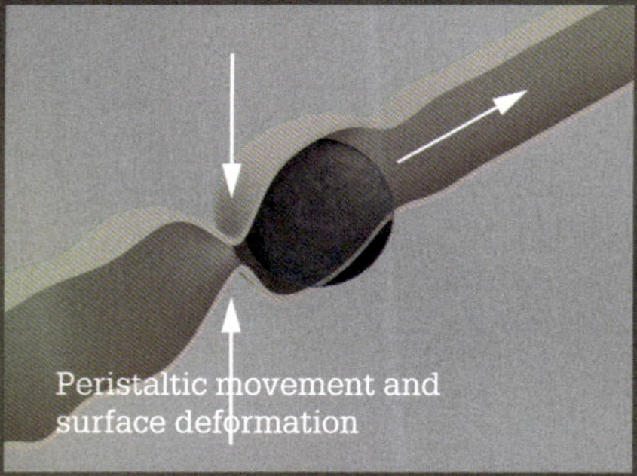

Peristaltic movement and surface deformation

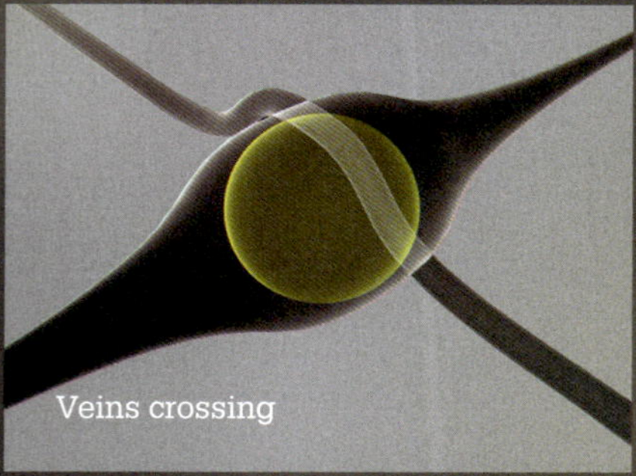

Veins crossing

Water would be stored in equal distribution between units. At given times, the collectors could expand in volume in order to hold enough water in stock for a bath or for cooking. Material flexibility could create peristaltic force to move objects and people through veins. By extending and compressing, it could pump whatever it is transporting. Alternatively, it could produce small ripples on its inner surface, gently moving objects or people along as it did.

The 'Vein Maker' is based on the geometric principle of the shortest walk. The shortest walk is a Grasshopper plug-in that enables the calculation of the shortest connection between two points in a network of possible paths.
In this model we tested the efficiency of several network topologies, from regular grid to Delaunay networks. The model allows us to test different configurations and detours by inserting rigid elements.
The model also explores the potential of merging between curves under a specific threshold. It studies the overlapping and crossing issues between multidirectional curves and the deformation of pipes.

The installation uses Rhinoceros, Grasshopper, Kangaroo and Weaverbird.

Distribution network

'The Makers' show how Barba might interact and become intelligent. But Barba material and Barba space remain hypothetical. How could it become real? We need a material that cannot only create different shapes, but that can interact by sensing and reacting to its environment. It needs to be able to change from one shape to the other in minutes and it needs to be rigid enough to carry the weight of several persons at the same time. How to make such a thing? Several recent developments could lead to adaptable, programmable materials and could eventually make our vision a reality.

Material-ization

The Why Factory: Ulf Hackauf

Dynamic Building Parts

The hypothetical Barba material needed for the fully adaptable environment requires three qualities: it needs to be able to sense the environment, react to what it senses and manifest this reaction through a physical change within a load-bearing material. We began our search for such a material or structure within the framework of today's existing architectural technologies.

Movable Walls

Adaptable, automated building parts are not new. The Rietveld Schröder House is one early, and iconic, example of adaptive architecture.[1] Built in 1924, it features movable walls that allowed the inhabitants to combine separate bedrooms into one large living room. While the building itself does not meet the requirements we've set for our adaptable, programmable material – the house's flexibility is limited to the four rails of the moving walls, which have to be moved by hand – we imagine this building as a predecessor to our fully adaptable environment.

Sensing Buildings

Since 1924, sensor technologies have been introduced in architecture, allowing spaces and materials to react to human needs and desires. The two main areas of application

Rietveld Schröder House, Utrecht

Jean Nouvel, Institute du Monde Arab

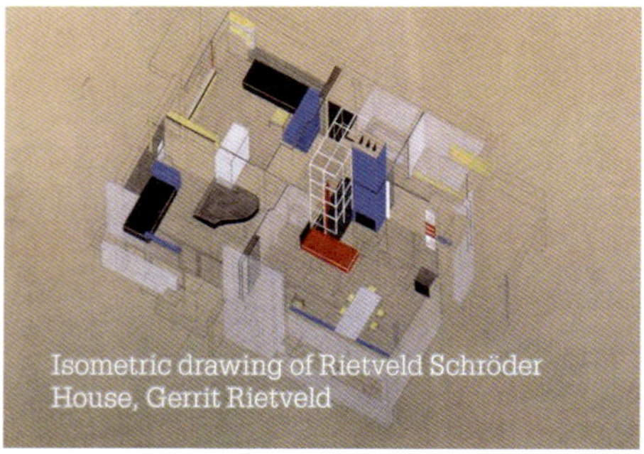
Isometric drawing of Rietveld Schröder House, Gerrit Rietveld

Jean Nouvel, Institute du Monde Arab

are typically comfort and safety. In regards to comfort: sensors can be used to control light, air, temperature or access. Motion sensors are used to turn lights on and off; proximity sensors can open and close sliding doors; and sensors can monitor air quality and adjust a room's air circulation system accordingly. Jean Nouvel's Institute du Monde d'Arab, in Paris, is a beautiful example in which light sensors modulate the entire façade's transparency depending on sunlight. [2] Hidden – and usually less oriented towards beauty – safety sensors make up the other significant group of sensors employed in contemporary architecture. They're typically used to detect smoke, fire or burglars.

Muscle Tower

Sensors and movable building parts already exist in contemporary architecture. But we aspire to more than changes in light, temperature and air quality and to more than walls or doors that move along single axes. Our Barba material will need to be more vital, more changeable, will need to move in multiple directions.

Muscle Tower, TU Delft

We found examples of developments towards these visions in different research groups. One such research team – the Hyperbody group – is a neighbour at Delft University of Technology. The team has been at working on complex transformations of load-bearing structures for years. Its Muscle Tower project demonstrates how a simple load-bearing framework can start to bend and transform with the use of actuators – in this case artificial muscles that can extend and shorten members of the framework. [3] These developments suggest that we may already have the ingredients for our fully adaptable environment at hand. We would need to shrink the actuators while keeping enough strength to carry the weight of a person. And we would need to build the structure in such a way that a flat wall could support openings at one moment and transform into a chair at another.

Automated Assembly

Instead of changing a structure from within, we could also manipulate it from the outside. Think of a chair: to move it automatically, we could add motors to its wheels and program it to roll towards us. But we could also program a small robot to push the chair to where we wanted it. This would allow for more flexibility in the actual material: the fully adaptable environment could be made from traditional bricks, from wooden cubes or even LEGO blocks, if we could find a tool to move these elements quickly, creating different arrangements and configurations in real time.

Flight-assembled Architecture

Flight-assembled Architecture, ETH Zürich

We found the first steps in this direction in the Flight-assembled Architecture project, presented in 2011 at the ETH Zürich following a collaboration between the chairs of Gramazio & Kohler and Raffaello D'Andrea. [4] In this project, drones were programmed to collect bricks and build a tower from them. Tireless swarms of drones flew back and forth, placing layer after layer of polystyrene blocks and eventually finishing a 6-m-high tower!

Aerial Construction

In a later project, completed in 2013, the same team presented their Aerial Construction project, in which drones were equipped with a coil of thread. [5] The drones release thread from the coil while flying around. The team's experiments show how a

Aerial Construction, ETH Zürich

drone can span a wrap and fasten one end of the thread to a pole and then span the thread between this and a second pole. Another drone was programmed to wrap thread around two parallel ropes, creating a loosely woven surface. Eventually, two drones flying around each other were able to weave shapes in space.

In these examples, drones act like small flying builders and are able to create structures. It would also be possible to destroy these structures in the same way. If we imagine this happening at high speed, we could imagine the structures becoming animated; how a wall could be built, unbuilt and rebuilt further. Done with the painstaking patience of a stop-motion movie, the wall could begin to move. Working with threads and woven nets could be a quicker and more elegant way; the drones would appear like sailors on a ship who keep adjusting the sails by pulling ropes. If we imagine transformable rooms made from suspended textile planes, the drones could pull and hold different ropes to adjust the room to a user's individual needs. We could see a solution for the fully transformable environment here. But this strategy would include having a buzzing swarm of flying drones around you at all times. It might be more elegant – and more calm – if the material itself would consist of small activated elements. In other words, if the material itself could be composed of small robots.

Modular Robotic

Claytronics

Claytronics, Carnegie Mellon University

Claytronics, Carnegie Mellon University

This magic material is composed of many small round balls that can be programmed to connect and disconnect. In this way, the material can change colour and create different shapes. In 2005, a research team led by Seth Copen Goldstein at Carnegie Mellon University in Pittsburgh presented this vision in a movie about their Claytronics project. [6] The material illustrated in this movie seems exactly the thing with which to create our fully adaptable environment. Thus far, Claytronics have not been built, or at least not on the scale presented in the movie. But the team at Carnegie Mellon is working on making this vision a reality. They have produced a number of larger prototypes and have simulated the required behaviour in several studies.

Smart Sand

Smart Sand, Distributed Robotics Lab MIT

Smart Sand, Distributed Robotics Lab MIT

The Distributed Robotics Lab – led by Daniela Rus at the Massachusetts Institute of Technology (MIT) – is at work on a similar concept. In their 'Smart Sand' project, they have managed to manufacture small cubes of 10 mm edge length that can connect and disconnect. [7] These cubes are programmed to execute different tasks. One experiment shows how they spread out a number of the cubes on a vibrating surface. They placed a simple, pixelated figure between the cubes, and showed how the cubes were able to sense the shape and convey that information to their neighbours. Following the script, some cubes started connecting to each other, while others remained loose. The result was a replica of the fixed shape, recreated by the programmed cubes. Daniela Rus's team aims next to shrink the cubes further, until they form the grains of this new kind of 'Smart Sand'. They imagine that you could fill a bag with thousands of grains of this material and that if you put an object inside the sand could sense and replicate that object, to scale or even larger or smaller.

M-Blocks

M-Blocks, Distributed Robotics Lab MIT

M-Blocks, Distributed Robotics Lab MIT

While Smart Sand passively reacts to given objects, M-Blocks, developed by the same research team at MIT, are more proactive. [8] The cubes resemble those used in the Smart Sand project. But in addition to the connectors that allow them to attach themselves to other cubes, the M-Blocks have motors inside that allow them to move. There are no moving parts on the outside but the inside contains a flywheel that can spin at 20,000 revolutions per minute. Stopping the flywheel abruptly creates a momentum that flips the cube to one side. This simple technology allows the M-Block to move around, to climb on other M-Blocks and even to create small towers.

These are only three projects from within the large field of distributed robotics. And at this point, all material examples remain much larger than the term 'sand' would suggest. But if these mini-bots keep shrinking, and if their ability to connect to one another remains strong, we can imagine that these prototypes of programmable matter could eventually form a material that would make a fully adaptable environment real.

Molecular Manipulation

With these developments in modular robotics, we may have found the means to create the material we need. But we wonder if we might go further still to achieve a more integrated solution. Could we work with much smaller components, components smaller, for example, than a grain of sand? This leads us to nanotechnology, a field in which we could manipulate matter on a molecular or even smaller scale.

Molecular Assembler

In his 1986 book *Engines of Creation: The Coming Era of Nanotechnology*, K. Eric Drexler describes the 'idea of a nanoscale 'assembler', which would be able to build and copy itself or other items of arbitrary complexity with atomic control'. [9] Since then, nanotechnology has found its way into many products in daily use in the form of nanomaterials. These engineered materials are used in cosmetics, in paint, in clothing and in bandages. They are 'passive' materials that have specific qualities but are far from the vision of small active machines that can create objects. The man-made nanoscale assembler, which Drexler described, thus far only exists in science fiction. Whether the concept could ever become real has been questioned. But Drexler's visions have inspired many speculative projects.

Utility Fog

Researcher and author Josh Storrs Hall further expanded the concept of self-replicating robots with his idea of 'Utility Fog'. [10] He proposed large numbers of 'foglets', 'tiny robots linked with their arms together to form a solid mass in the shape of the object you want'. The robots would be abundant and would fill the surrounding space. They would be able to simulate the properties of air. If you moved your arm through this fog of robots, you would not be able to feel them unless they locked their little arms and became an object or a wall. As Stoors Hall describes:

If you filled your house in with Fog this way, furniture no longer need be extruded from the floor; it can appear instantly as a pattern formed

out of the 'air' robots. Non-Fog objects can float around at will the way you did in your 'car'. But what's more, your surroundings can take on the appearance, and feel, of any other environment they can communicate with. Say you want to visit a friend; you both set your houses to an identical pattern. Then a Fog replica of him appears in your house, and one of you appears in his. The 'air' fog around you can measure your actions so your simulacrum copies them exactly. The pattern you both set your houses to could be anything, including a computer-generated illusion. In this way, Utility Fog can act as a transparent interface between 'cyberspace' and physical reality. [11]

Organic Production

Since we are willing to speculate, why not take it all even further? On the molecular level, we could cross the border between physics and biology and look for our adaptable material in the organic realm.

Nature's Assembly

In the previous chapter we expressed some doubt about the feasibility of molecular assembly. But if we look to living cells we find things arguably similar to that concept. Ribosomes, which construct proteins in cells, could be described as nanoscale assemblers. Manipulating code, they can change the outcomes of this biological process. This is a form of genetic modification and is a controversial yet widely applied technology. Its applications include ways to enhance plants in agriculture and produce insulin or vaccines. But there are also examples in which genetically modified organisms are used to produce building material.

Biomolecular Materials

Angela Belcher leads the Biomolecular Materials group at MIT. She and her team use genetically modified simple organisms to produce materials and devices. [12] In 2009, they presented a lithium ion battery built by genetically modified viruses. [13] A year later, they showed how genetically engineered baker's yeast can convert carbon dioxides into carbonates that can be used as a building material. [14]

Living Matter

What if such organically produced materials could be responsive and life-like? Rather than being merely programmed, the material could evolve to adapt to our desires. It could grow and heal and reproduce itself. Instead of giving orders, we might have to train our environment, to teach it what we want. This vision is both beautiful and scary. Scary, as the material might not always do what is best for us. In *Engines of Creation*, Eric Drexler speculates on how molecular replicators could eventually turn against nature and humans:

Early assembler-based replicators could beat the most advanced modern organisms. 'Plants' with 'leaves' no more efficient than today's solar cells could out-compete real plants, crowding the biosphere with an inedible foliage. Tough, omnivorous 'bacteria' could out-compete real bacteria: they could spread like blowing pollen, replicate swiftly, and reduce the biosphere to dust in a matter of days. Dangerous replicators could easily be too tough, small, and rapidly spreading to stop – at least if we made no preparation. [15]

The more advanced the material became, the more attention we would have to give to ensuring its safe use. But, then again, couldn't that be said of every technological innovation?

1 The Rietveld Schröder House, in Utrecht, the Netherlands, was designed by Gerrit Rietveld for Truus Schröder-Schräder and her three children. The house is an important example of De Stijl architecture.

2 The building, located in Paris's 5th arrondissement, was designed between 1981 and 1987 in a collaboration between architect Jean Nouvel and the firm Architecture-Studio.

3 Muscle Tower II, 2004 by Hyperbody/chair Kas Oosterhuis, Delft University of Technology.

4 Flight-Assembled Architecture by Gramazio & Kohler and Raffaello D'Andrea in collaboration with ETH Zürich, exhibition 2012 at FRAC Centre Paris, online at http://www.gramaziokohler.com/ and http://raffaello.name/.

5 Aerial Construction, 2013–2015, by Federico Augugliaro and Raffaello D'Andrea, IDSC/ETH Zürich and Gramazio & Kohler, dArch, ETH Zürich.

6 Claytronics at Carnegie Mellon University, http://www.cs.cmu.edu/~./claytronics/.

7, 8 Smart Sand, M-Blocks and Robotic Pebbles are projects by the Distributed Robotics Lab, CSAIL, MIT, http://groups.csail.mit.edu/drl/.

9 K. Eric Drexler, *Engines of Creation: The Coming Era of Nanotechnology* (New York: Doubleday Publishers, 1986).

10 John Storrs Hall, 'Utility Fog: A Universal Physical Substance', in *Vision-21: Interdisciplinary Science and Engineering in the Era of Cyberspace*, G. A. Landis, ed., NASA Publication CP-10129, 115–126 (1993).

11 John Storrs Hall, 'Utility Fog: The Stuff that Dreams are Made Of'. See www.autogeny.com, as accessed on 11 June 2015.

12 The Biomolecular Materials Group at MIT. See: http://belcherlab.mit.edu/.

13 'New virus-built batteries could power cars and electronic devices', MIT News Office, 2 April 2009, online at http://newsoffice.mit.edu/.

14 'Putting carbon dioxide to good use, MIT biological engineers have found a way to convert carbon dioxide emissions to useful building materials, using genetically altered yeast.' MIT News Office, 22 September 2010, online at http://newsoffice.mit.edu/.

15 See Drexler, *Engines of Creation*, op. cit. (note 9).

Image from the "Grey Goo" video game

Although the previous chapter presents some promising developments, the omnipotent Barba material we've imagined remains unavailable. We aim to approach these visions. Different existing materials are used to simulate the experience of Barba spaces. Life-size installations test possible interactions. Architectural projects apply our concept using today's technologies. And these material applications – and the future visions from which they arise – inform, challenge and inspire one another.

Tests

How might we bring Barba and his fully adaptable city to physical reality? This first test is a very direct attempt. Our vision is remade with analogue materials and rendered in film. While jumps in scale are allowed, virtual models are prohibited in this exercise. The aim is twofold, directed towards experience and experimentation. The intention is to simulate the experience of how the smoothly rendered surfaces of the previous chapters could play out in real life, to simulate the texture of the material and how Barba would feel. The test is simultaneously an experiment with simple and available materials. Could these materials themselves hold the key to new building methods? And is it possible that the advanced building materials about which we've speculated are already known to us? That they already exist, are hidden in our kitchens, say?

The results presented are deliberately light and playful. Some work with materials that can easily be scaled up to room size, while others are bound to remain conceptual. The aim of this exercise is, above all, to inspire.

Material Experiments

The Why Factory with Buckylab: Building Barbapapa workshop, TU Delft

Building Barbapapa workshop, TU Delft

Callum Andrews, Laura Ospina Botero, Gloria Chen, Maosen Geng, Ching He, Aleksander Jensen, Mark de Klijn, Niek van Laere, Chen Li, Marie d'Oncieu de la Batie, Guendalina Rocchi, Ansis Sinke, Ugur Sütcü, Fei Wu

Teachers: Marcel Bilow (Buckylab, TU Delft), Ulf Hackauf, Adrien Ravon

Guest critics: Ulrich Knaack (Façade Research Group, TU Delft), Winy Maas

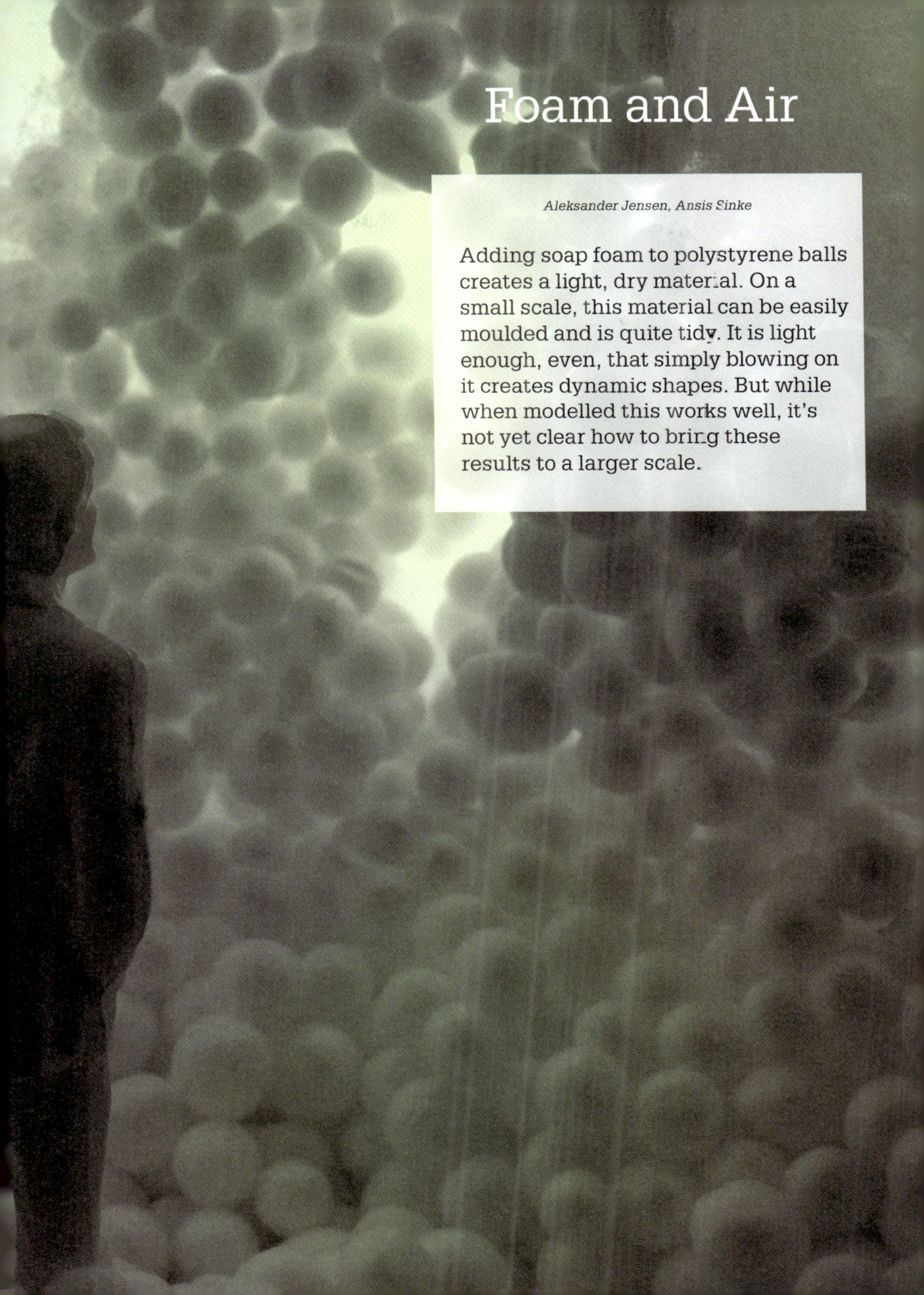

Foam and Air

Aleksander Jensen, Ansis Sinke

Adding soap foam to polystyrene balls creates a light, dry material. On a small scale, this material can be easily moulded and is quite tidy. It is light enough, even, that simply blowing on it creates dynamic shapes. But while when modelled this works well, it's not yet clear how to bring these results to a larger scale.

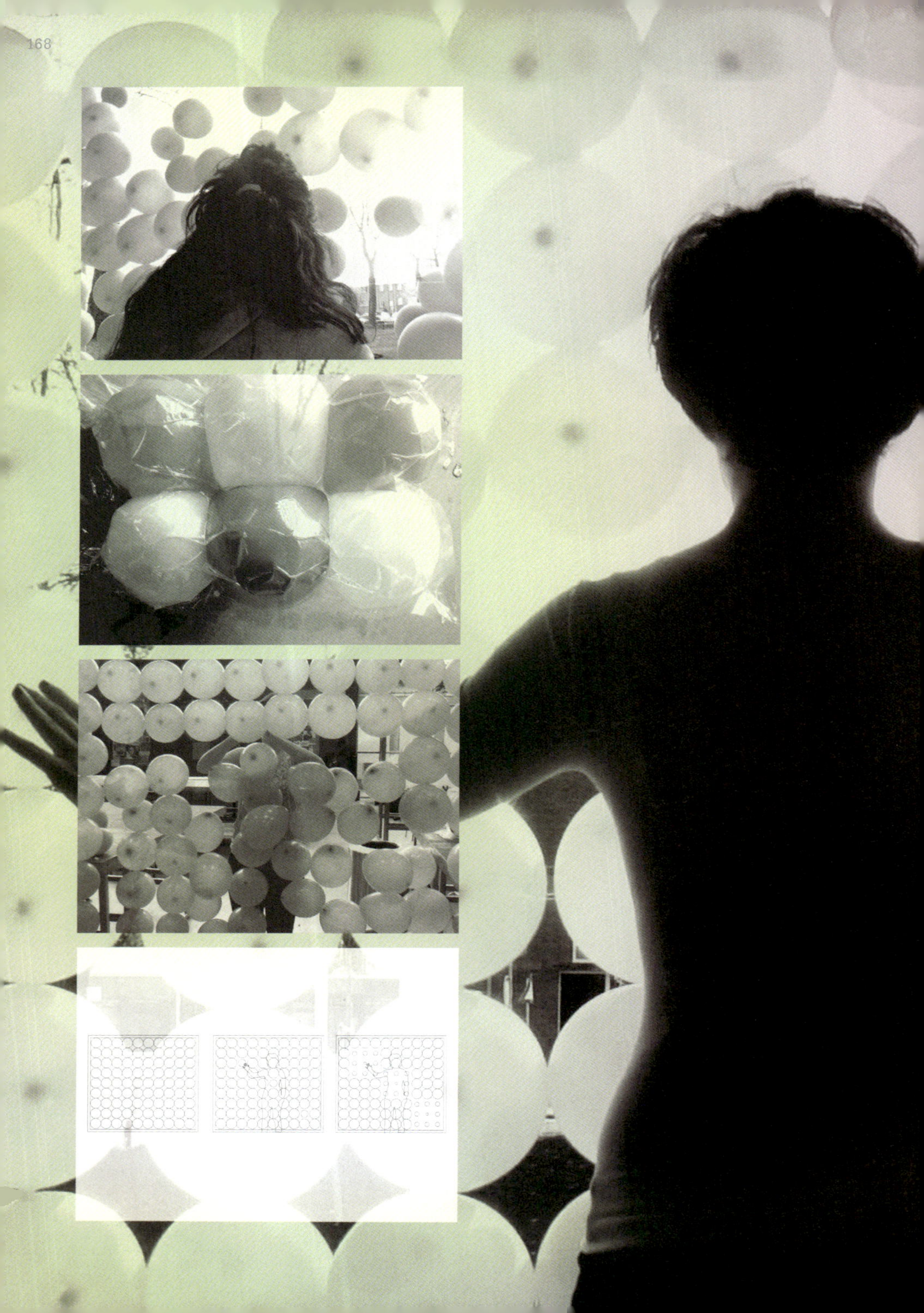

Balloon Transparencies

Marie d'Oncieu de la Batie, Guendalina Rocchi

Inflatable balloons seem a good base for flexible building elements. In this experiment, they form a façade. Changing the size of the balloons, such that their skin is stretched and compressed, adjusts the surface's transparency. Small ventilators and valves could automate this process. In another step, inflated balloons were vacuum-sealed in an airtight envelope, creating a structure that could be adjusted to different sizes and then frozen in a rigid state.

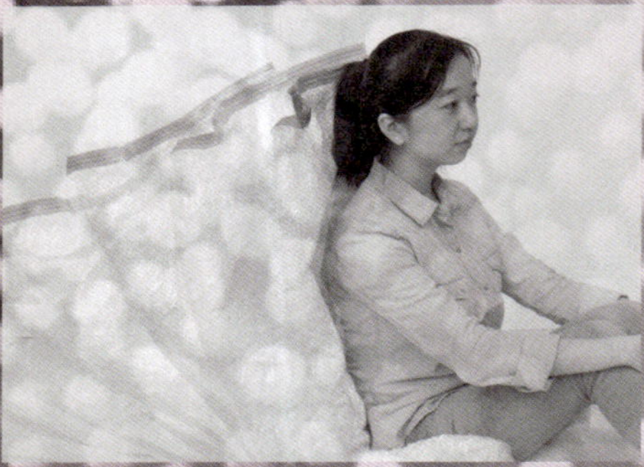

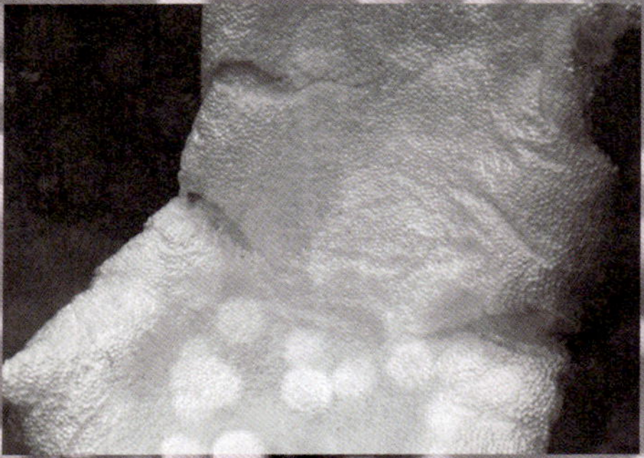

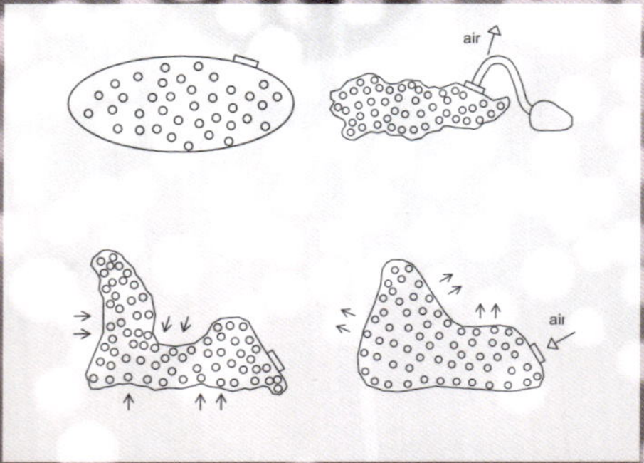
air
air

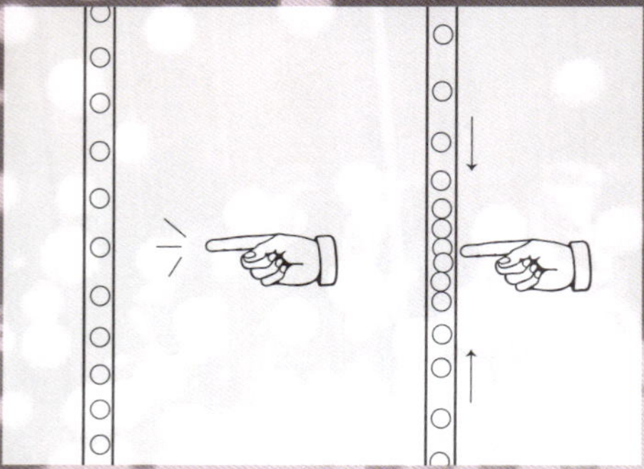

Vacuumed Grain

Gloria Chen, Maosen Geng, Ching He, Chen Li, Fei Wu

Polystyrene balls and a vacuum, elements taken from the previous two experiments, are combined here. Attaching a vacuum cleaner to a bag filled with foam balls allows for the transformation of free shapes into fixed forms. This technology enables the construction of both adaptable furniture and adaptable walls with relative ease. The experiment then explores another quality of polystyrene: its potential electrical conductivity. Rubbing the plastic creates static electricity, which attracts the balls. Could this property be applied to façade design as well?

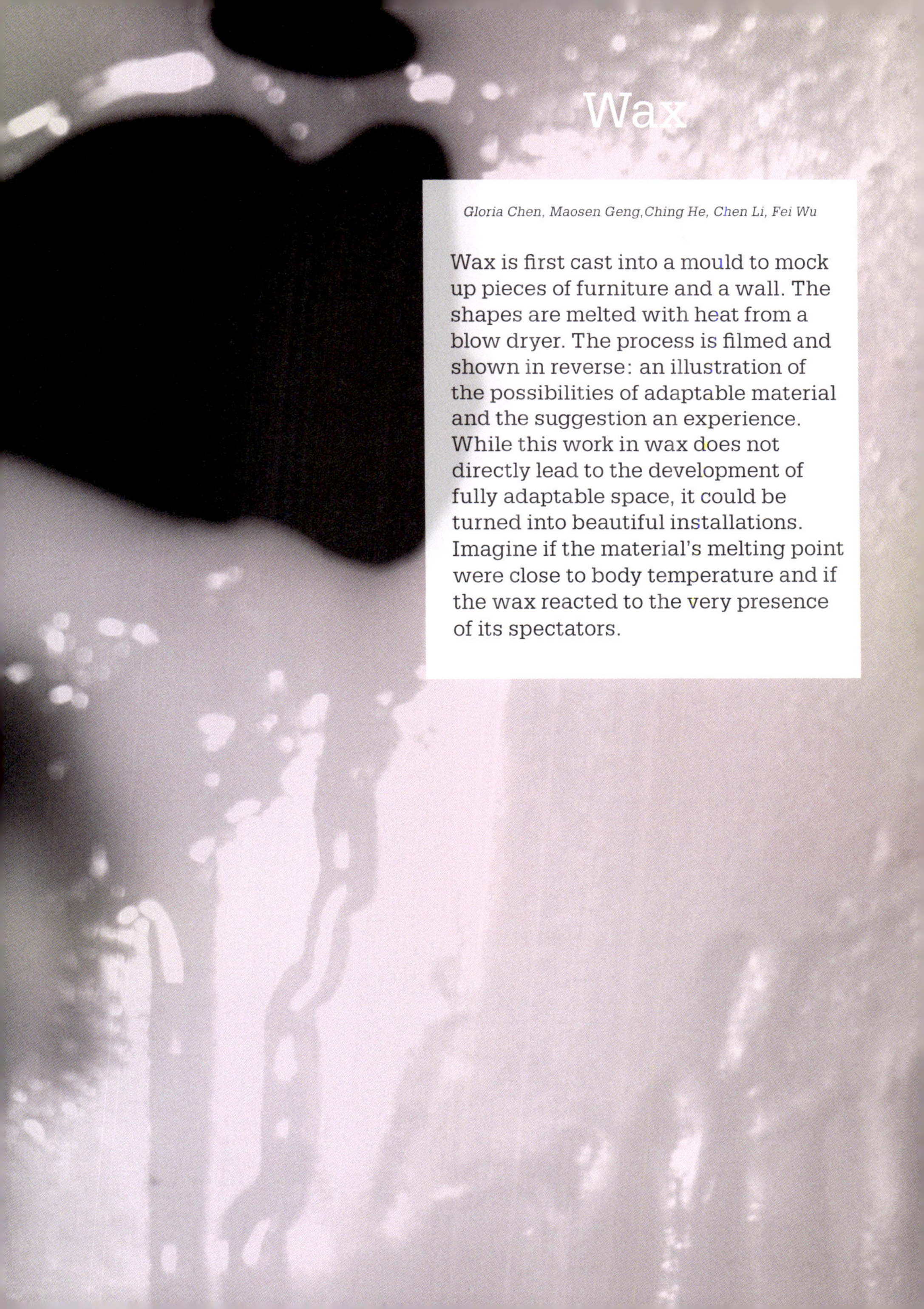

Wax

Gloria Chen, Maosen Geng,Ching He, Chen Li, Fei Wu

Wax is first cast into a mould to mock up pieces of furniture and a wall. The shapes are melted with heat from a blow dryer. The process is filmed and shown in reverse: an illustration of the possibilities of adaptable material and the suggestion an experience. While this work in wax does not directly lead to the development of fully adaptable space, it could be turned into beautiful installations. Imagine if the material's melting point were close to body temperature and if the wax reacted to the very presence of its spectators.

Skin and Pistons

Mark de Klijn, Ugur Sütcü

There are precedents for the exploration and successful use of pistons hidden behind a membrane surface. The challenge here remains the membrane. How far can it stretch? And this system requires a deep floor or wall, behind or in which the pistons would be hidden. As a correction here, we propose a membrane with inflatable balls, rather than pistons. We imagine that, if these can be steered individually, the membrane could bend by itself.

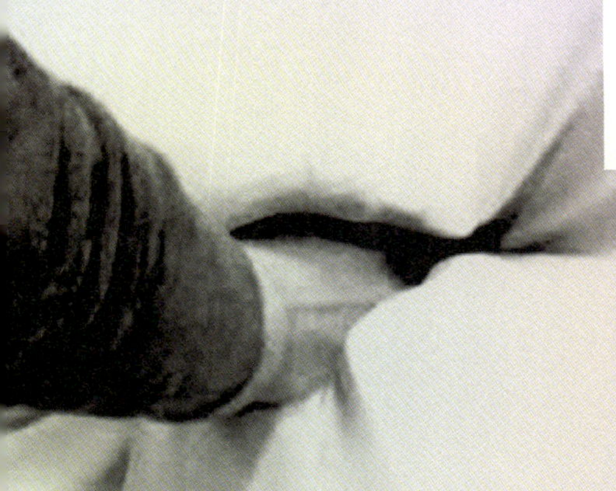

Membranes

Callum Andrews, Niek van Laere

Simple plastic planes are welded together to create a large, airtight cocoon. An air blower is used to inflate it. You can step inside, walk around and have the real feeling of minimum space. Using strings, you can adjust the size and shape of the cocoon, exploring, as you modulate its form, the limits of your comfort and claustrophobia.

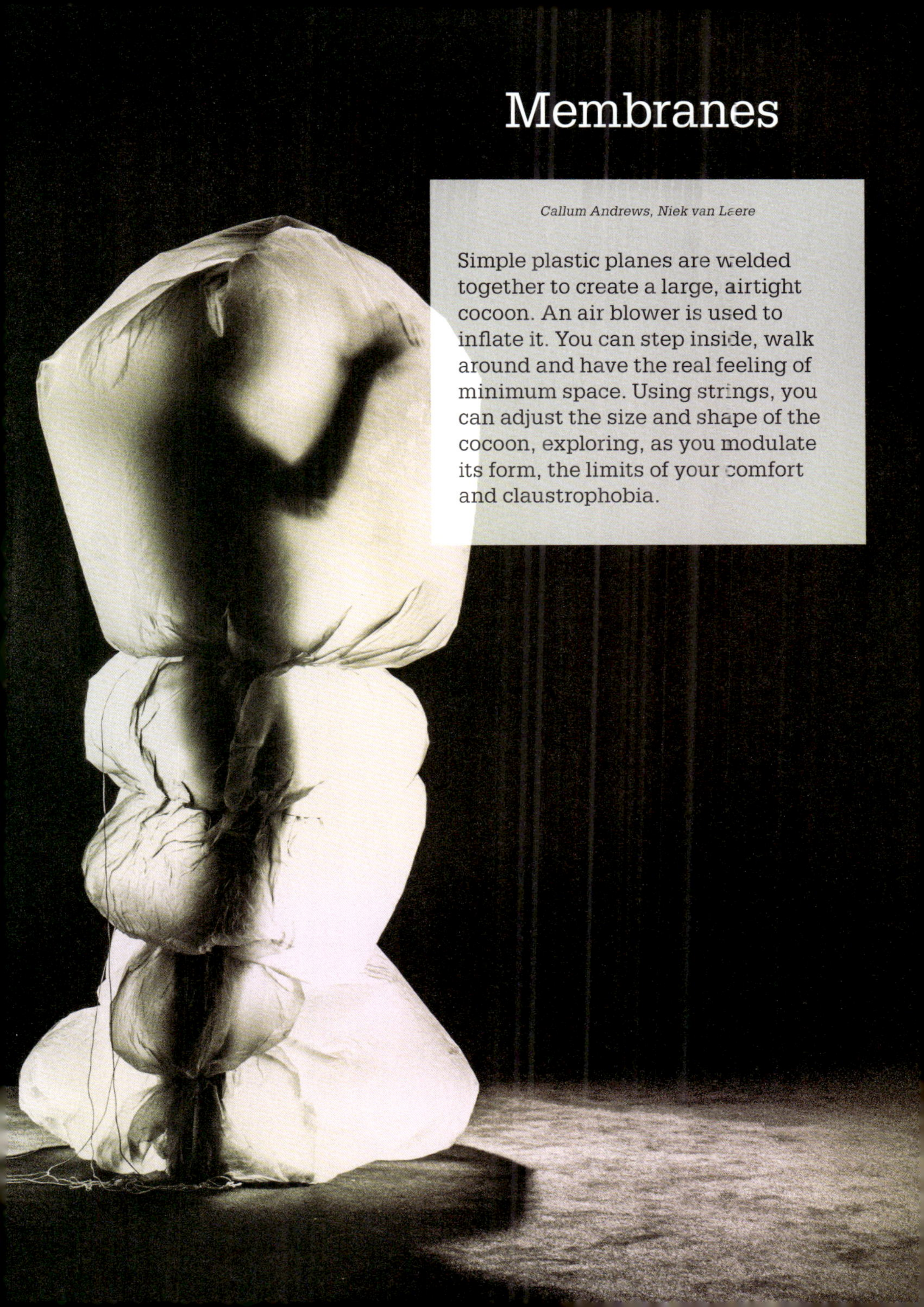

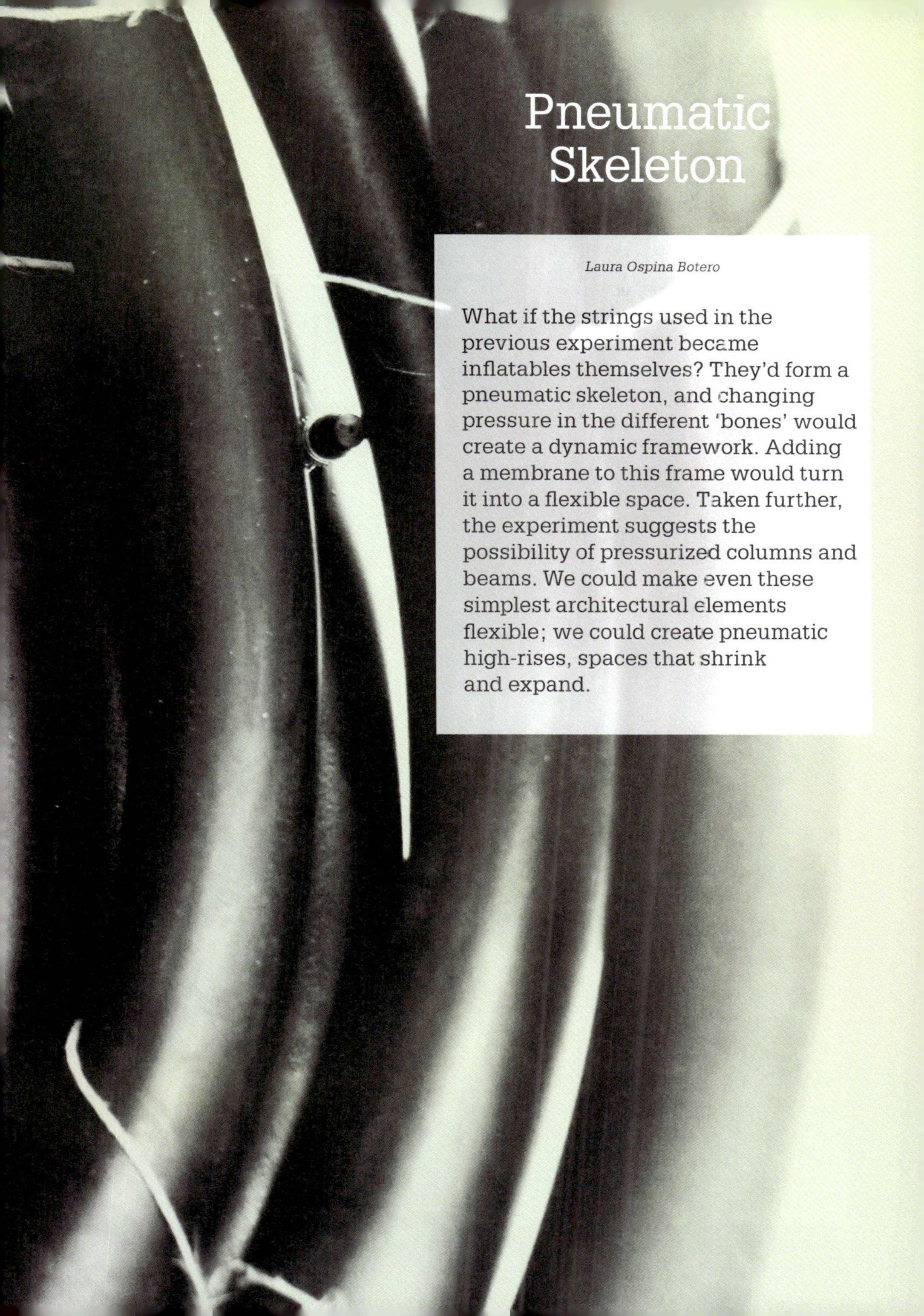

Pneumatic Skeleton

Laura Ospina Botero

What if the strings used in the previous experiment became inflatables themselves? They'd form a pneumatic skeleton, and changing pressure in the different 'bones' would create a dynamic framework. Adding a membrane to this frame would turn it into a flexible space. Taken further, the experiment suggests the possibility of pressurized columns and beams. We could make even these simplest architectural elements flexible; we could create pneumatic high-rises, spaces that shrink and expand.

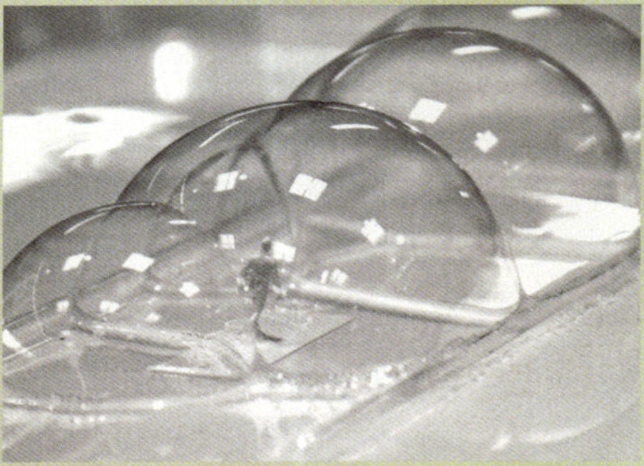

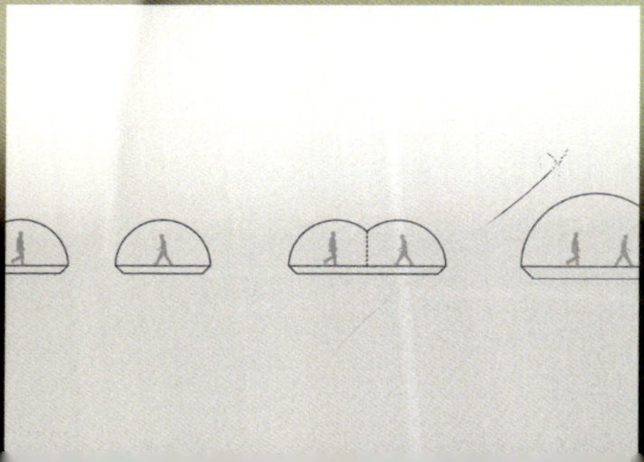

Bubbles

Callum Andrews, Niek van Laere

The previous experiment points to a promising approach to creating a fully adaptable city using today's existing technologies. But here we'll move back from reality and return to the visionary and speculative. Imagine soap bubbles that form ephemeral spaces around humans. The spaces could be touched, pushed, connected. Enacting this on a larger scale would be hard, nearly impossible. And while bubbles are too fragile to provide any shelter, they visualize personal space and human interaction in a most poetic way.

The previous chapter's explorations serve as starting points for a longer, more thorough set of experiments. These experiments are directed towards a fully adaptable city achievable with today's existing technologies. They result in a series of physical and virtual prototypes, each one using a different material as point of departure. These investigations are both experimentation and simulated experience towards the vision of Barba space. They lead us to ask: How can we make any of this, today or in the near future? And how, then, would it feel?

Models

The Why Factory: Transformer Installations studio, TU Delft

Transformer Installations studio, TU Delft
Roxane Belot, Arslane Benamar, Nicola Campri, Sam Chia, Federico Gobbato, Nadia El Hakim, Ruth Hoogenraad, Cathrina Lee, Lex te Loo, Marine Manchon, Vincent Marchetto, Andrea Migotto, Rob Moors, Piet de Reuver, Dominik Saitl, Robbert Verheij, Hans Vlaskamp, Jordan Yerbury
Teachers: Ulf Hackauf, Adrien Ravon, Sander Mulders, Huib Plomp
Guest critics: Winy Maas, Marcel Bilow, Agata Jaworska

Inflatables

Andrea Migotto and Nicola Campri

We imagined the space surrounding a person as pure space, a space without any material. We imagined it as an ephemeral volume able to expand and contract with the movement of the person. If we pull a very thin layer around this volume, we reveal its shape. This thin layer would be mutable, composed so it could expand its surface and then contract again, like a balloon. Increasing the pressure on the inside, the material expands and the space increases; applying pressure on the outside shrinks both material and space. We used latex, a milky emulsion that can be stretched like rubber once dried, as the material with which to begin. In our work we asked: Could fully adaptable space be curved and pressurized? Could its material be thin and translucent?

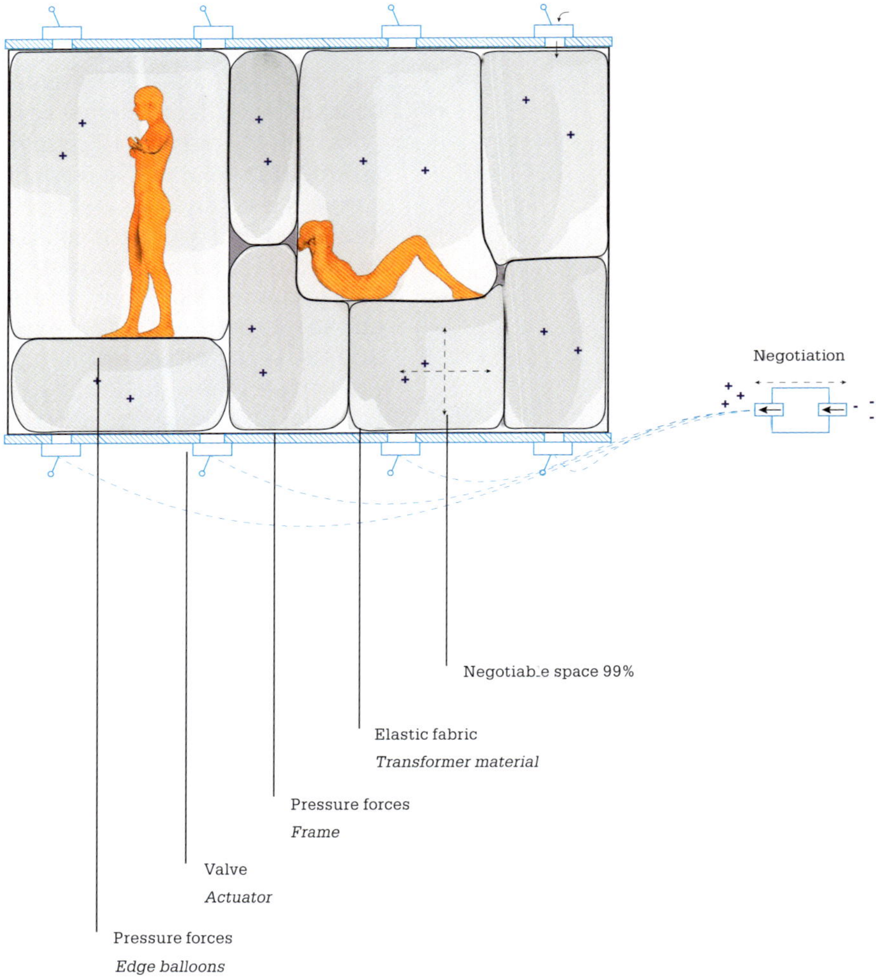

Negotiation
Negotiable space 99%
Elastic fabric
Transformer material
Pressure forces
Frame
Valve
Actuator
Pressure forces
Edge balloons

Making Latex

We used fluid latex to create custom-made membranes. This allowed us to play with different shapes and to make skins of varying thicknesses. Once the latex dried, we were able to stretch it in all directions. We imagined a room made of this membrane. If deflated, the material would collapse to a minimum. What if a completely transformable house could be stored in the space of a single wardrobe?

Prefab Inflatables

We first covered the walls of a room with a latex membrane. Balloons placed behind the membranes allowed the walls to expand and the room to shrink. But this new space remained curved and organic in form and it lacked an architectural articulation. We explored the possibility of shaping of basic architectural elements, such as the column and wall. We covered moulds with latex to create these shapes. Without these moulds, the shapes lay deflated at the bottom of the model room. When we inflated them with little pressure, their shapes appeared. Applying excessive pressure returned the shape to a curvy spherical form.

Curvy space with inflated membrane walls

Creating membrane from liquid latex

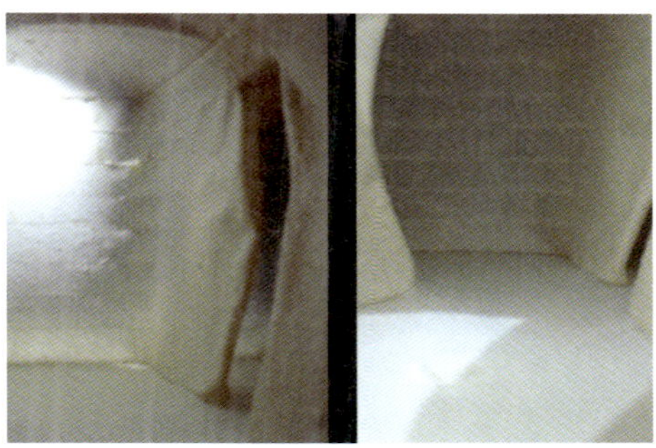

Interior of inflatable wall space

Space Variety

These prefabricated shapes were limited. We could define a shape once and then turn it on or off. But we needed a different approach in order to make more customizable and adjustable shapes. We tried different ideas: plates, for example, which reinforced the latex membrane, or multiple balloons used to shape the membrane. None of these allowed us to modify spaces in a convincing way.

Supporting Rooms

To achieve better control over the shape, we placed eight inflatable volumes in one rigid container. If all eight shapes are inflated with the same pressure, they assume the same size and shape. If the pressure increases in one shape, it grows and pushes the others to the side. Carefully manipulating the pressure in all eight shapes allows us to create precise configurations of smaller and larger rooms and to adjust the shape of each room.

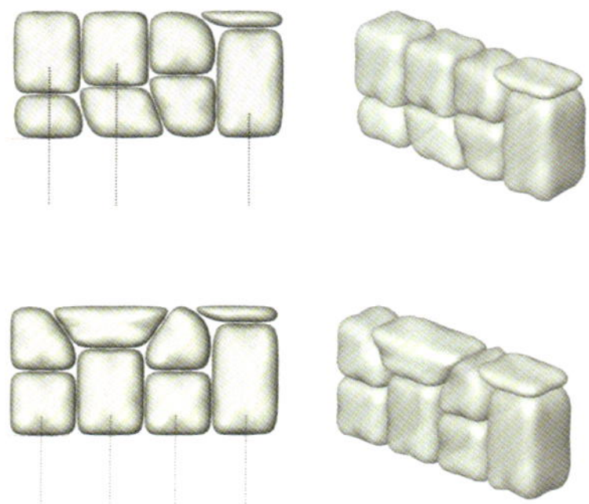

Supporting rooms concept

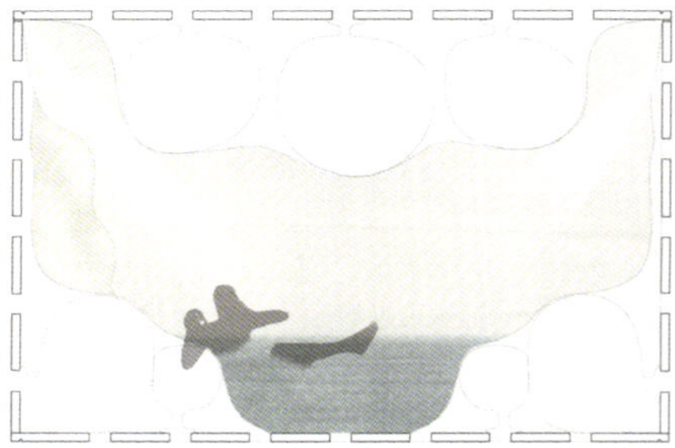

Inflatable volumes control interior space

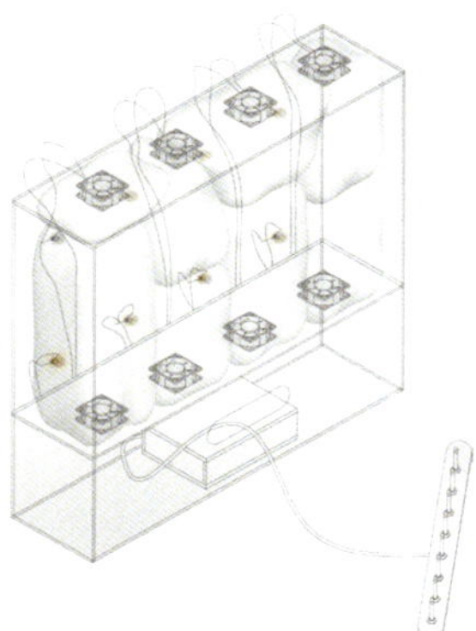

Model set-up for supporting rooms experiments

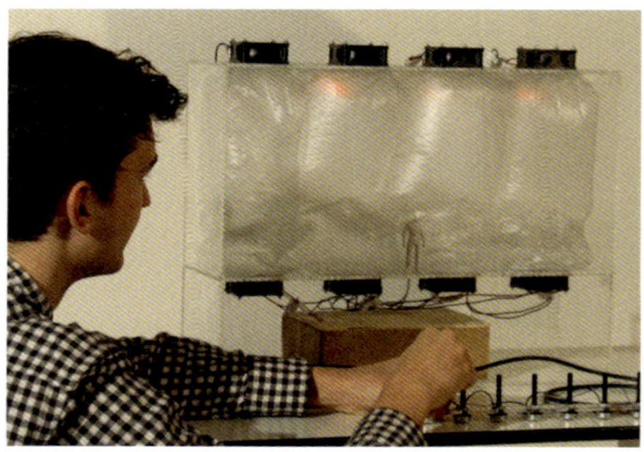

Installation for supporting rooms experiments

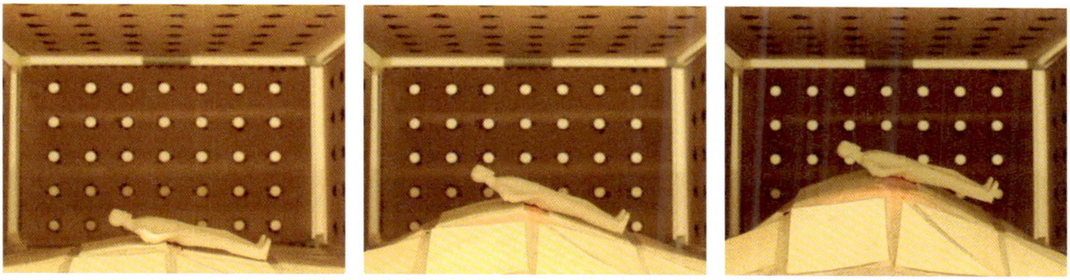

Test box model: Latex reinforced with solid plates, trying to control the shape of the inflatable

Test box model: Testing a column shape

Test box model: multiple balloons behind a latex surface

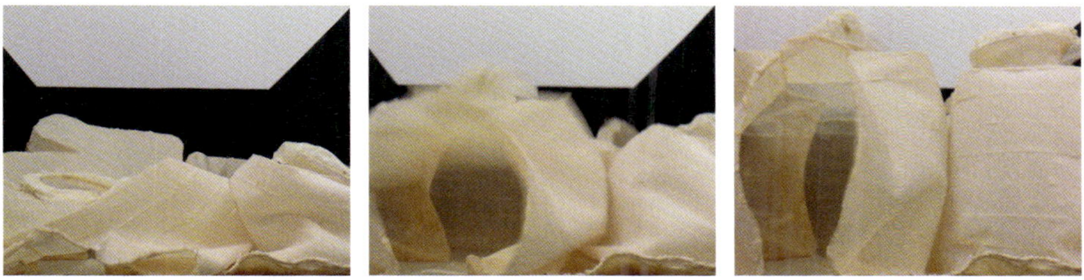

Test box model: preshaped latex shapes. Air, introduced by a pump, flows through the space from below

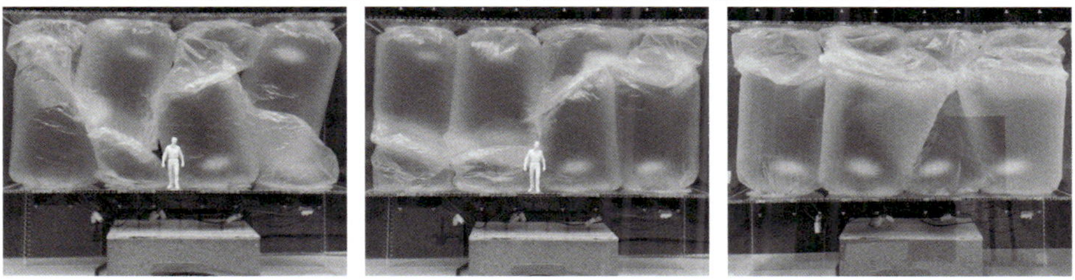

Test box model: voronoi concept tested with eight fans and small plastic bags
The light corresponds with the amount of pressure

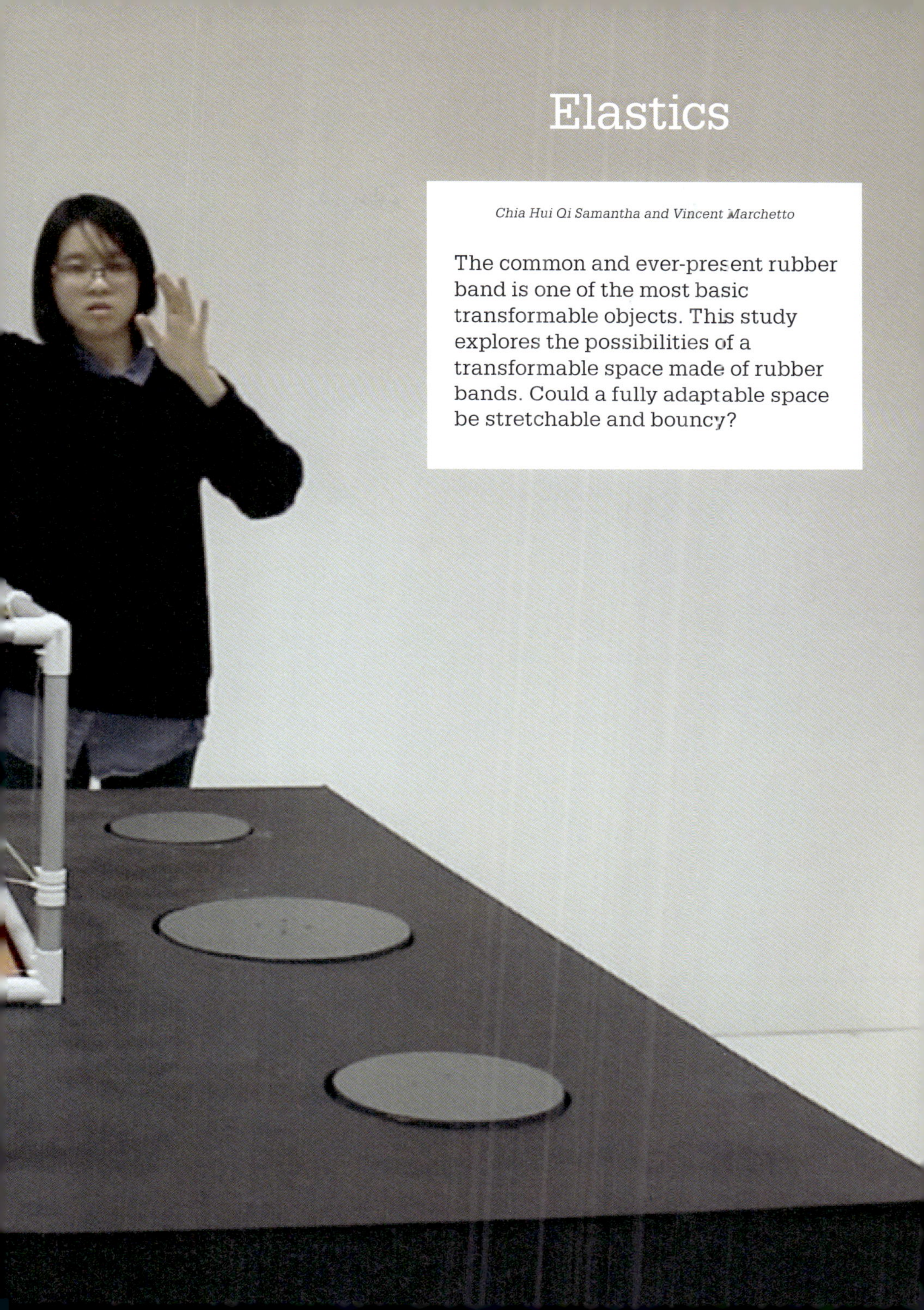

Elastics

Chia Hui Qi Samantha and Vincent Marchetto

The common and ever-present rubber band is one of the most basic transformable objects. This study explores the possibilities of a transformable space made of rubber bands. Could a fully adaptable space be stretchable and bouncy?

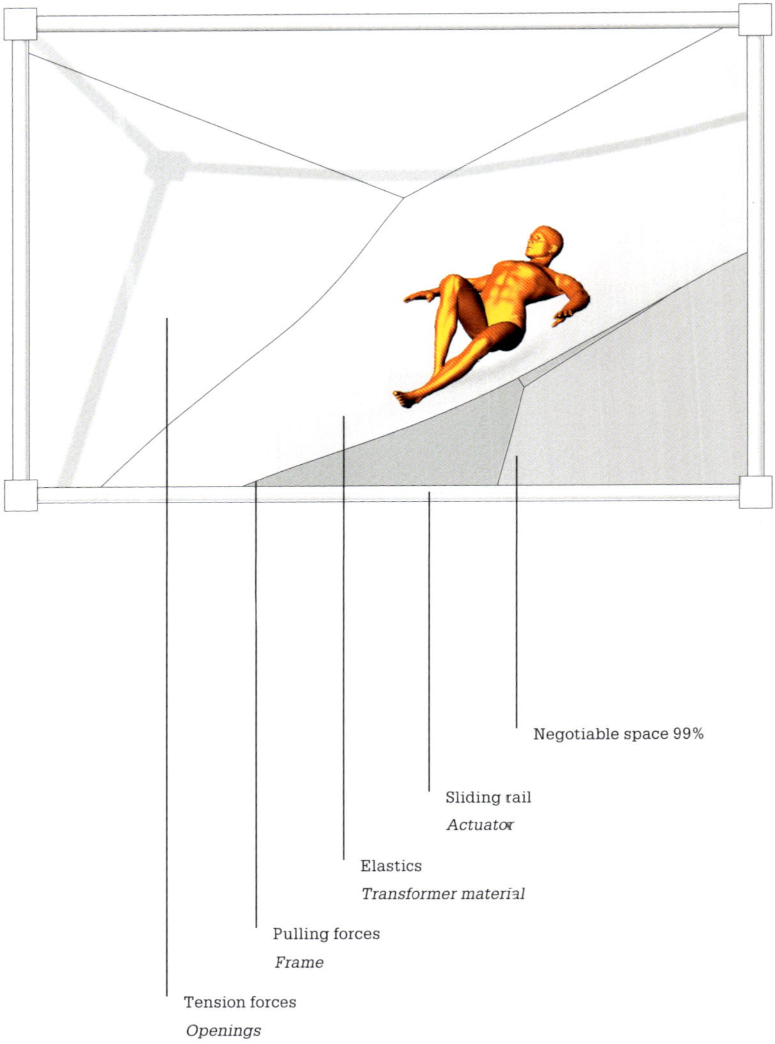
Negotiable space 99%
Sliding rail
Actuator
Elastics
Transformer material
Pulling forces
Frame
Tension forces
Openings

Stretchable Space

To create an elastic wall, we began by placing some wooden sticks in a foam board. We stretched one rubber band around them to create a boundary. Then we added more rubber bands. One atop the other, they created a wall that enclosed an angulated space. Pulling one side of the elastic wall allowed us to change the shape of the enclosed space.
One elastic band is easy to stretch. Pulling a whole row of bands requires more force.
We experimented with different pulling systems in order to create different shapes of the enclosed room.

Cocoon Space

In the first experiment, we used one rubber band stretched around four anchor points to create a trapezium shape. The space becomes narrower and narrower towards the top, until it is nearly closed. We call this model 'the cocoon'. We use a single activation method to stimulate a series of anchor points in the z-axis in order to transform the space. With less tension, the cocoon is small and its surface is closed; by increasing the tension, we transform the space, enlarging it to reveal the space behind. At night, the cocoon can be a bed and, during the day, the same form might expand to become a large room.

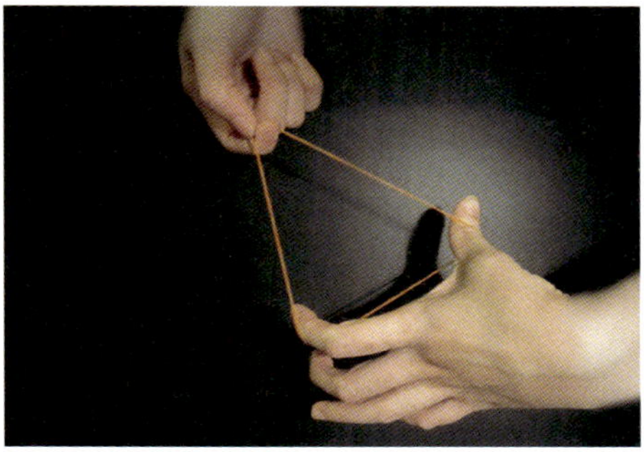

Rubber band

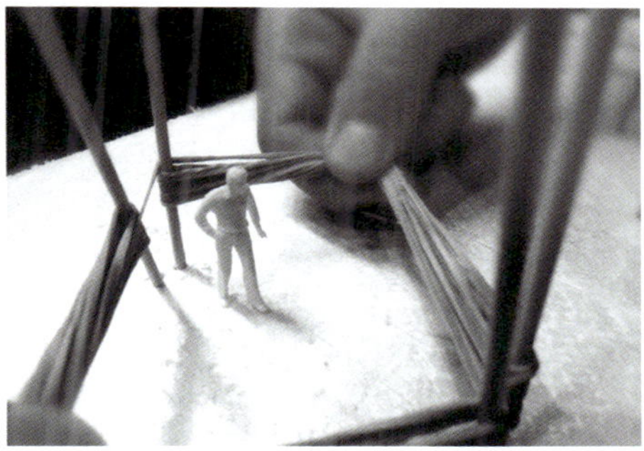

The first elastic space

Cocoon space

Cocoon space at night

Dead Space

We used strings to suspend and stretch the rubber bands inside of a rigid frame. Pulling and releasing the strings allowed us to adjust the shape of each rubber band. The frame had to be strong enough to resist the forces of the rubber bands. We've called the space between the frame and the actual usable cocoon space 'dead space'. We did not use this dead space; it functions as technical space, as storage for unused volume. This is not very efficient. Therefore, the next step was to find a way to use both sides of a rubber band wall.

The Elastic Chair

To test how our elastic space would feel, we built a chair as a one-on-one prototype. The chair was made from a wooden frame with regular rubber bands knotted together, stretched and attached to the frame. You could sit in this chair quite comfortably! By turning it, it could become a lounge chair or even a bed. We then replaced some of the rubber bands with adjustable strings. This allowed us to change the resistance of the rubber bands to match different parts of the body.

Voronoi

How can we have more rooms, in one frame, divided by elastic walls? For this set of experiments, the rubber bands represented the edges of an elastic wall, leaving the actual surface of the wall open. The edges of the rubber band were connected to sliders on the rigid frame. Moving the sliders allowed us to shift, stretch and distort the rubber bands and the rooms they defined. We called the rooms 'voronoi spaces', referring here to the mathematical model for space division.
To test user interaction, we created an installation in which the model was combined with a motion-tracking device. With different gestures, we were able to change the size of the 'voronoi' rooms in real time.

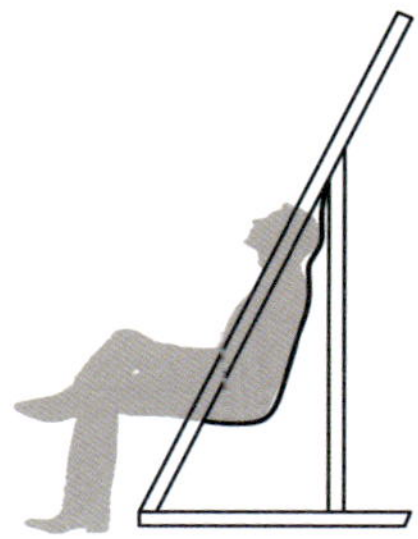

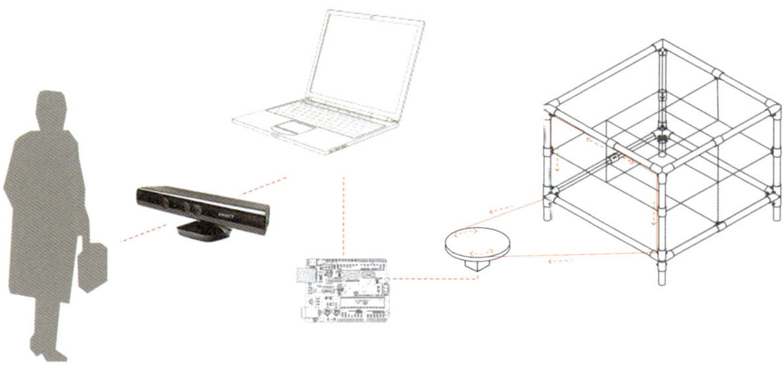

Experiments with one elastic wall

The cocoon model: from triangular to multi-layered room

Elastic chair studies

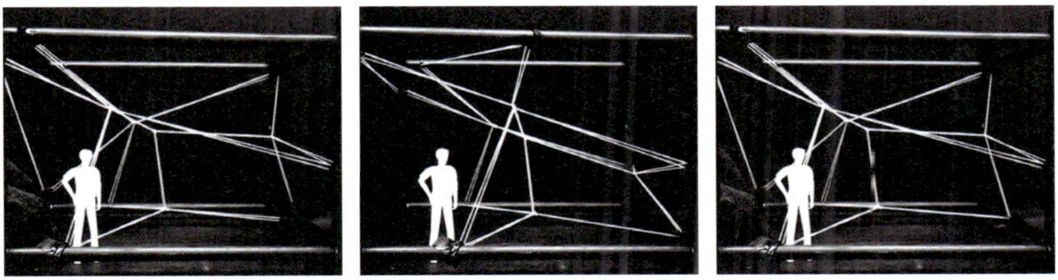

The voronoi room

Elastic installation

Drones

Arslane Benamar and Federico Gobbato

Arslane once had a dream. A dream about flying particles that he could control. The particles could be assembled into movable walls and flying chairs. They could form any object he wanted. Could he create transformable space from these flying particles? We found today's particles in the form of drones. And we wondered: Could we use these drones to make Arslane's dream a reality?

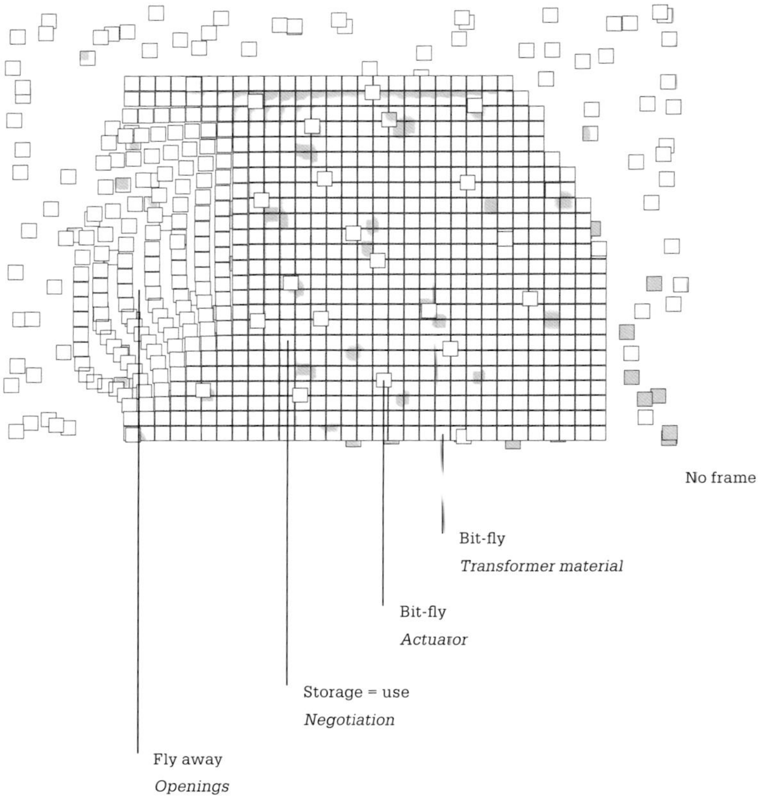
No frame
Bit-fly
Transformer material
Bit-fly
Actuaror
Storage = use
Negotiation
Fly away
Openings

Fans and Foam Balls

We first tested this idea in a model. Instead of drones, we used small foam balls. We built a box with two walls lined with fans and placed the foam balls inside. When we turned on the fans, the foam balls were blown from one side to the other, just like the flying particles we imagined. The movement of the foam balls was chaotic, like an angry swarm gone wild. To increase their controllability, we mixed the foam balls with bath foam, which gave us some control over the shape of the swarm. Then we connected the balls to a cloth, to simulate a stable surface formed by drones. In these experiments, the control we had over the formation of the particles was minimal. In order to achieve greater precision, we needed the fans to be located inside the particles. Within the scope of this study, it was not realistic to work with a large number of miniaturized drones. We therefore chose to describe the ideal drone set-up rather than build it in model form.

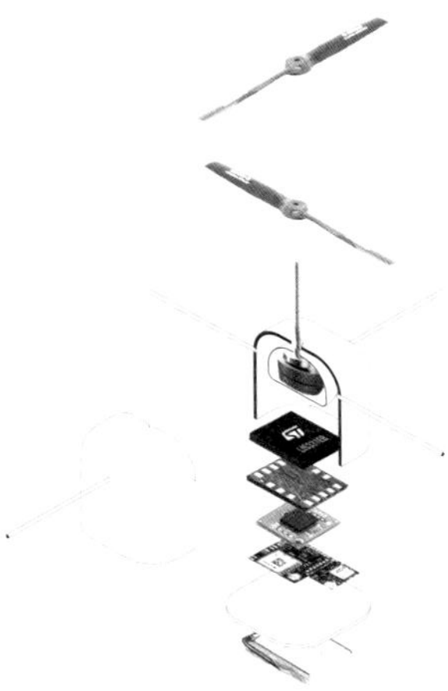

The Drone in a Cage

We put the drone in a cage to give it volume and, enclosed like this, it became a flying building brick, a transformable building material. We can stack one on top of the other so that the cages connect when they touch. The shape of this cage must address three main issues: it has to be modular and allow for a diversity of shapes, it has to be ergonomic and fit the human body, and the gaps between two cages must be as small as possible, so that they are able to support water- and windproof shapes. Even though cubes are not ergonomic, we chose to stick with this shape, as it could allow for both wind- and water-resistant structures.

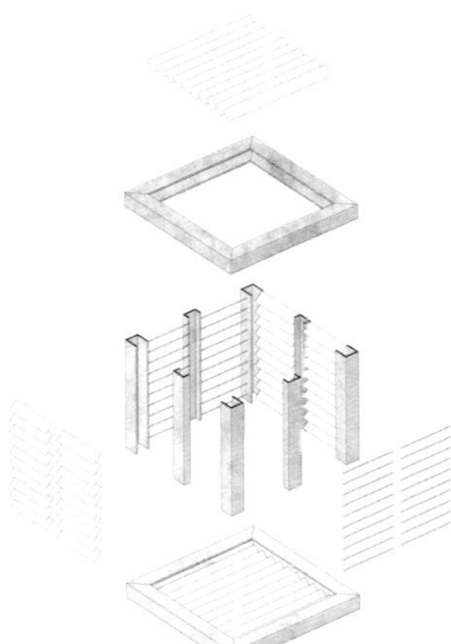

Grain

Discovering the optimal size of the drones was the next step. The smaller the drone, the more detailed the spaces we are able to build with them. But reducing the size of the drones also increases the number of drones needed. The larger they are, the easier it would be to build the drones and control the swarm with the technology currently available. To find the right balance between the two, we simulated a 24-hour use cycle of a living and a dining room composed of different particle sizes. This helped us to understand how to transform a space from one configuration to another, and to determine how to store these particles efficiently.

Swarm

Every time a room or a piece of furniture was reconfigured, a few drones would change positions. The transformable space would constantly behave like a living, busy, vociferous swarm. We tried to imagine some of the consequences of this strategy. One would be noise. The drones have turning rotor blades used to keep them in the air. This sound might be disturbing. We could reduce it by limiting the movement of the drones. As long as one piece of furniture or the walls of one room stay in position, the drones could connect to each other and lock. Or we would need to find other techniques for levering the drones such as, for example, magnetic force. We also considered the idea of ownership in this scenario, and asked questions like: How many drones would I need? Would I take some with me when I went to work? Would I share them with my neighbours? Or would they be provided by some central authority, as is water, gas and electricity? Would there be a citizen's right to drones?

Swarm shapes by moving foamballs with two walls with 4x4 computer fans, adjustable in speed

Swarm shapes by moving foamballs with nylon wire

Simulating drone behavior with stacked wooden boxes and digital drones

Simulation of different drone sizes: 20 cm, 5 cm and 0.1 cm. On the right a dining room,
on the left the volume of the required drones

Digital simulation:Drone behaviour relating to a person

Wood

Rob Moors and Jordan Yerbury

Wood has been used throughout history to create objects and structures. It is a material that has come to be intimately related to human life. It allows us to carve objects of beauty and utility. Inspired by these qualities, we asked: Can transformable space be both beautiful and precisely crafted?

Objects Form a Surface

While reducing the material – carving it down – can modify wood, this is not typically a reversible process. Our focus in this project is to turn traditional wooden structures into transformable systems. In our first experiment, we bundle small skewers and examine the surface created by the top part of the skewers. Pushing and pulling the skewers in and out, the surface begins to create depth that can form objects in space.

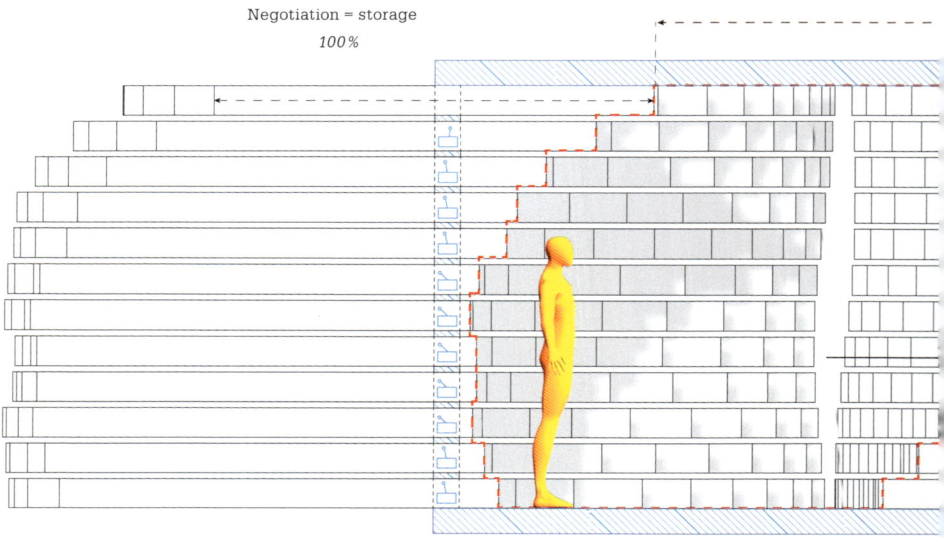

Shifting Wood Wall

We translated this idea into a room of shifting wooden beams. Double layers of grid formed the walls of the room. Wooden sticks were pushed through the grid. They could move back and forth. If all sticks were aligned, they formed an almost massive wall. If some were pushed back, an indent appeared. Pulling sticks out created seating areas, floors or stairs. An animated log hut appeared.

The Backside

When all of the wooden sticks were pushed outwards, they occupied a large space outside of the actual room, a negative of the inhabited space. This was not efficient. To improve this, we imagined telescopic beams. But these would also likely be less rigid.

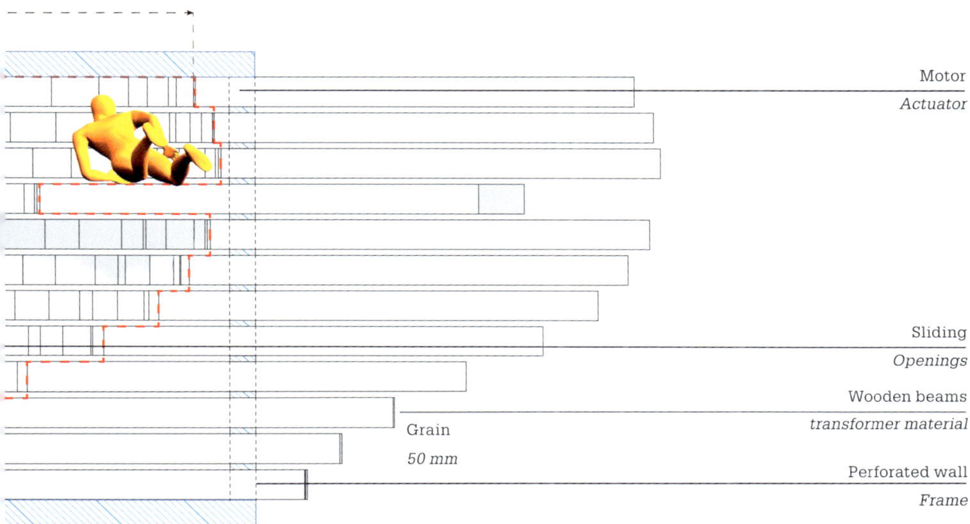

Animated Blocks

The problem lay in the grid that supported the beams. Instead, we tried wooden blocks, one lying on top of the other. We added a system in which the blocks could glide over each other. The blocks can cantilever and create arches or bent walls. The length of the block and the degree to which it could be set back are related. Longer blocks can create larger horizontal planes but also necessarily create thicker and thus less efficient horizontal walls. To keep the blocks in position, we proposed a simple locking system. The blocks could be pushed and pulled into position by hand to create different spatial configurations in the room. Motors inside of the block could make the system automated.

Beauty

By simulating different grain sizes of the push-pull system with rendered images, we demonstrated the feeling and beauty of organic life inside this wooden, fully adaptable space.

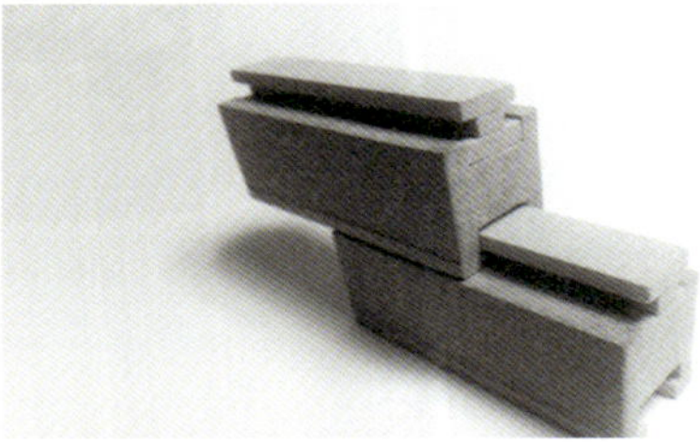

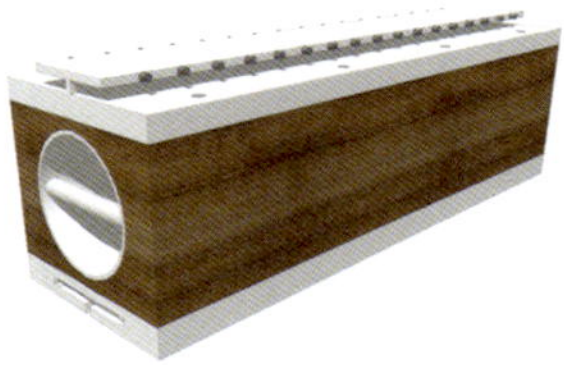

Mock up: A pack of skewers, twisted by two hands transforming the image

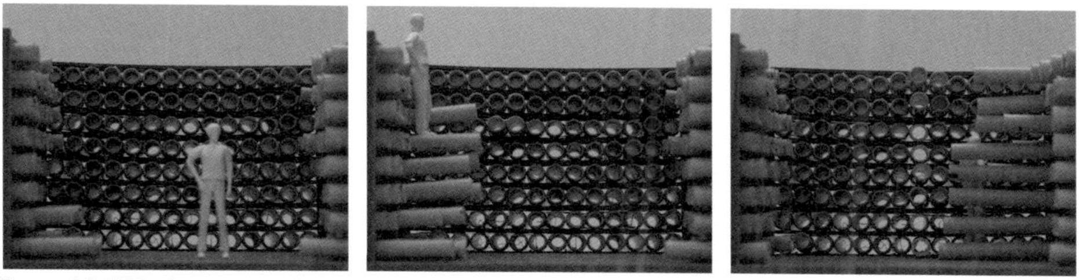

Box model: PVC tubes representing wooden beams, arranged along three walls

Laser-cut frame holding wooden sticks from two sides sliding back and forth

Box model: Stacked wooden profiles sliding between two rails

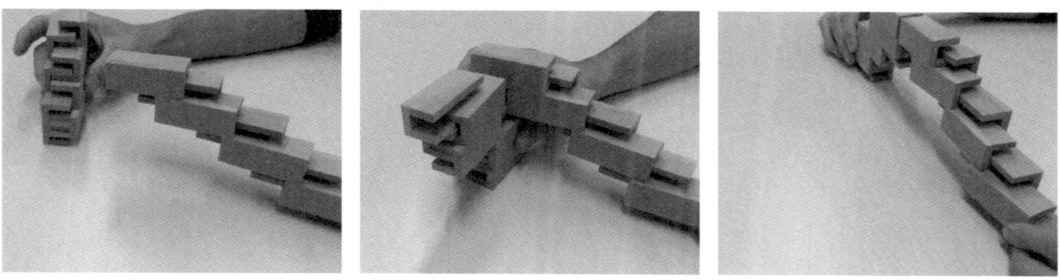

Mock up model: wooden profiles sliding and twisting, creating more freedom than the non-rotating box model

Mesh

Ruth Hoogenraad and Hans Vlaskamp

Metal is a stable and rigid material. It can provide the rigidity and security that a transformable room might need. But can it also be flexible? We took our inspiration from medieval chain mail. This armour's metal mesh is smooth and fits the wearer's body, but is hard and resistant when struck by a weapon. Can we create a mesh that is simultaneously soft and hard? Can we build a room from such a transformable mesh?

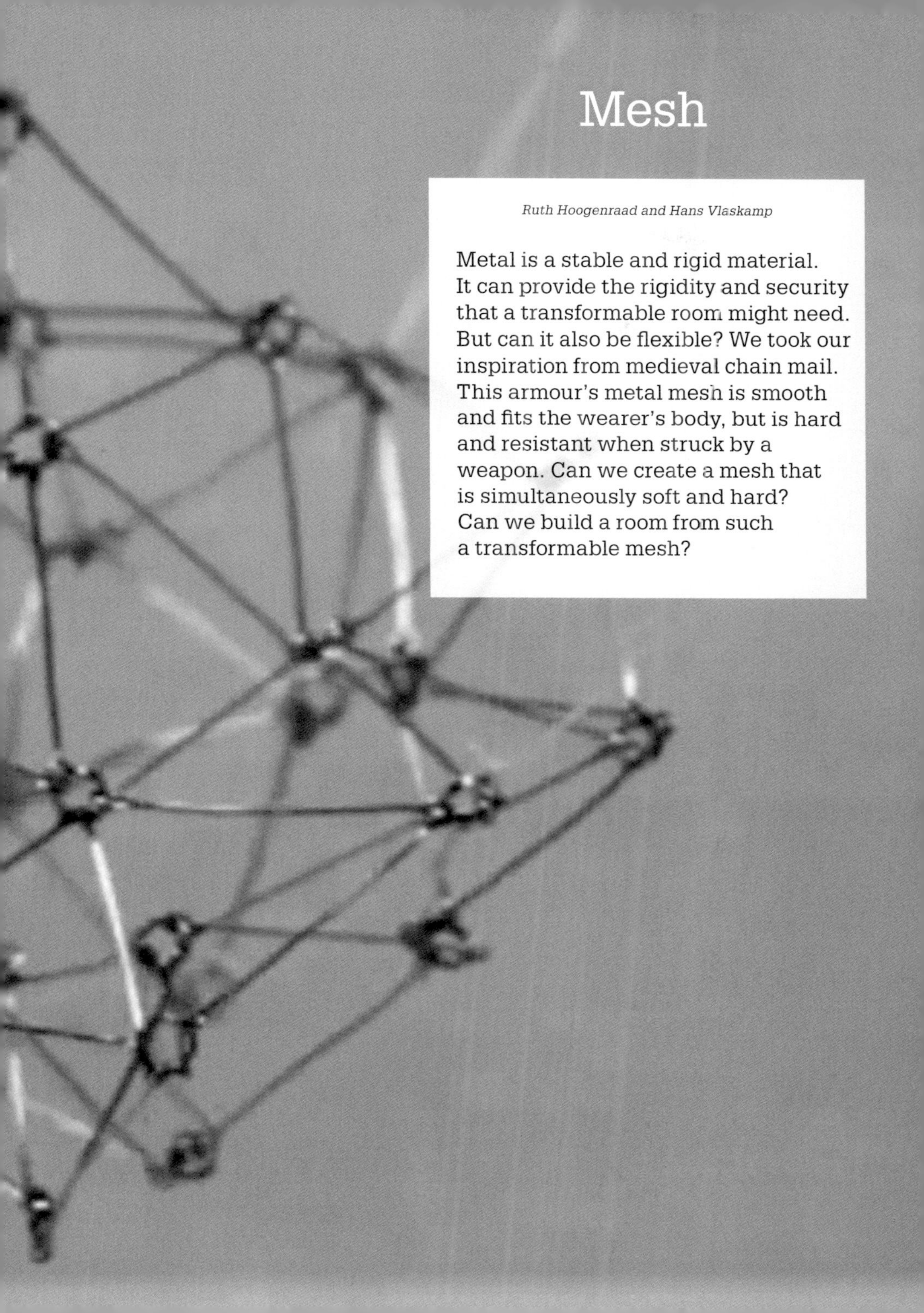

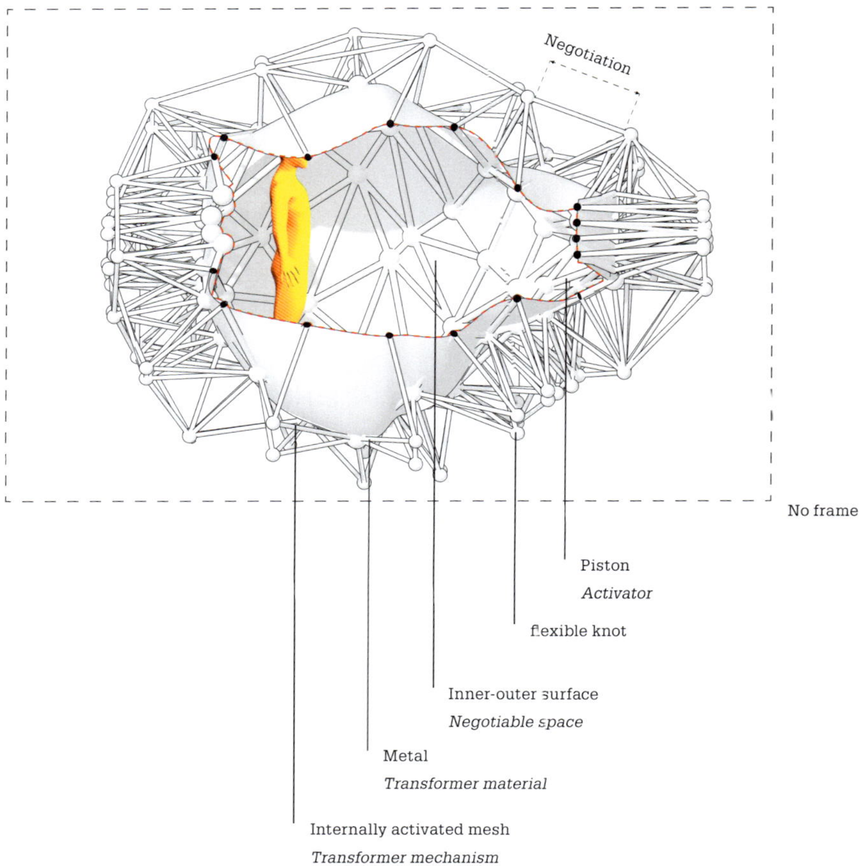
Negotiation
No frame
Piston
Activator
flexible knot
Inner-outer surface
Negotiable space
Metal
Transformer material
Internally activated mesh
Transformer mechanism

Mesh Options

We connect metal sticks with metal rings to create a flexible mesh. The mesh can be crunched into a small heap or stretched out to form a large surface. Different shapes were possible. We then covered the walls of a model room with this mesh. Pushing the mesh from behind modified the room's shape. To add further flexibility, we reviewed the connecting points. We replaced the metal sticks with hollow, plastic tubes and used rubber bands to connect them. This gave us a light yet rigid mesh.

Manipulating Mesh Surfaces

Thus far, the fabricated meshes were flat and had to be controlled from the outside. To introduce activation within the mesh itself, we use a double-layered system with two skins and a structure between them. As in the system of double musculature inside the human arm, here both surfaces can contract or extend in order to bend the surface. We first used wires to pull the three-dimensional mesh. To vary the shape, we cut the mesh into lanes. This limited the transformations to one axis. We studied origami, and paper-folding techniques, to find inspiration as to how we might add transformation in the second axis.

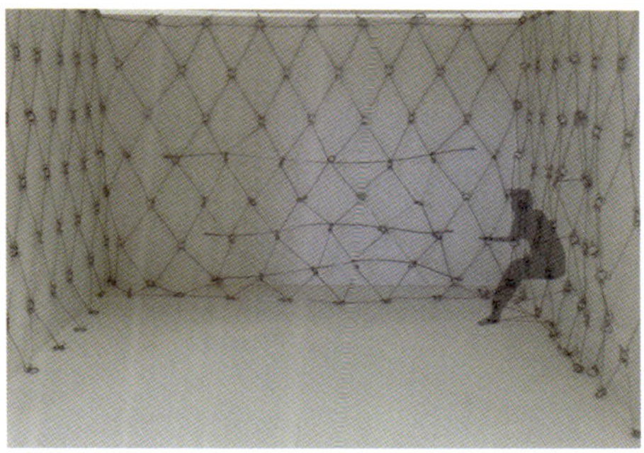

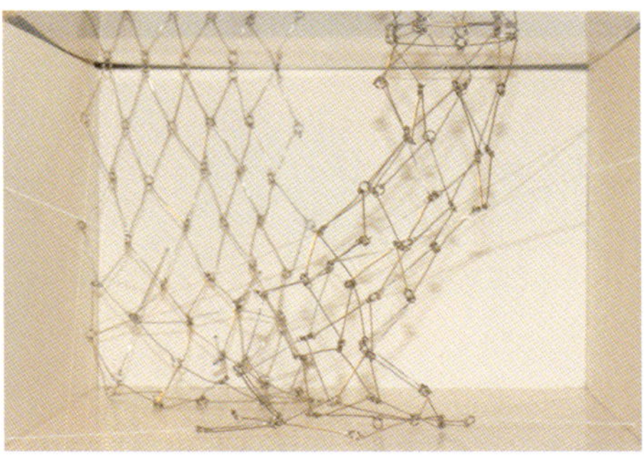

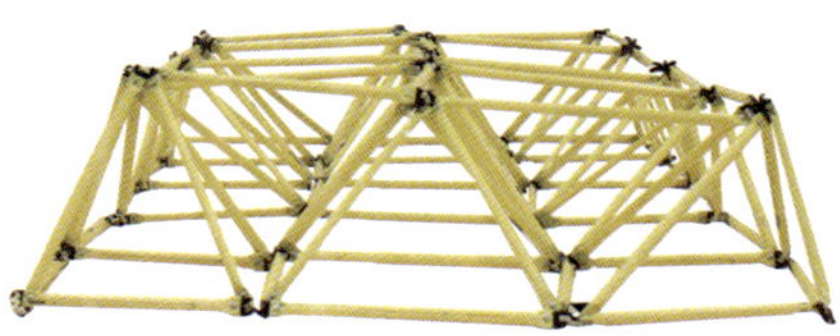

Activated Mesh

To activate the mesh, we had to make the rods of the grid flexible. We used plastic pipes with slightly different diameters and pushed them into each other. This gave us rods that could extend and shrink. All rods parallel to the 'active' rod had to be flexible too, in order to allow the whole grid to react. As a 'motor' for this adjustment, we created a hydraulic system. We used syringes filled with water as actuators. Each syringe was connected to a second syringe outside of the mesh with a flexible plastic tube. Pushing that second tube added water to the actuator and stretched it, which in turn transformed the mesh. To reverse the movement, we used the syringe to suck the water back out of the actuator. Throughout these tests, we saw that the mesh's rigidity can be alternately favourable and challenging in the fabrication of a flexible, transformable space. We created a load-bearing system actually useful for walls, roofs and furniture. At the same time, the shapes we created were mainly large curves. Detailed, refined shapes were difficult to achieve.

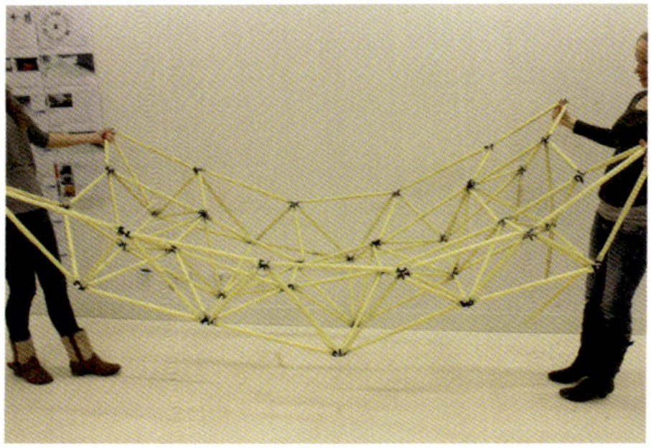

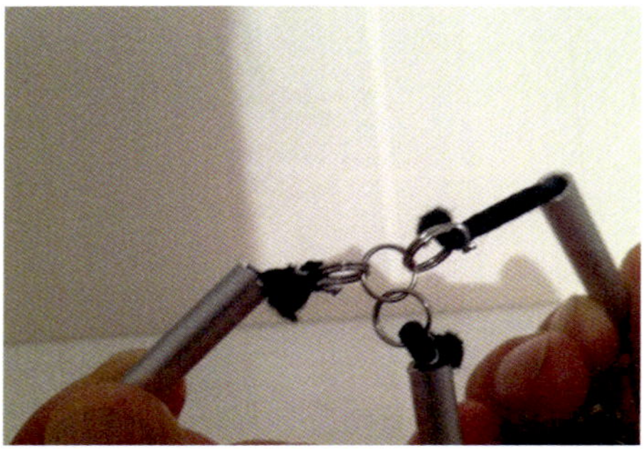

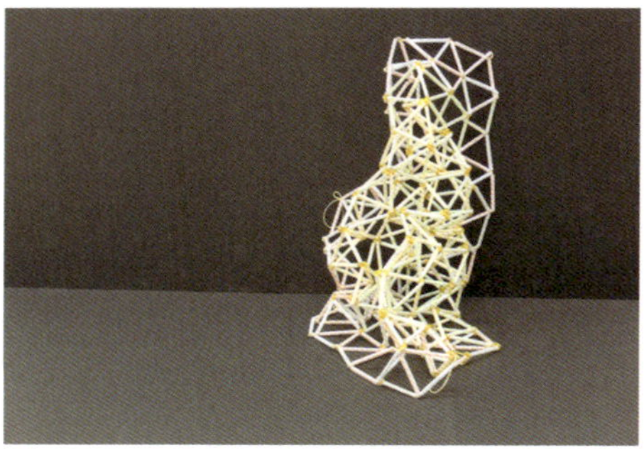

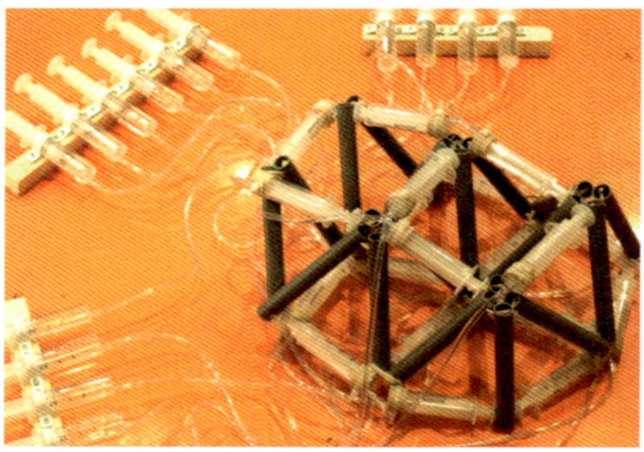

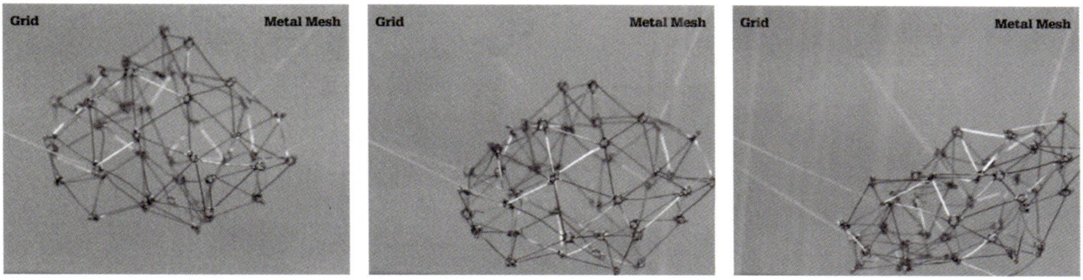

Steel sticks and metal connections

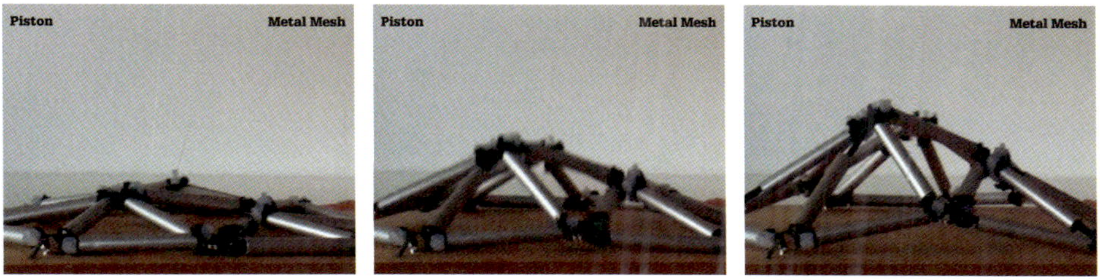

Hollow steel tubes and elastic connections

Flexible space frame

Origami studies

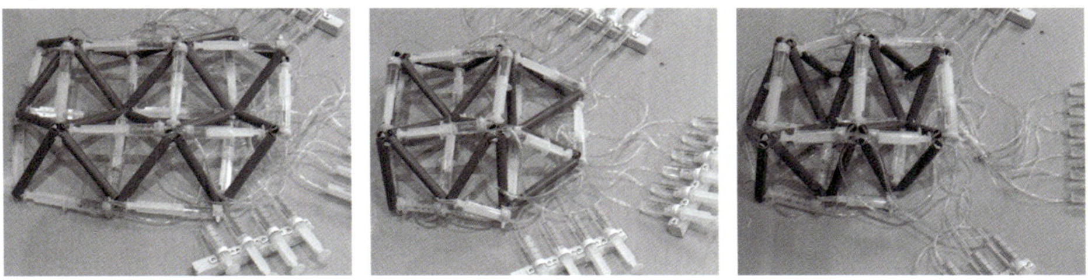
Hydraulic spaceframe

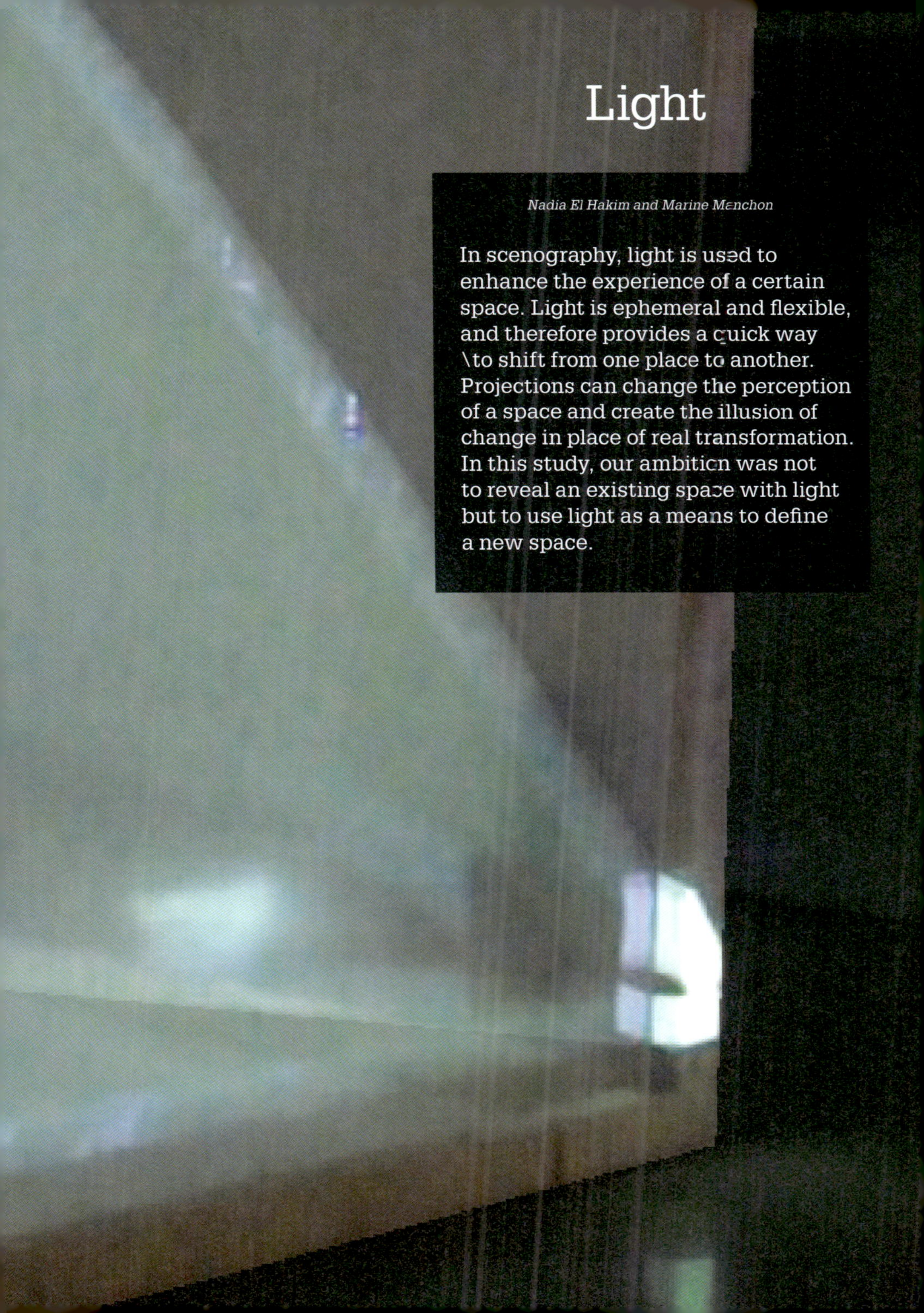

Light

Nadia El Hakim and Marine Menchon

In scenography, light is used to enhance the experience of a certain space. Light is ephemeral and flexible, and therefore provides a quick way to shift from one place to another. Projections can change the perception of a space and create the illusion of change in place of real transformation. In this study, our ambition was not to reveal an existing space with light but to use light as a means to define a new space.

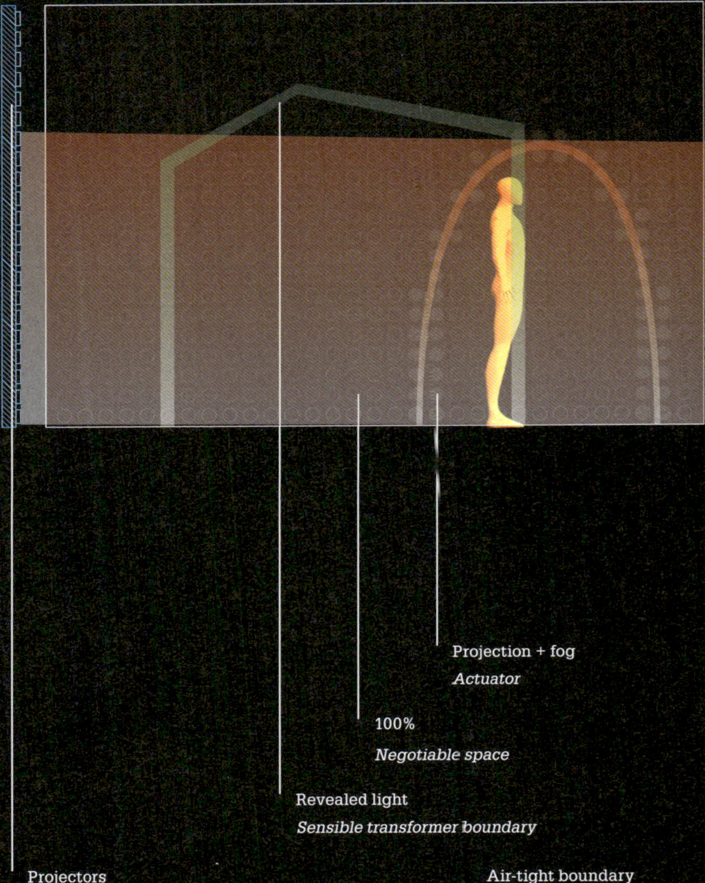
Projection + fog
Actuator
100%
Negotiable space
Revealed light
Sensible transformer boundary
Projectors
Air-tight boundary

Light without a Screen

A beam of light becomes visible when it reflects on a surface. We don't see light in an empty space until it hits some kind of edge. To create space with light, we experimented with fog. Low-density fog or smoke particles revealed light beams in a clear and sharp way. They allowed us to make opaque planes in space without relying on objects.

Test Light Box

We started with a box made of transparent PVC planes. We blew smoke from a smoke machine into this sealed box. To control the light, we projected different shapes of light with a projector into a small, fog-filled plastic cube. To control the shapes, we ran a script on a smart phone. The script allowed us to draw simple shapes with a finger, shapes that would then appear as if 'drawn' by light in the fog. Where the beam of light hits the fog, it is revealed, so the flat image is extruded into a three-dimensional light element. A line becomes a plane and a dot becomes a beam.

Two Planes

In the first box, the shapes were all extruded from the bottom of the box. In the next experiment, two projectors were placed around the box. One projected light from below and the other from the side. We created a catalogue of two-dimensional images of architectural elements – including wall, column, arch, roof, door and window – that could be projected onto the test box from two sides. Projecting with two beamers and allowing the light elements to overlap resulted in more complex elements, such as the dome and niche. These forms, entirely composed of light, created beautiful boundaries and spaces, and gave the impression of enclosure.

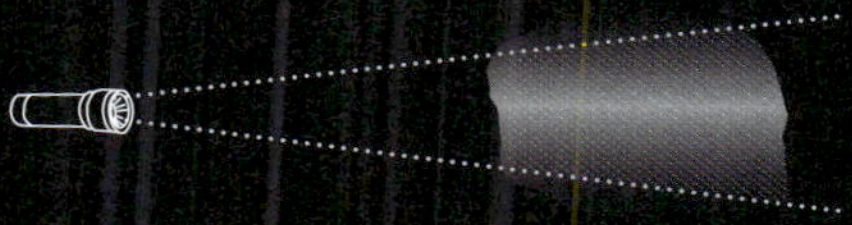

Light Room Installation

To experience the shapes in life scale, we built a fog room. We covered a wooden frame with transparent PVC sheets and filled it with fog from a smoke machine used for stage set designs or at parties. We replaced the smart phone control with a motion-tracking device. Standing in the fog room, we were able to draw shapes of light with our hand or body and to experience vaults, roofs, planes and other architectural elements made from light.

What If . . .

While these experiments created impressions of spaces and provided interesting experiences, they as yet were not able to create physical, usable separations of rooms. The value of our experiments clearly lay in the experience they created rather than in any practical use – unless we were suddenly able to use the light to materialize forms in space. We found one promising research project that might point in this direction.

In 2012, a researcher from the National Institute of Advanced Industrial Science and Technology (AIST) in Japan managed to change the physical state of a material from liquid to solid and back at constant room temperature, by projecting UV light on it. [1] If we imagined that our fog could be made of liquid droplets of that same material, we could imagine our light-generated shapes actually materializing and becoming solid. While it is a long way from the small-scale experiments at AIST to the materialization of a chair from fog and projected light, we find this development promising.

1 'Materials Able to be Repeatedly Liquefied and Solidified at Room Temperature by Light Irradiation', press release by AIST National Institute of Advanced Industrial Science and Technology, 6 April 2012, online at http://www.aist.go.jp/.

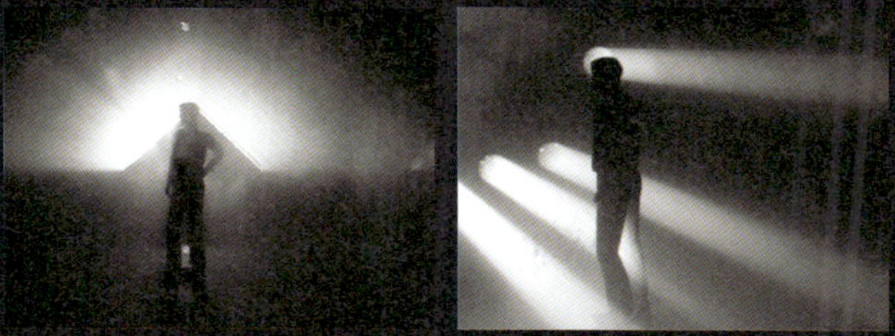

Projecting shapes into a fog-filled light box

Interactive box model, with shapes drawn on a smartphone

Architectural spaces drawn with light

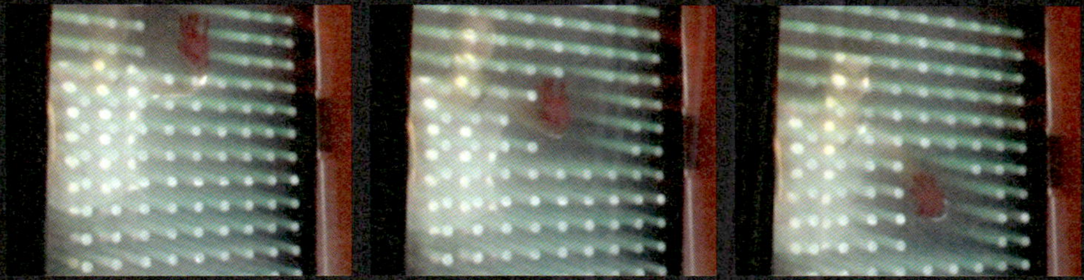

Deactivating light with your hand, using motion tracking

Life scale fog room

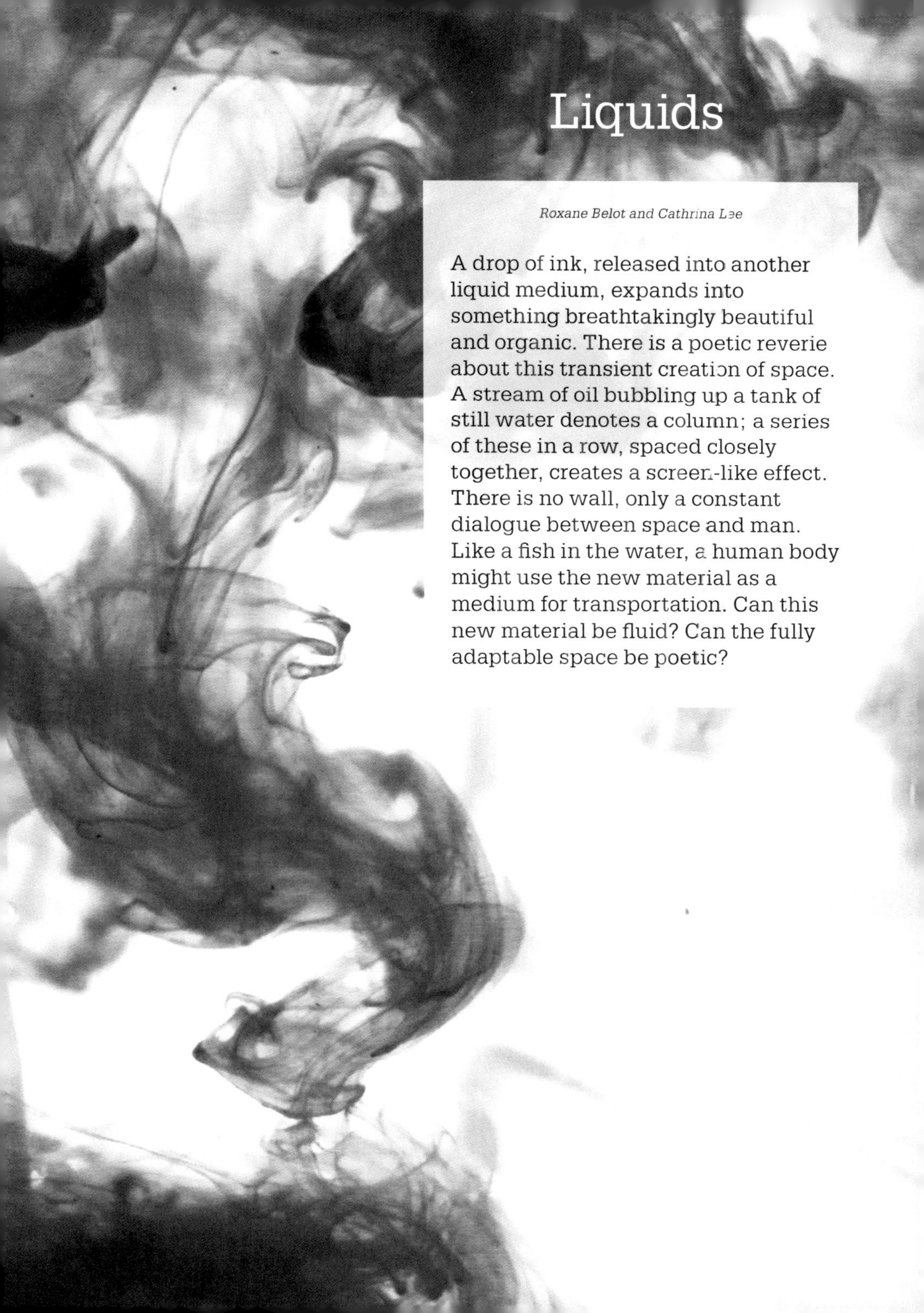

Liquids

Roxane Belot and Cathrina Lee

A drop of ink, released into another liquid medium, expands into something breathtakingly beautiful and organic. There is a poetic reverie about this transient creation of space. A stream of oil bubbling up a tank of still water denotes a column; a series of these in a row, spaced closely together, creates a screen-like effect. There is no wall, only a constant dialogue between space and man. Like a fish in the water, a human body might use the new material as a medium for transportation. Can this new material be fluid? Can the fully adaptable space be poetic?

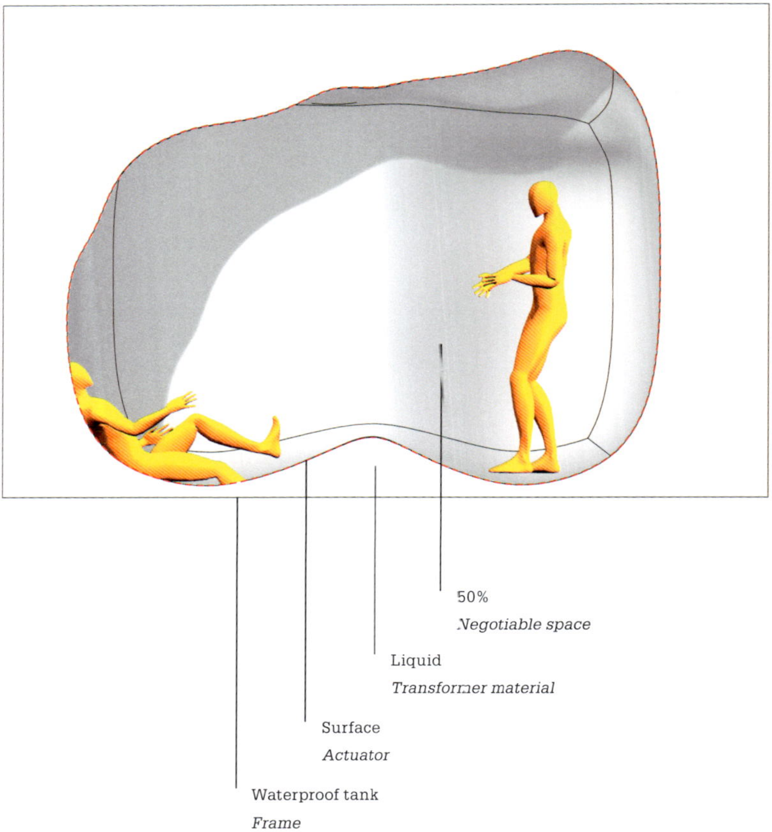
50%
Negotiable space
Liquid
Transformer material
Surface
Actuator
Waterproof tank
Frame

Suspended Liquids

Working with liquids did not seem to be a very practical way to create transformable spaces. But the schlieren created by the free flows of ink in water were so suggestive that we wanted to explore this technique further. We built a double-walled tank and tested the behaviour of different liquids mixed with each other. Our aim was to control these shapes. Ink dropped in water quickly dispersed and mixed. Oil in water did not mix, but rather gathered on the surface of the water. Hair gel in water remained suspended and did not mix with the water. We injected the gel with a syringe. This allowed us not only to 'draw' shapes but also to suck the gel back in and, in a way, to 'undraw' them. An interesting thought: if we could print and erase shapes and reuse the material, we would have a new way of creating transformable spaces.

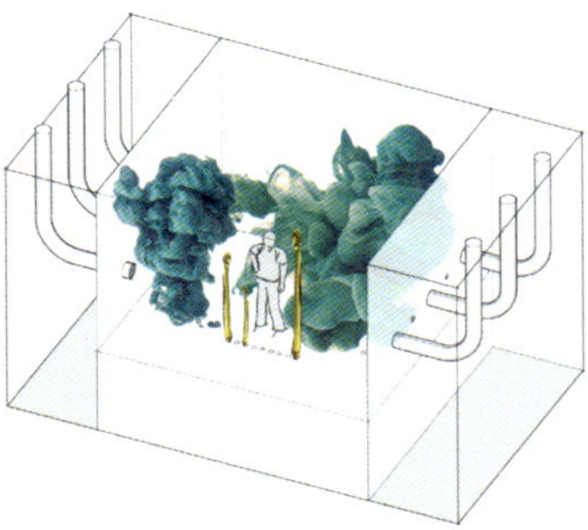

Erasable 3D Printing

We used the first experiments as an inspiration and moved on to more speculative studies. The first borrowed from the concept of 3D printing: A liquid material is deposited by a device and solidifies as it is built up. But here's the twist – the process would be reversible. An external stimulus would be used to change the phase of the material and return it to its liquid form. Our references here were photopolymers, which change property when exposed to light. Providing the polymers could alternate between solid and liquid states, the device could absorb them again, resulting in reversible 3D printing. To simulate how one would use this device to print different types of furniture, we made an installation. A motion-tracking device records your movements. Your hand acts as a kind of brush and is able suddenly to 'paint' in space.

Liquid Envelope

The installation showed that printing and erasing with your hand worked very intuitively. But we lacked a strategy through which the material could be stored when not in use. In the next step, we therefore imagined that we moulded a constant volume of material rather than having it appear and disappear. The reference here is an envelope filled with foam balls. To make this concept usable for our purpose, we want to be able to control the rigidity. We refer to the existing technology of Electrorheological fluids, which can change from liquid to gel and back when an electric field is applied. We imagined these fluids contained in a very flexible envelope. The electric field would be activated when you touch the envelope with your hand. We simulated this model with an interactive installation.

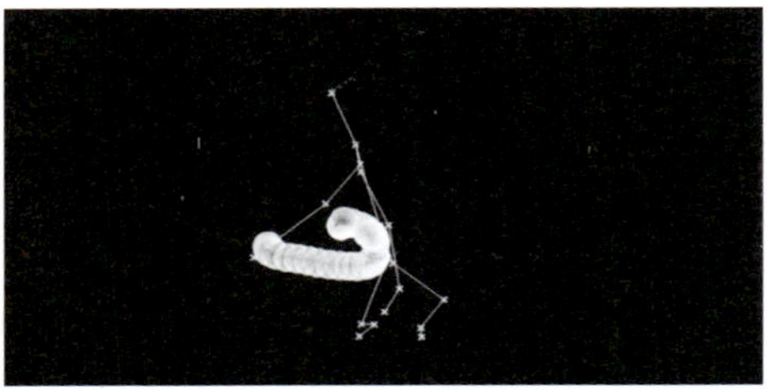

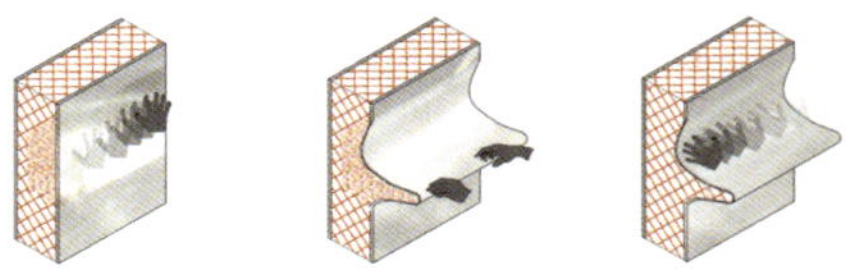

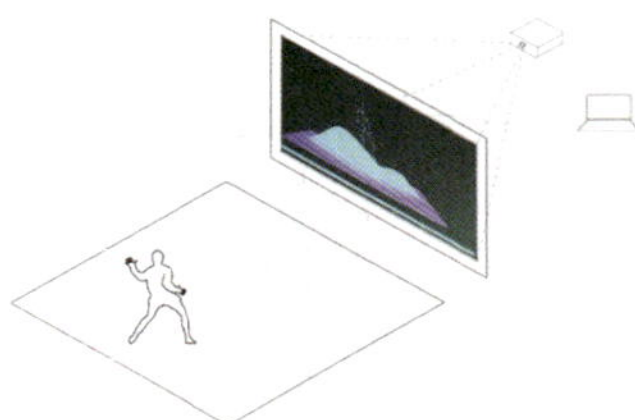

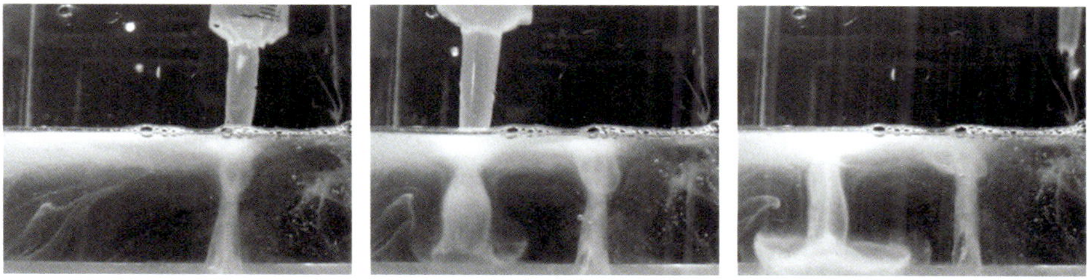

Glycerol mixed with ink injected in water

Oil mixed with water injected in water

Hair gel mixed with ink injected in water

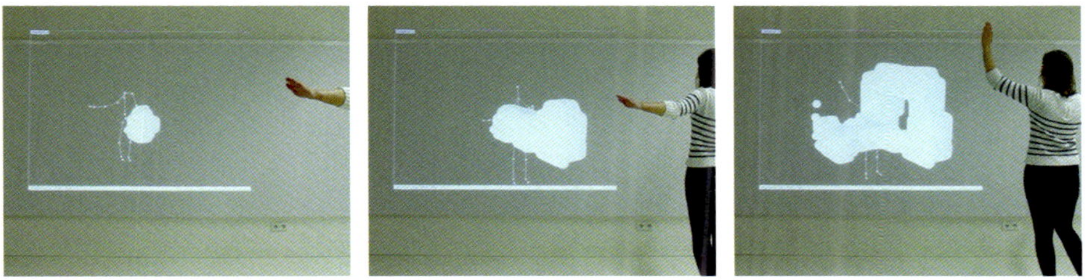

Installation: Erasable 3D printing with gestures

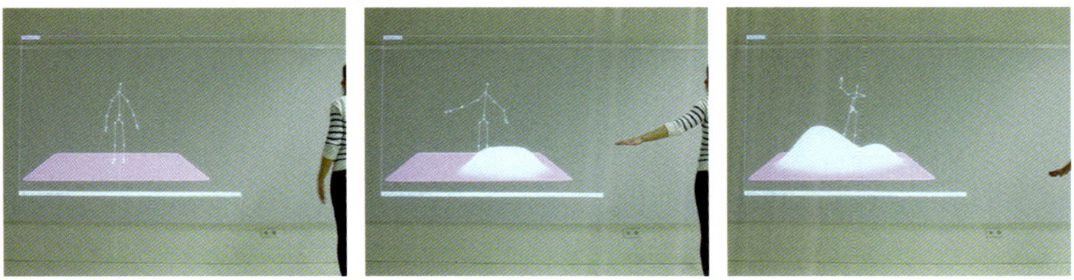

Installation: liquid envelope with gestures

Lex te Loo and Dominik Sartl

This project begins with a yellow cleaning sponge. The kind of thing that can be found in almost every kitchen. It has a clear, square shape, but if you squeeze it, it can easily be deformed. If you roll it up, you can reduce the size of the sponge to a minimum. The same happens if you wrap the sponge in an airtight plastic bag and suck out the air: it shrinks to a flat volume. We wonder if we can control this process in order to create different shapes. Could we build a room from transforming sponges? And how would it be to live in such a soft spongy space?

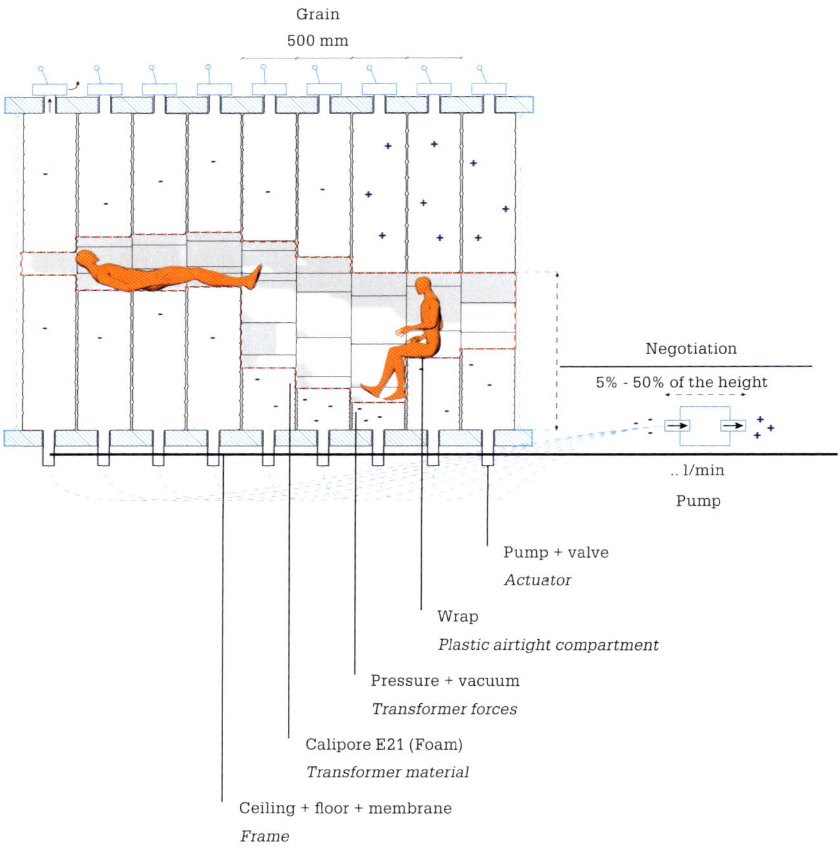

Grain
500 mm
Negotiation
5% - 50% of the height
.. l/min
Pump
Pump + valve
Actuator
Wrap
Plastic airtight compartment
Pressure + vacuum
Transformer forces
Calipore E21 (Foam)
Transformer material
Ceiling + floor + membrane
Frame

Deformation by Vacuum Force

We started with simple yellow household sponges, which we stacked to simulate a column or a wall section. We wrapped the sponges with plastic foil and attached a pump and sucked out the air. While the sponges shrunk, they created a crumpled shape that did not look very practical. We wanted the sponges to maintain their straight edges even after shrinking.

The Accordion Sponge

To achieve this, we first cut holes in the sponges, while keeping their outer surfaces intact. When applying negative pressure, the sponges now retained their straight edges until they were about halfway shrunken and then started crumpling again. To further improve the shape, we placed sheets of perforated plastic between the sponges. The deformation of a sponge between two plates would be mostly in the direction perpendicular to the plane of the plates. By combining multiple layers of sponges and plates, a column-shaped element could move up and down like an accordion.

Conducting Spaces

We then made a scale model of a room and filled the ceiling and floor with a grid of four-by-four sponge columns. All columns were sealed in plastic. We attached a valve to each column and connected it to a pump. We attached each of the 32 valves to a servomotor and the servomotors to a microcontroller board. We connected the microcontroller to a laptop and used visual scripting software to control the valves. This allowed us to program and activate different spaces. If no air pressure was applied, all columns were fully extended and touched each other. The space was completely filled by the sponges. Deflating one column, which would shrink back towards ceiling or floor respectively, created a small room. Applying more or less negative air pressure allowed us to control the height of each column. In the model, we were able to create all kind of spaces. To further improve the model, we used smart phones as an interface to control the individual columns.

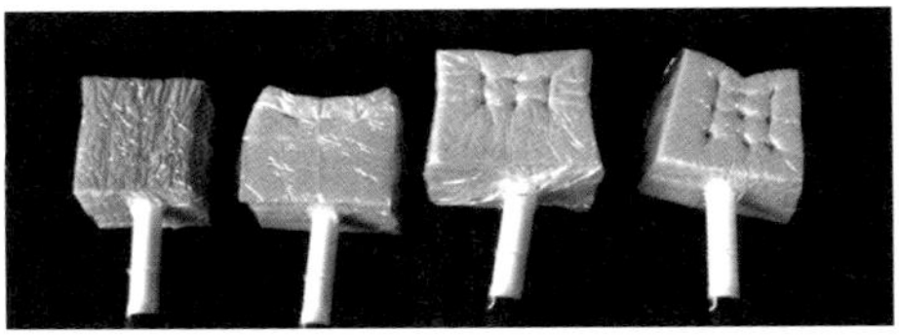

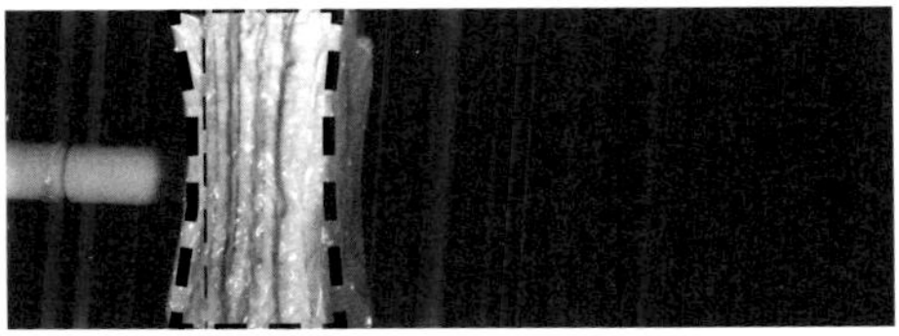

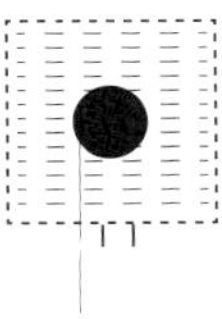

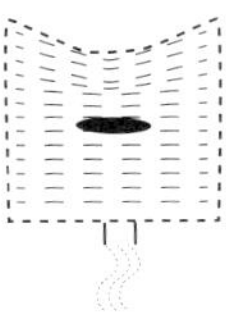

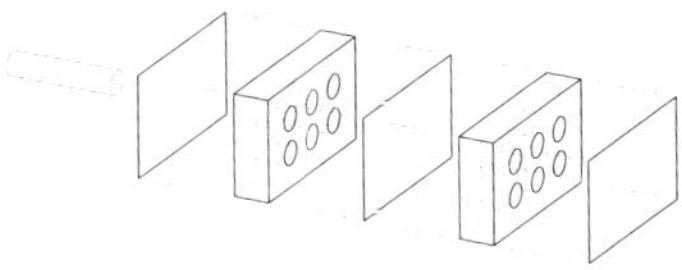

Bending and Complexity

To combine both space and ergonomics in a model, we proposed a layer of thinner columnar forms on top of a layer of thicker columns. The thicker columns provide stability and large spatial changes, while the smaller columns on top allow for more refined shapes and edges and thereby provide detail and comfort. Another study involved wrapping four columns to form one trunk. Deflating only one column made the trunk bend. This would make it possible to create curved shapes.

Soft Space Installation

The next step was to fabricate this on a one-to-one scale. We built large versions of the accordion columns: 50 x 50 cm wide and 1 m high. At the bottom of each column, we added a wooden box that was 40 cm high. When completely deflated, the sponge column is reduced to a fraction of its size and fits into this box. The box also holds the required valve and the connection to the pump. We created a grid of 16 – 4 x 4 m^3 – of these columns. Making use of the previous experiments, we were able to choose the right density of foam for this installation. The columns were easy to deflate, yet sturdy enough to carry the weight of a person. We were able to create different configurations that could function as stairs, chairs, beds or as new, diverse sitting landscapes intended for use by several people.

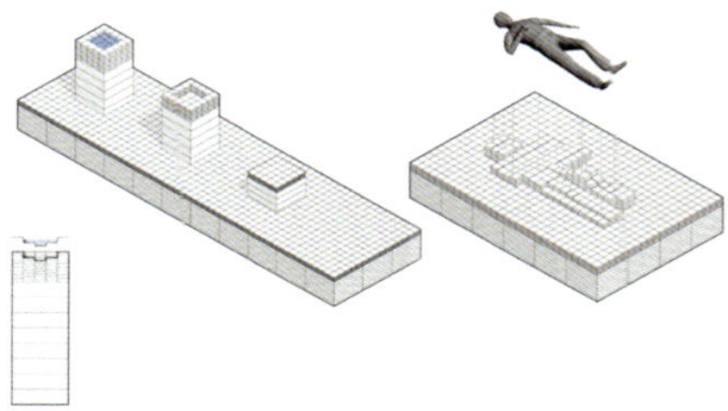

Box model: 4 x 4 grid of the layered sponge, making spaces

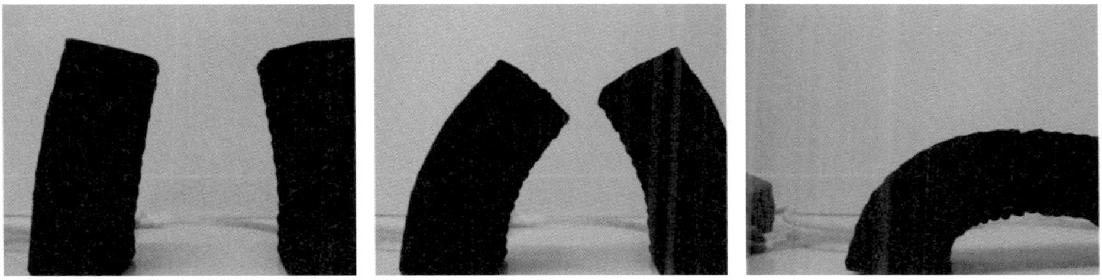

1:1 scale sponge installation: Creating a meeting space

1:1 trunk model: Bending up when user sits down

1:1 scale sponge installation: Basic shelter in a floor

1:1 scale sponge installation: Creating a meeting space

Felt

Piet de Reuver and Robbert Verheij

Textiles are close to the skin. We dress in textiles. They can protect and comfort us. When folded, they can be efficiently stored. Light clcth can dance in the wind and heavy cloth can keep us warm. Can textiles be our transformable material? Might a fully adaptable space keep us warm? We explored textiles as the basis for transformable space.

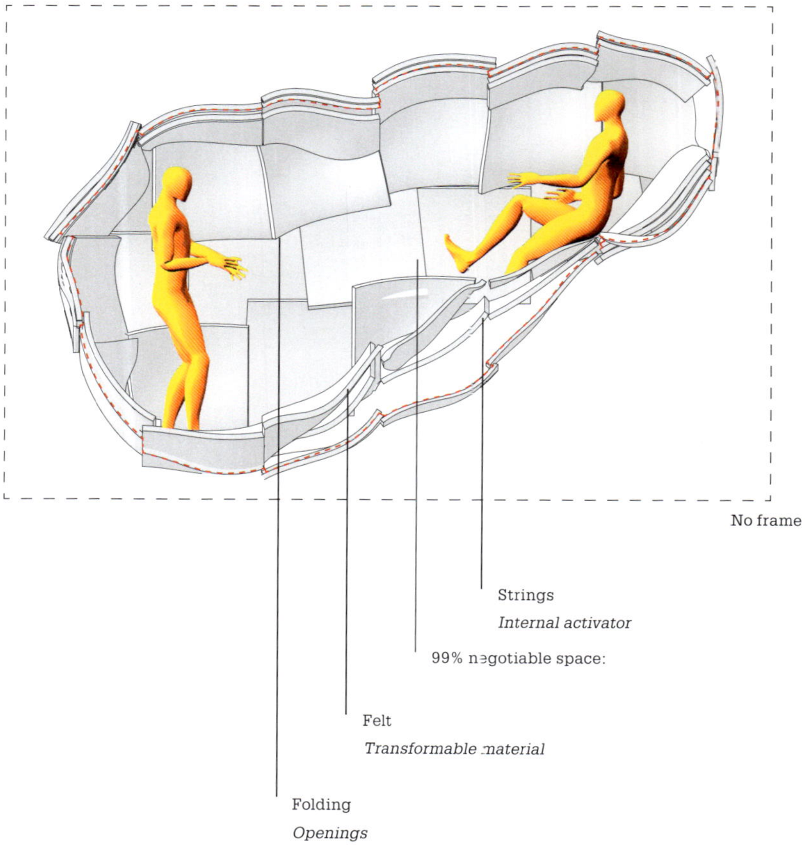
No frame
Strings
Internal activator
99% negotiable space:
Felt
Transformable material
Folding
Openings

Stiff Materials

We began our study with a series of tests using different kinds of textiles. We folded, bent and hung cloth in very flexible ways. Our experiments showed the potential of beauty of such soft spaces: we saw differences in opacity and stiffness that proved relevant to the way textiles could be applied. We chose to continue our experiments with felt. Thicker and stiffer than other textiles, felt gave us controlled motions and beautiful curved spaces. Making use of the material's stiffness, we created geometric shapes; we composed arches, peaks and ridges.

Muscle Fabric

The felt's stiffness did have disadvantages. For example, while it was easy to bend the felt in one direction, it was hard to manipulate it in two directions. To solve this, we cut the felt into strips. To control each strip, we sewed a tunnel to it and added a plastic tube. Pushing it in and out seemed to function as might an artificial muscle, bending and stretching the felt strip.

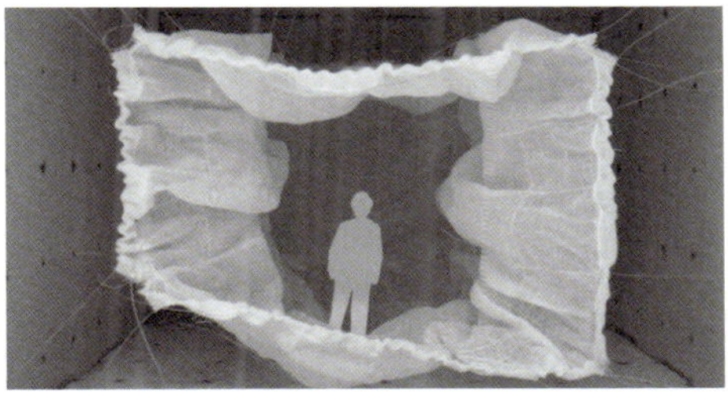

Pattern Furniture

To create more complex shapes – such as, for example, a table – we used two generally different approaches. We first cut a piece of felt and folded it to form a table. We attached strings to hold the shape in place. When releasing the strings, the stiffness of the felt pushed it to the sides until it was a flat piece of felt. In this way we were able to activate a table – to 'turn it on' when needed and to let it fall flat and become part of the floor when unused. This approach would require custom-made felt pieces for each piece of furniture and for each room.

Smart Felt

A modular system provided more freedom. One piece of felt could, at one moment, become a chair. At another moment it could become a table and then, later, part of the wall again. To achieve this, we designed a generic felt panel. The panel had a number of activation points capable of bending and stretching the panel. On its surface, the panel had a number of Velcro-like connection points that could be turned on and off. These points would allow one panel to attach itself to another panel. Turning the connectors on and off while bending and stretching would allow one panel to crawl over the other. In this way, several panels might stack themselves to form a wall, to create a ceiling, to become a room or to provide furniture. Felt repels water and insulates, so we could easily imagine such a cocoon becoming a soft and comfortable shelter.

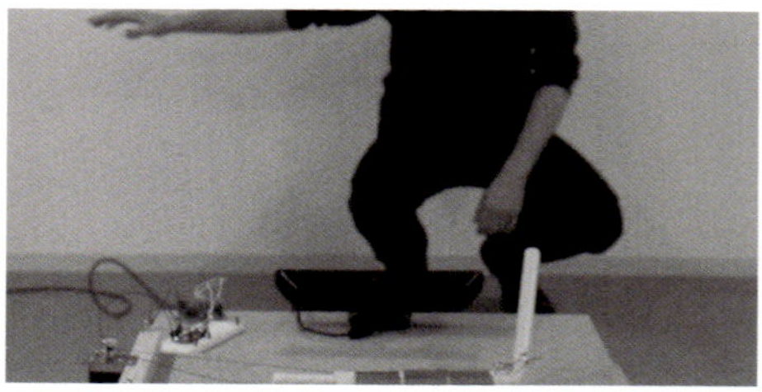

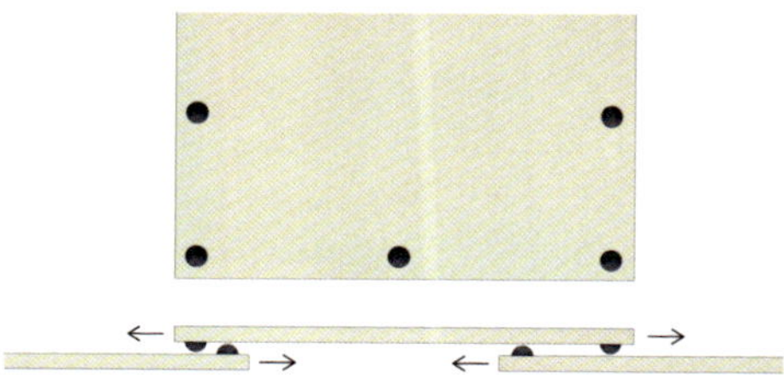

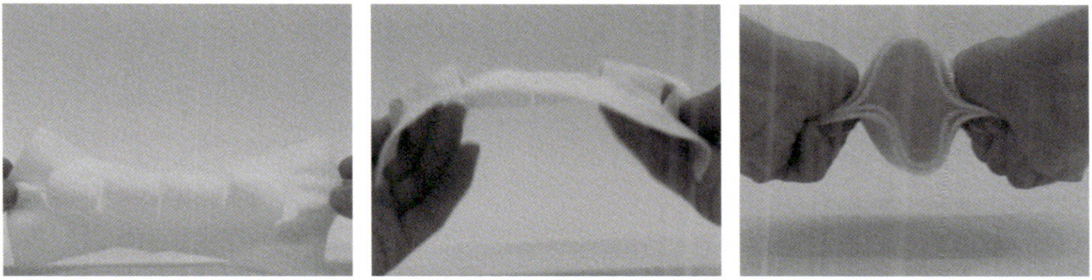

Testing felt properties

Simulating rigid felt

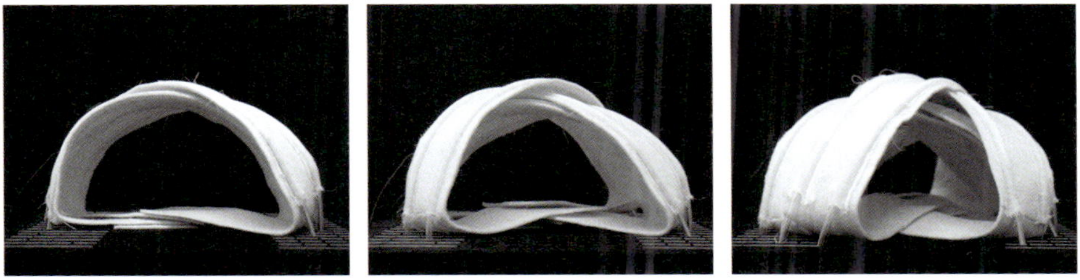

Onion model with tunnel activation

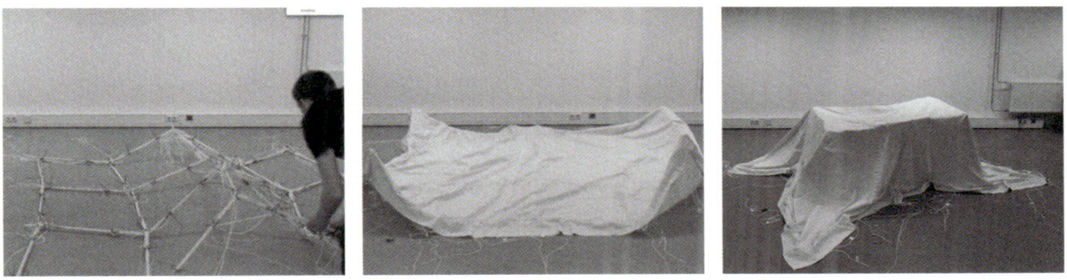

Skeleton with cover

Smart Felt simulation

The Rietveld Schröder House was built in Utrecht 90 years ago. The house became not only a beacon of the De Stijl movement, but also a forefather of adaptable architecture: its sliding walls allow the separation and connection of spaces and make it possible to adapt the space to different use configurations throughout the day. This adaptation is part of a three-dimensional composition that connects technology and art.

Rietveld Revisited

The Why Factory: Transformer Installations studio, TU Delft

Transformer Installations studio, TU Delft
Roxane Belot, Arslane Benamar, Nicola Campri, Sam Chia, Federico Gobbato, Nadia El Hakim, Ruth Hoogenraad, Cathrina Lee, Lex te Loo, Marine Manchon, Vincent Marchetto, Andrea Migotto, Rob Moors, Piet de Reuver, Dominik Saitl, Robbert Verheij, Hans Vlaskamp, Jordan Yerbury
Teachers: Ulf Hackauf, Adrien Ravon, Sander Mulders, Huib Plomp
Guest critics: Winy Maas, Marcel Bilow, Agata Jaworska

Mrs Schröder lived in the original Rietveld Schröder House with her three children and Gerrit Rietveld, who used the ground-floor as an office space.

We imagined a day in the life of this busy family. And then each research group used this script as a basis for imagining a day in the life of their version of 'Rietveld Revisited'.

06:00

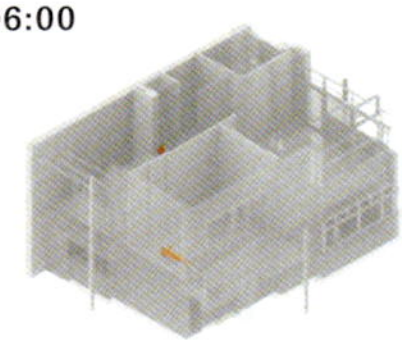

07:00

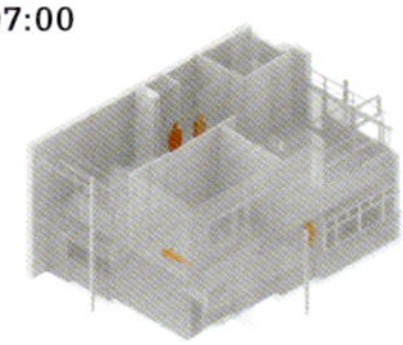

08:00

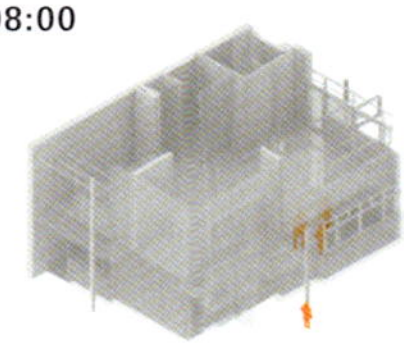

09:00

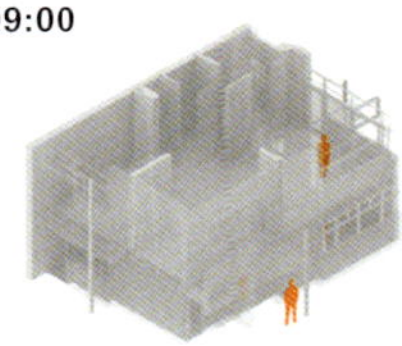

10:00

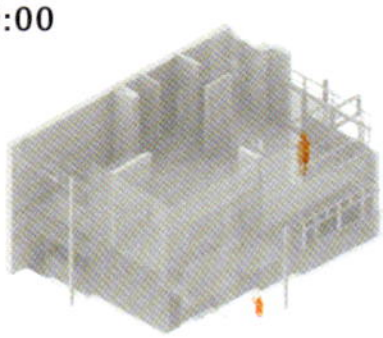

11:00

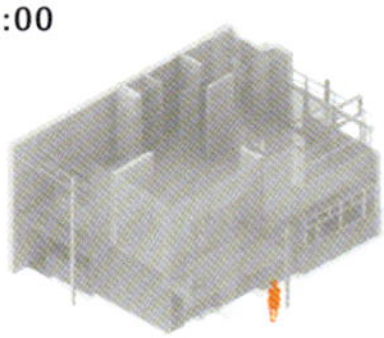

12:00

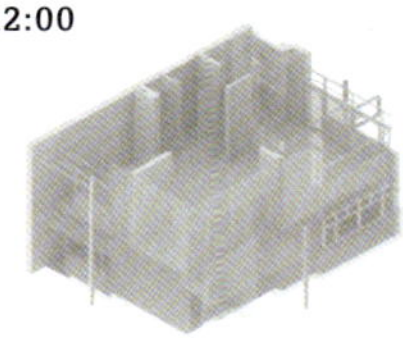

13:00

14:00

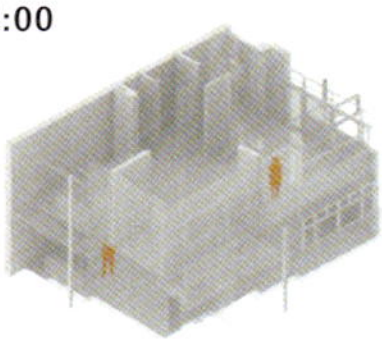

15:00

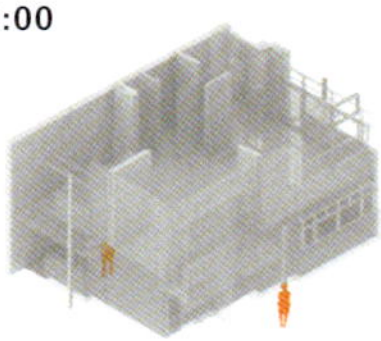

16:00

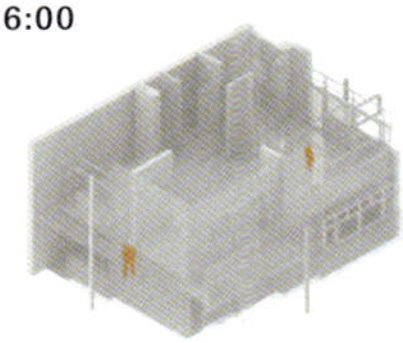

17:00

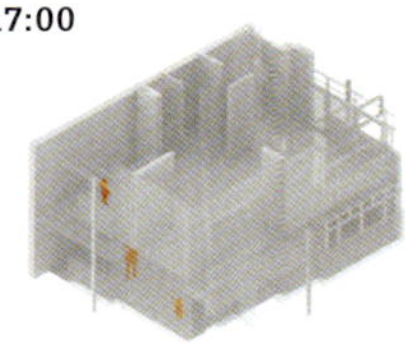

18:00

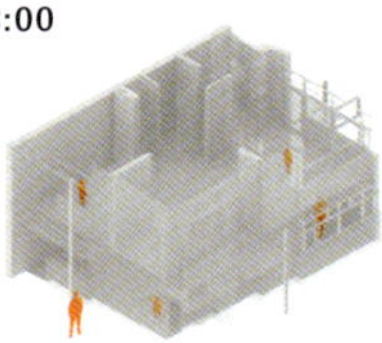

19:00

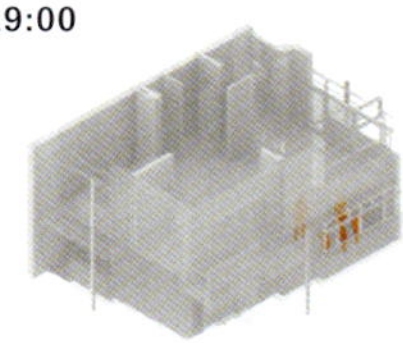

20:00

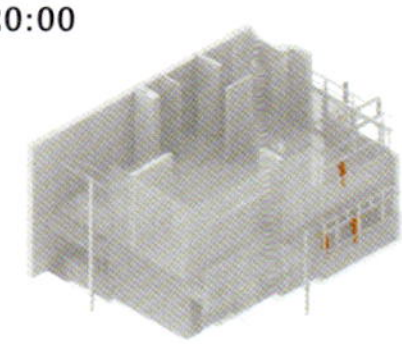

21:00

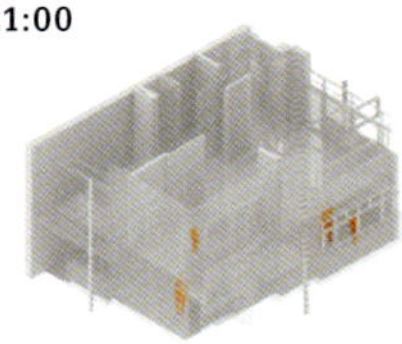

22:00

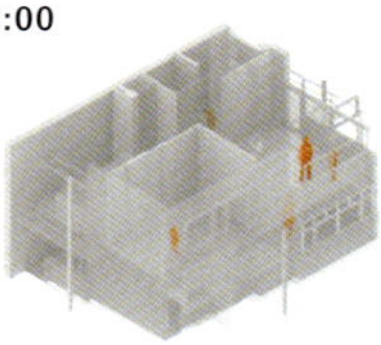

23:00

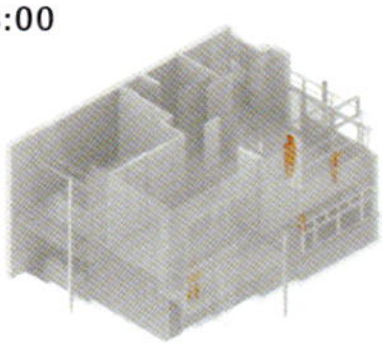

00:00

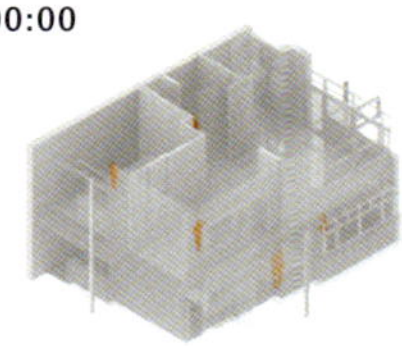

01:00

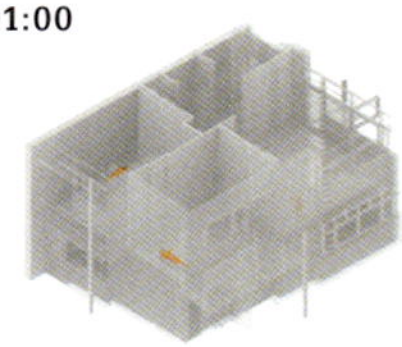

02:00

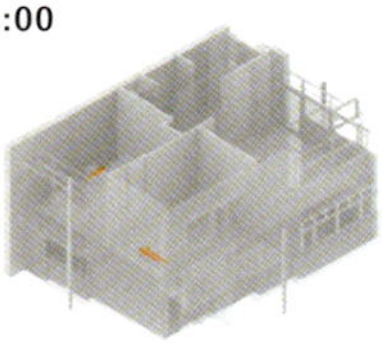

03:00

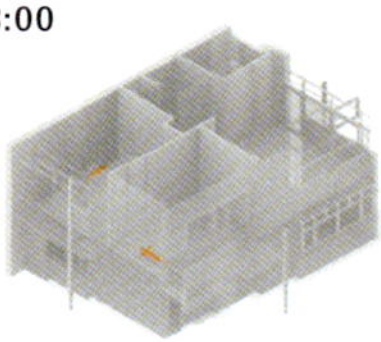

04:00

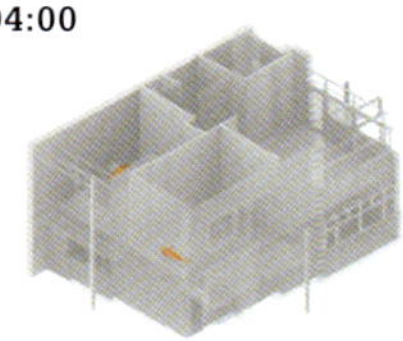

05:00

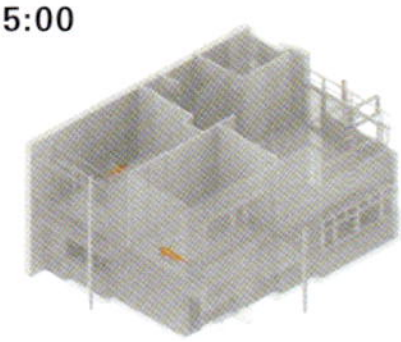

The Rietveld Schröder Schedule

An imagined day in the life of the Rietveld Schröder House

06:00
The man and the woman sleep. The daughter gets up and takes a shower. The other daughter and the son are both sleeping.

07:00
The man and the woman get up and use the bathroom. After a quick breakfast, the daughter leaves home. Her sister gets up. Their brother is still asleep.

08:00
The woman prepares breakfast and eats with the man and the second daughter. Their son is in the bathroom.

09:00
The man leaves for a meeting. The woman works at home. The daughter leaves for school. The son has breakfast.

10:00
The son leaves for school. The woman continues to work at home.

11:00
The woman leaves the house. The house is now empty.

12:00
Nobody is home and the house remains empty.

13:00
The man returns home and prepares lunch. The woman joins him and they have lunch together.

14:00
The man works in his office and the woman makes some phone calls.

15:00
The woman leaves the house and the man remains at work in his office.

16:00
The man continues to work in his office. The first daughter returns home. She does some fitness routines.

17:00
The man is still at work in his office. His son returns home. He and his sister do their homework.

18:00
The man does maintenance work in the garden. The woman prepares food. One daughter and the son both do their homework. The other daughter is reading.

19:00
The whole family has dinner together.

20:00
The woman and the man have a coffee together outside, in the garden. The first daughter and her brother clean up the table. Their sister is on the phone.

21:00
The man watches television with his daughter. The woman watches a different programme in another room. Her other daughter plays the guitar and her son does some exercise.

22:00
The man has a glass of wine and talks with his daughter, while his wife reads a book. The first daughter still practices the guitar. The son takes a bath.

23:00
The woman, her husband and their daughter all read. Their other daughter uses the bathroom and then goes to bed. The son watches television.

00:00
The woman, her husband and their daughter all use the bathroom and go to bed. The son continues to watch television.

01:00
The son watches television. The rest of the family is asleep.

02:00
Eventually, the son uses the bathroom and also goes to bed.

03:00
The woman wakes up and uses the bathroom.

04:00
Everybody is asleep.

05:00
Everybody is asleep.

To honour this inspirational project, we used the Rietveld Schröder House as a test case for the conceptual technologies discussed in the previous chapter. Each group shows how their innovations would play out if applied to the famous house in Utrecht. How well would the proposed technologies cope with the dense schedule of the house's inhabitants? And what kind of sculptural spaces and artistic languages do they create?

Inflatables

Chia Hui Qi Samantha and Vincent Marchetto

House under constant pressure

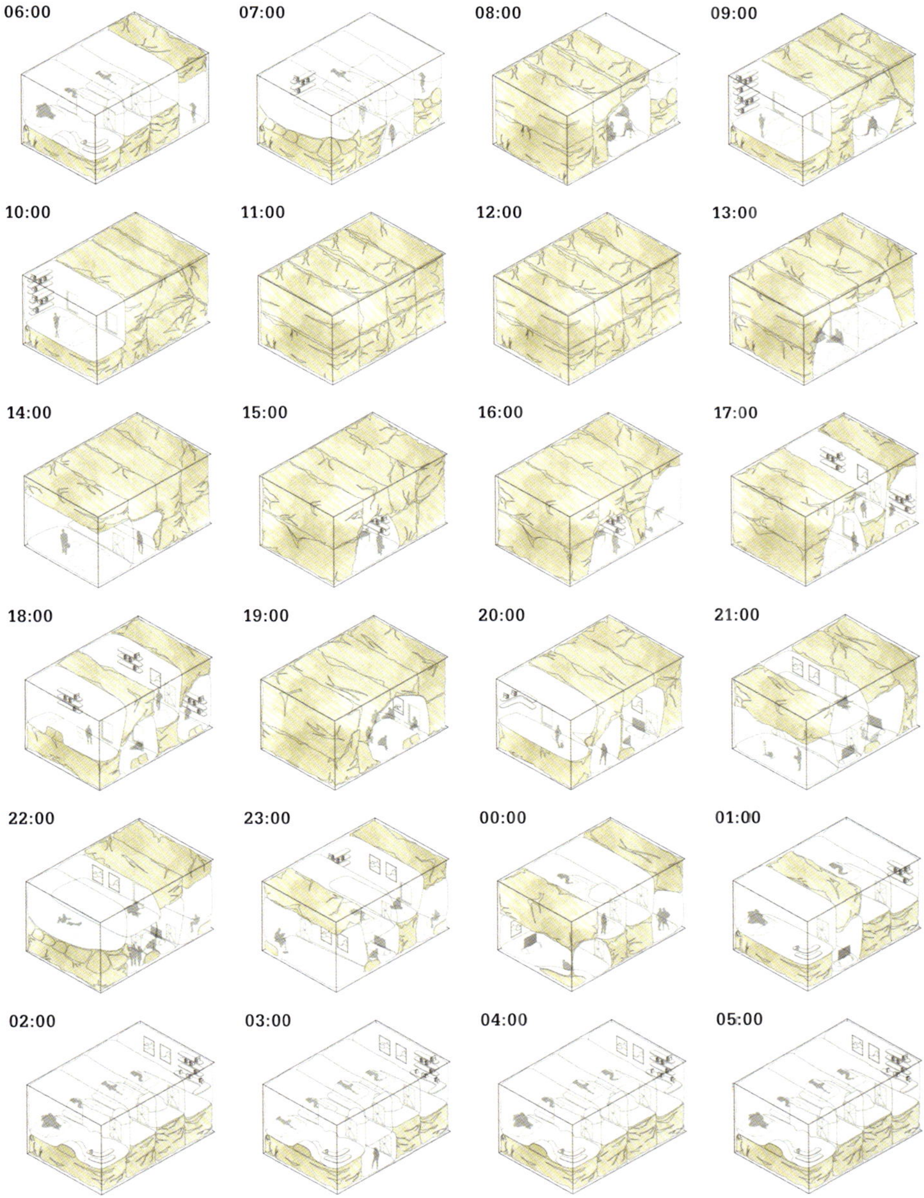

Elastics

Chia Hui Qi Samantha and Vincent Marchetto

Rigid framework with tensile planes

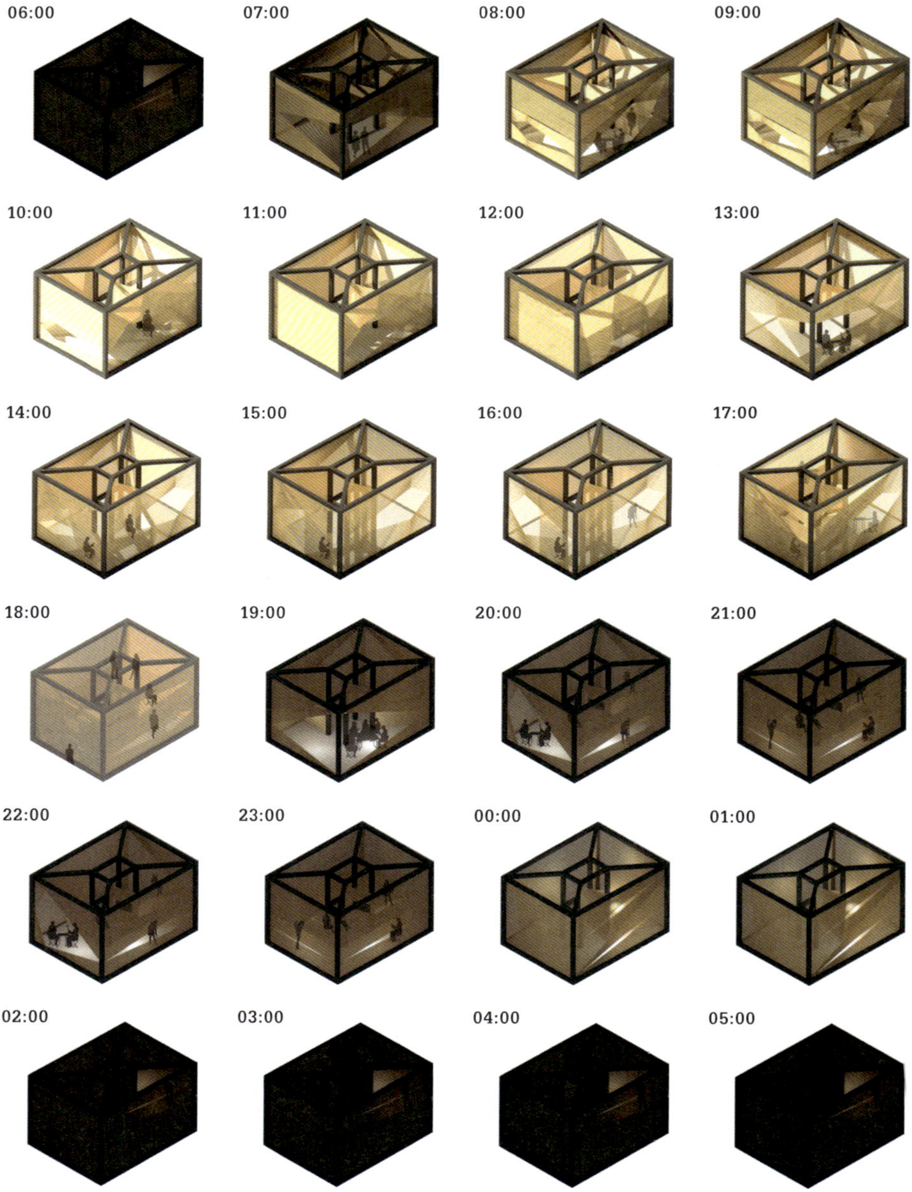

Drones

Arslane Benamar and Federico Gobbato

Swarm of bricks, constantly reorganizing

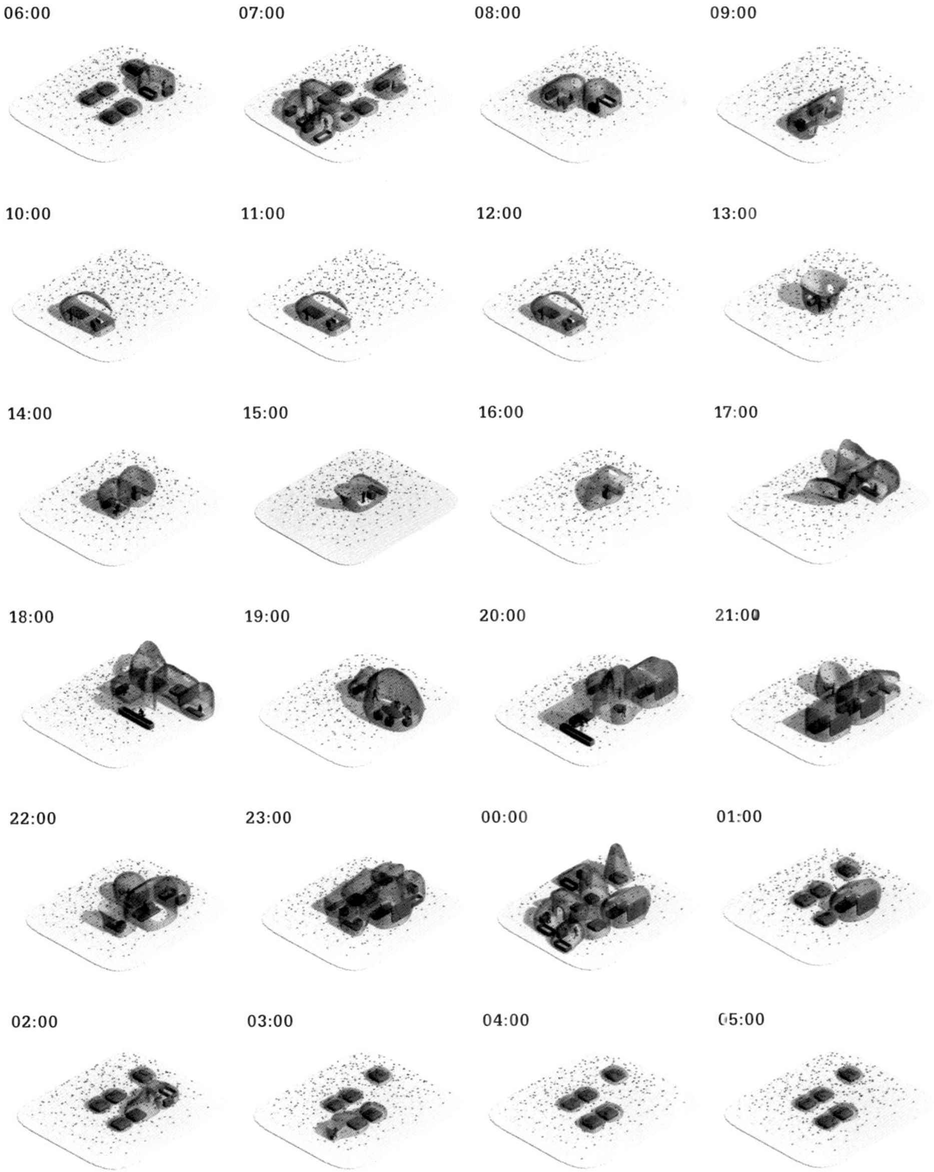

Wood

Rob Moors and Jordan Yerbury

Sturdy, animated log cabin

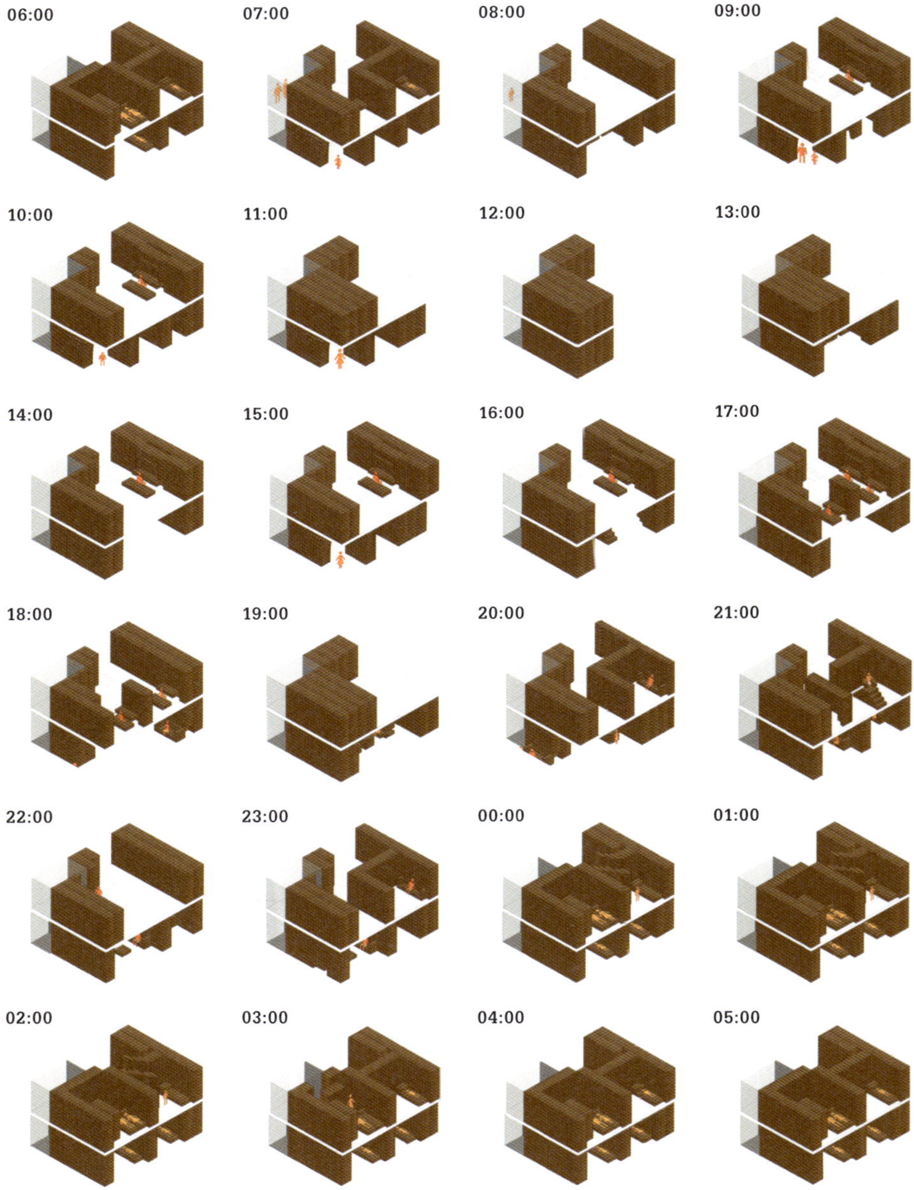

Mesh

Ruth Hoogenraad and Hans Vlaskamp

Steel waves, folding rooms

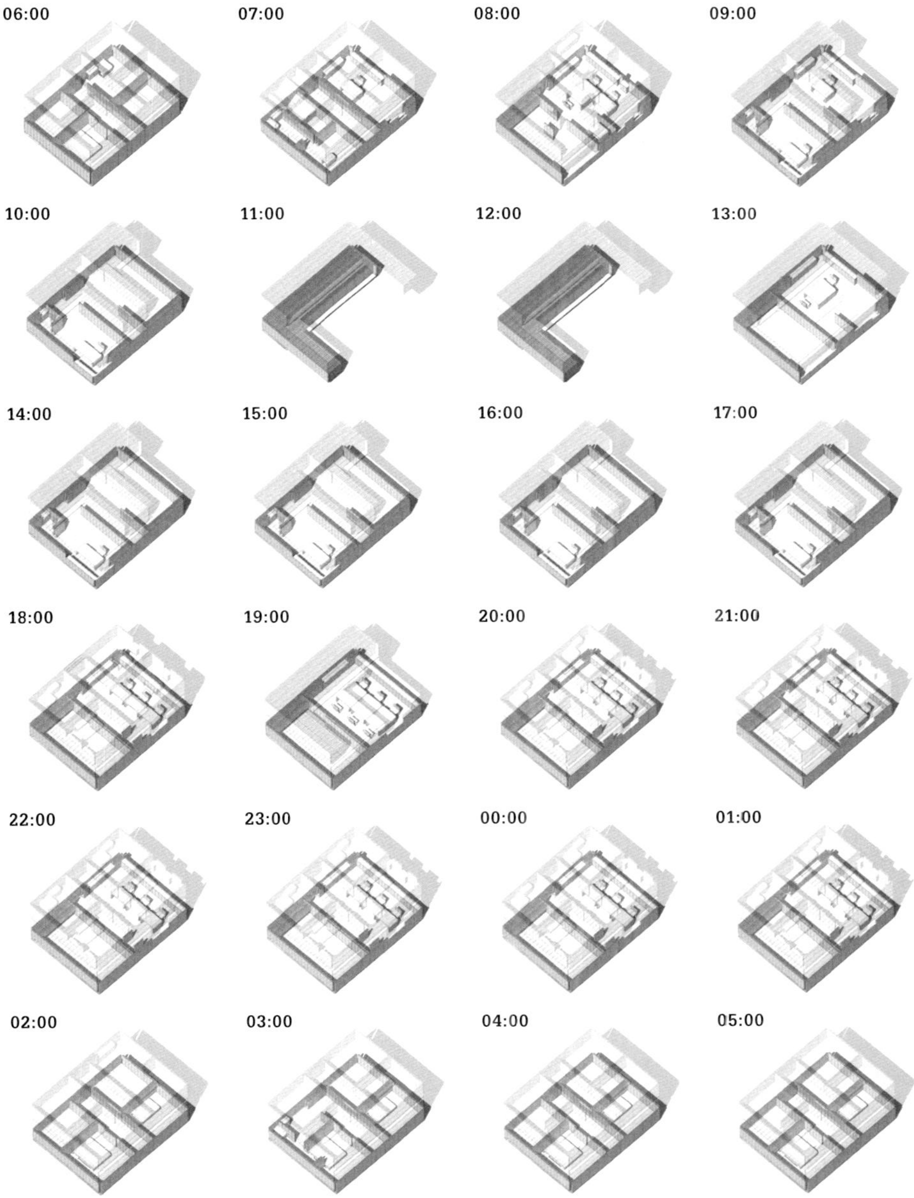

Light

Nadia El Hakim and Marine Manchon

Ephemeral spaces, appearing and vanishing

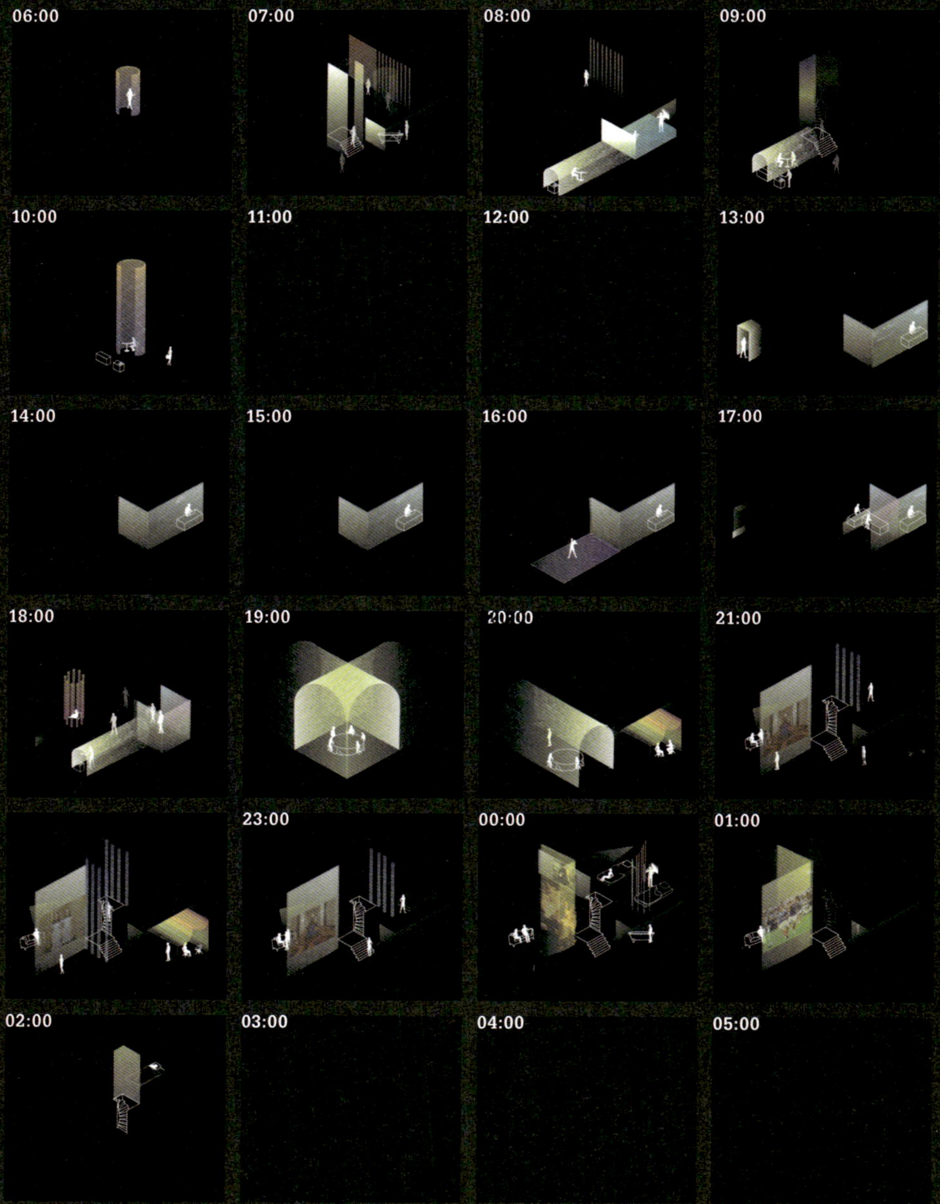

Liquids

Roxane Belot and Cathrina Lee

Constantly building and unbuilding your house

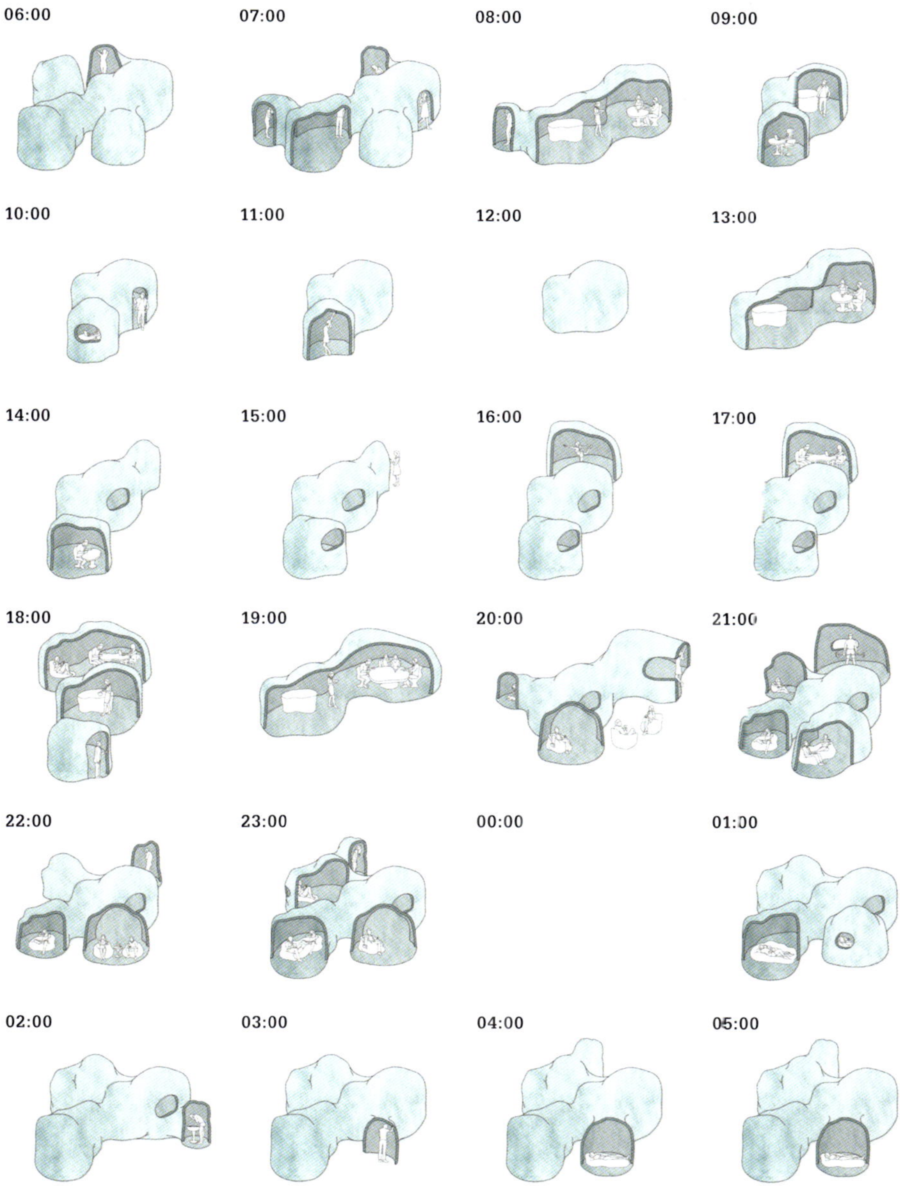

Sponge

Lex te Loo and Dominik Saitl

Massive, soft block with continuous caves

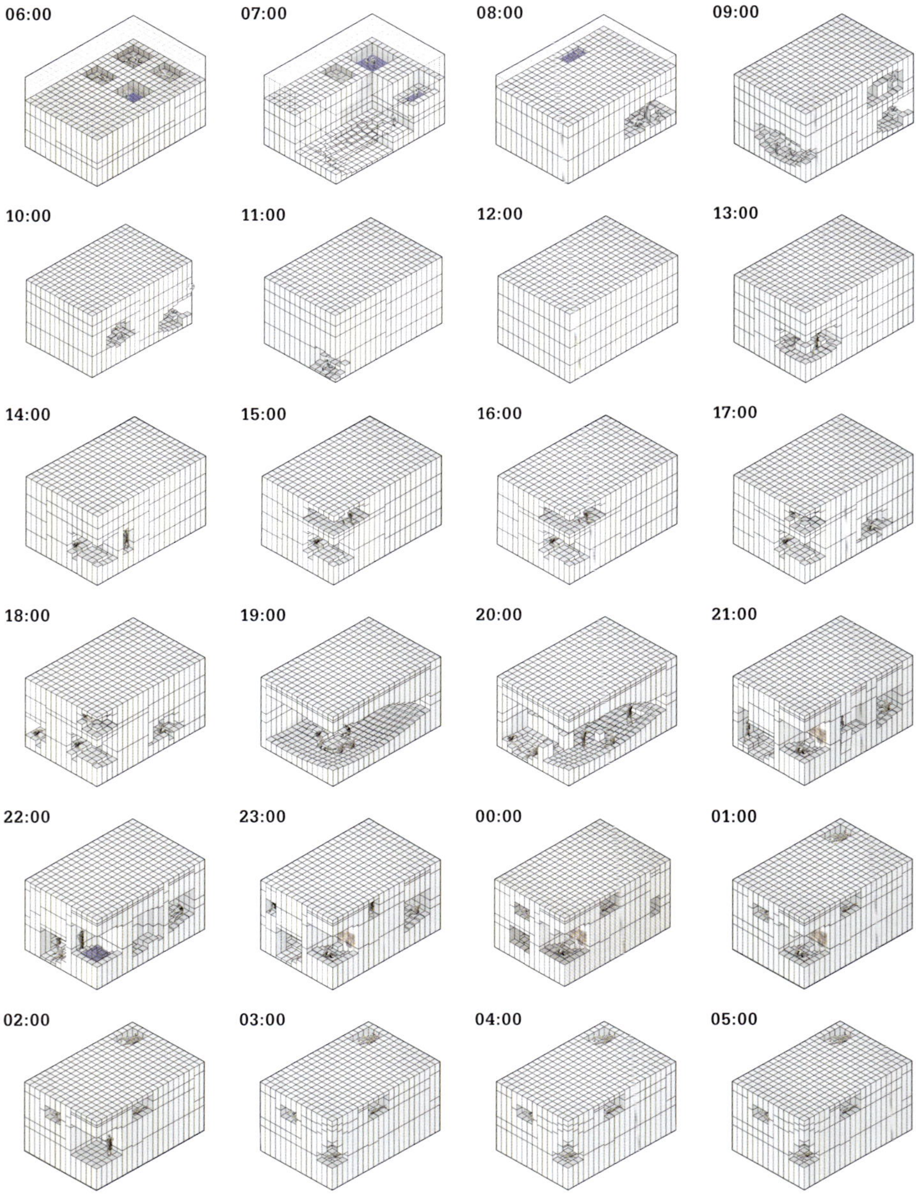

Felt

Piet de Reuver and Robbert Verheij

Soft caves and garment-like shelter

06:00

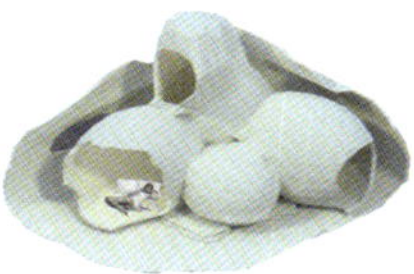

07:00

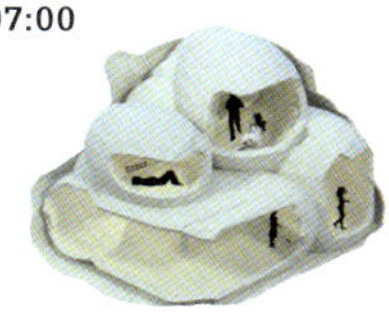

08:00

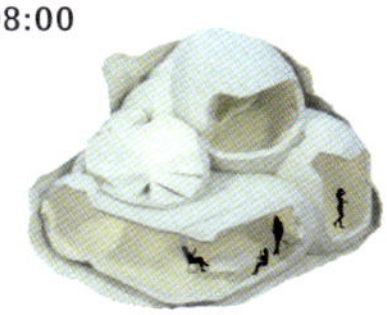

09:00

10:00

11:00

12:00

13:00

14:00

15:00

16:00

17:00

18:00

19:00

20:00

21:00

22:00

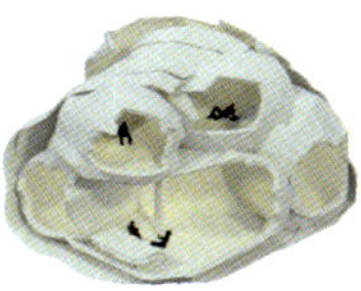

23:00

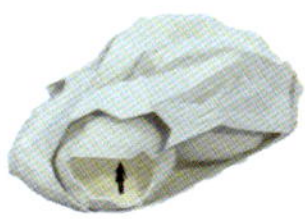

00:00

01:00

02:00

03:00

04:00

05:00

The three individual studies that follow consider visions of full adaptability in relation to the reality of daily needs and desires. All three were originally thesis projects conceived at The Why Factory.

The first project considers the smallest architectural scale: the human body and the space it needs and creates. A set of tools is developed to precisely notate the body's spatial requirements during different activities. Step by step, the bare minimum is translated into a usable space without ever generalizing or compromising. The challenging goal is the achievement of minimal yet livable space, space that is never larger than needed. The goal is inhabitation without wasted space.

Watch the movie:
www.thewhyfactory.com/barba

Inhabiting Zero Wasted Space

The Why Factory: Stavros Gargaretas

Adaptive City thesis studio, TU Delft
Teachers: Ulf Hackauf, Adrien Ravon, Huib Plomp with Winy Maas

Tracking and Logging Space Usage

The Halo Maker described in the Space Makers chapter records the movement of a body over time. This is done in two steps:

1 The Halo is intersected by a vertical plane defined by the position of the spine. The intersecting curve between the two describes the position of the body at any moment.

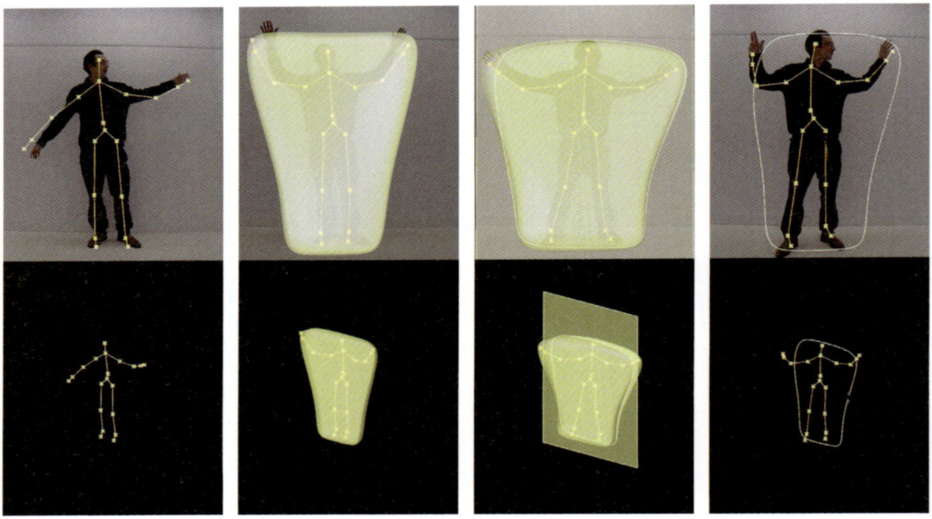

2 Every two seconds, this curve is frozen and moved in space. This creates a log of movement in space over time.

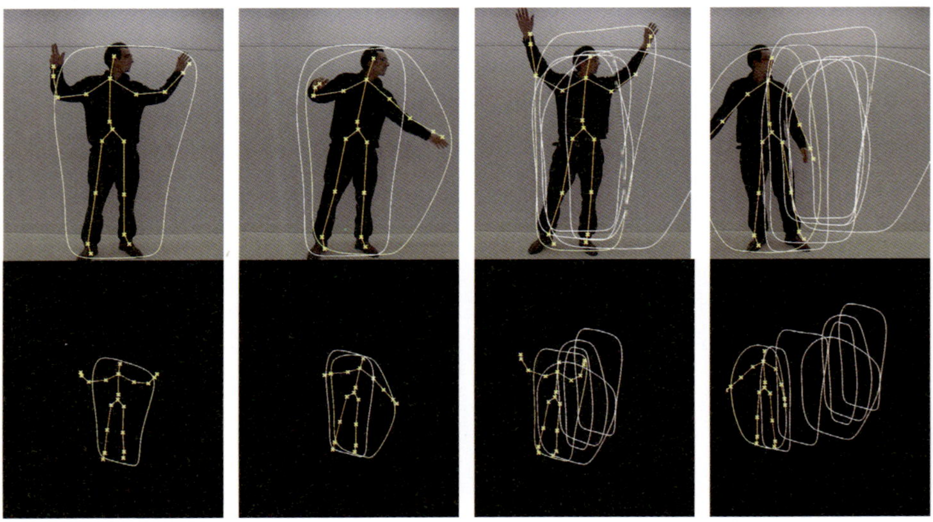

after 0 sec | 2 sec | 8 sec | 12 sec

Visualizing our life as a series
of two-second timeframes
timeframes, which we can access.

The Mapping Tool

The mapping tool uses these logged curves to create a spatial model of movement over time. This can be visualized in two ways:

1 Space over time: an endless worm – a whole life potentially – of space as actually used.

One person

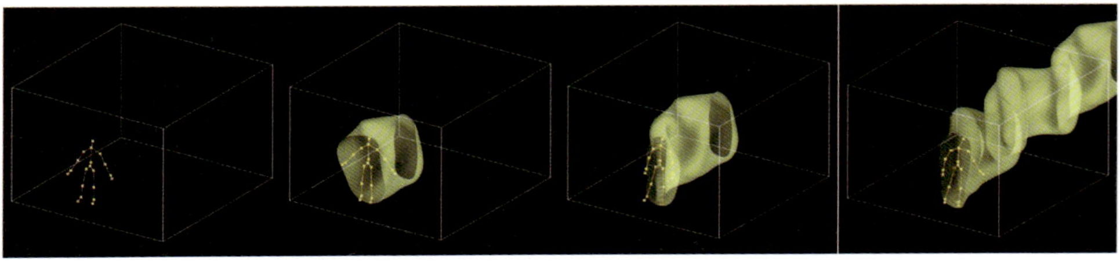

Two people

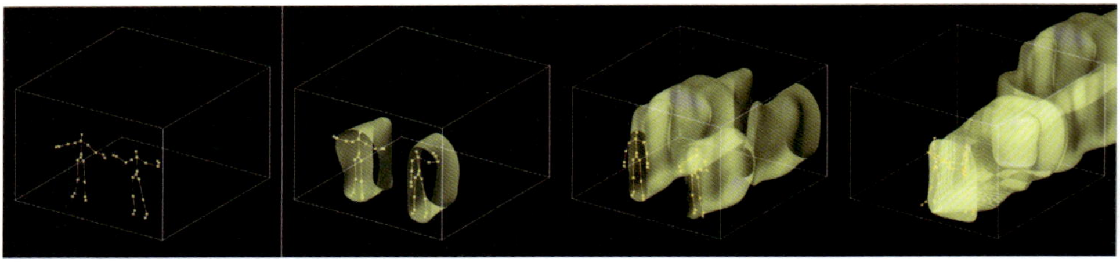

1 sec 4 sec 8 sec 24 sec

2 Space with time collapsed: the total space used during mapping.

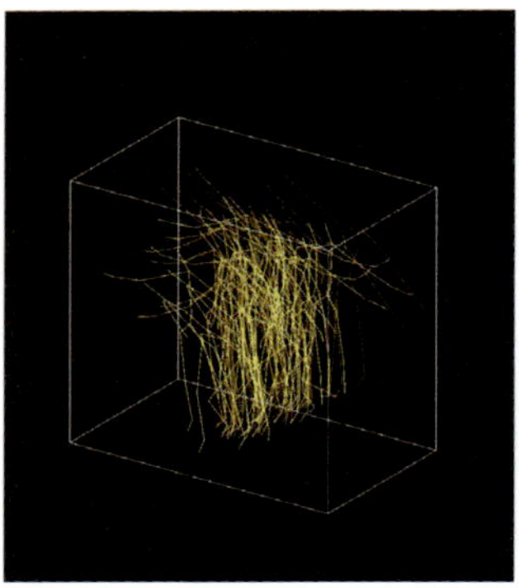

Skeleton log
catalogue of all the postures of the human during the time of mapping

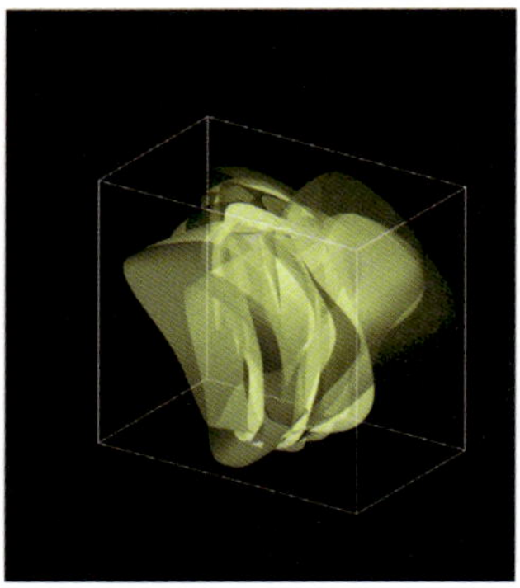

Space log
a catalogue of all the spaces that were required for the different activities of the human

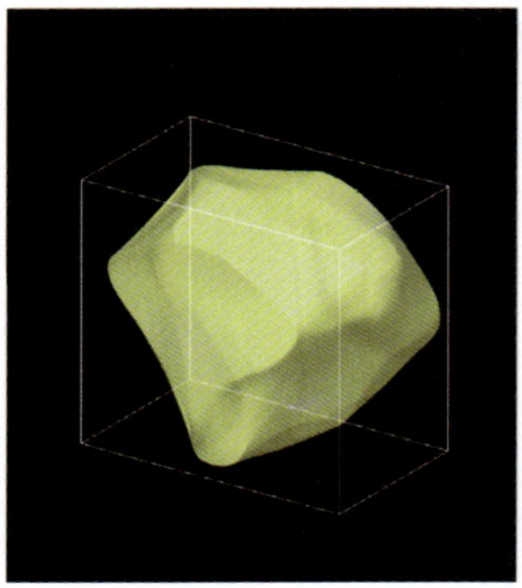

Volume log
a complete calculation of the volume used during the time of mapping

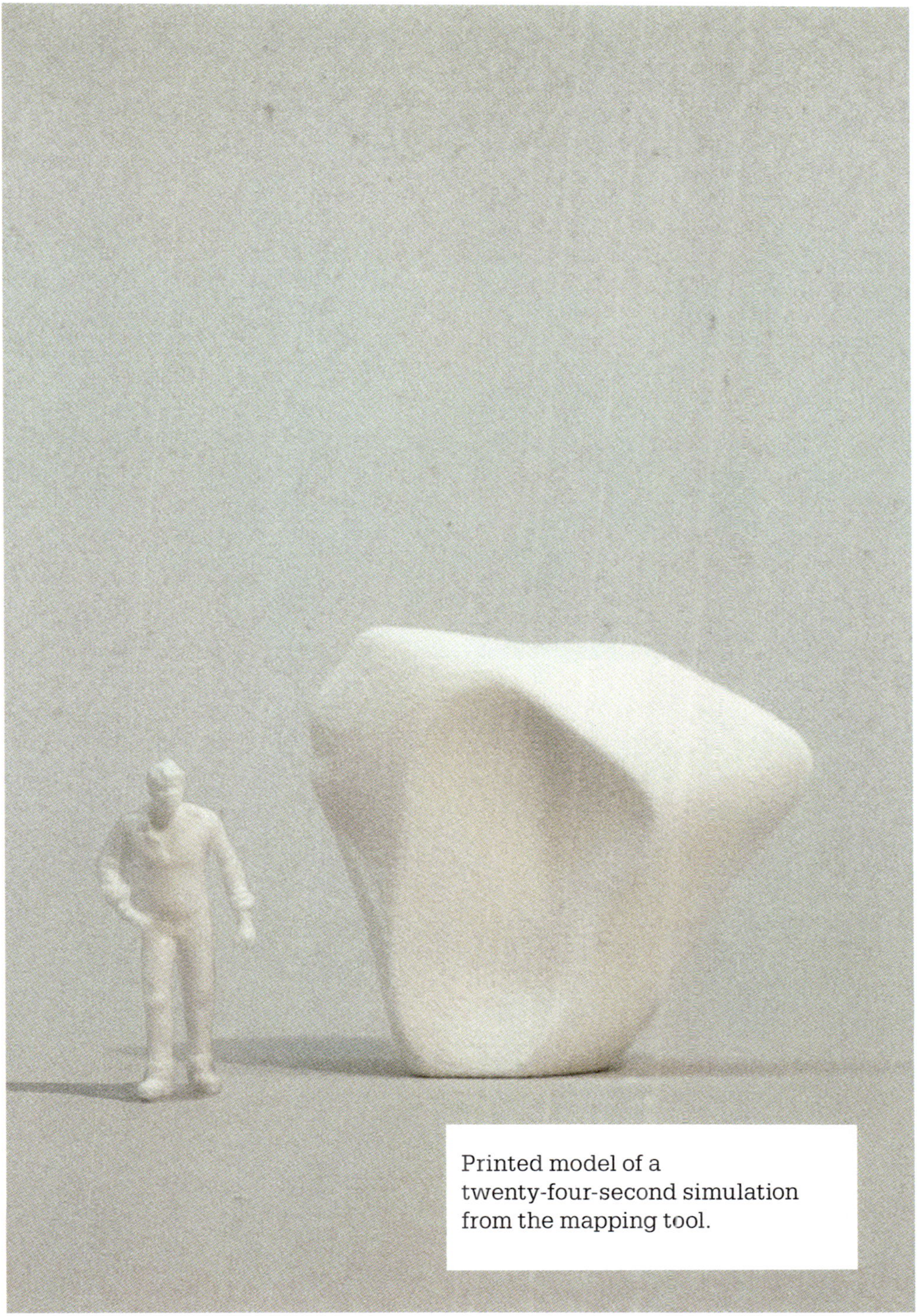

Printed model of a
twenty-four-second simulation
from the mapping tool.

Mapping Human Relations

The research continues as we test how the mapping tool might work in relation to different human interactions. Ettore Scolas's movie, *Una Giornata Particolare* (1977), is a good object for this study. Important scenes from the movie are acted out and mapped, demonstrating how the mapping tool reads the interaction between several people. When do two Halos merge, when do they stay separate and when do they collide?

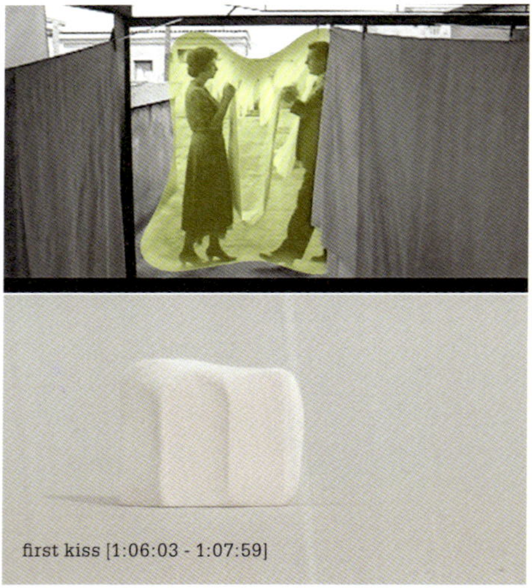

first kiss [1:06:03 - 1:07:59]

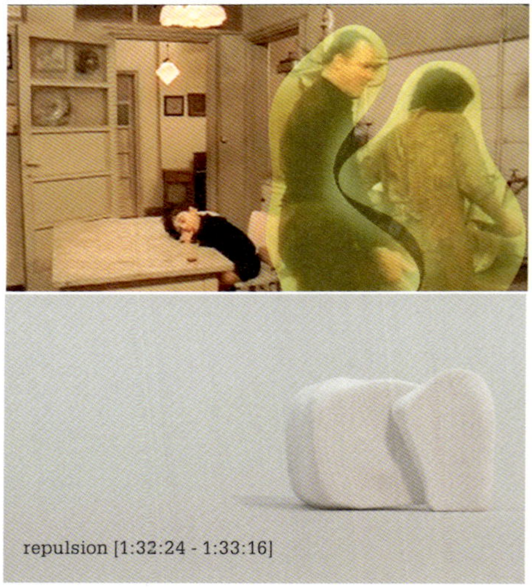

repulsion [1:32:24 - 1:33:16]

coexistence [1:13:02 - 1:14:25]

passion [1:20:20 - 1:24:25]

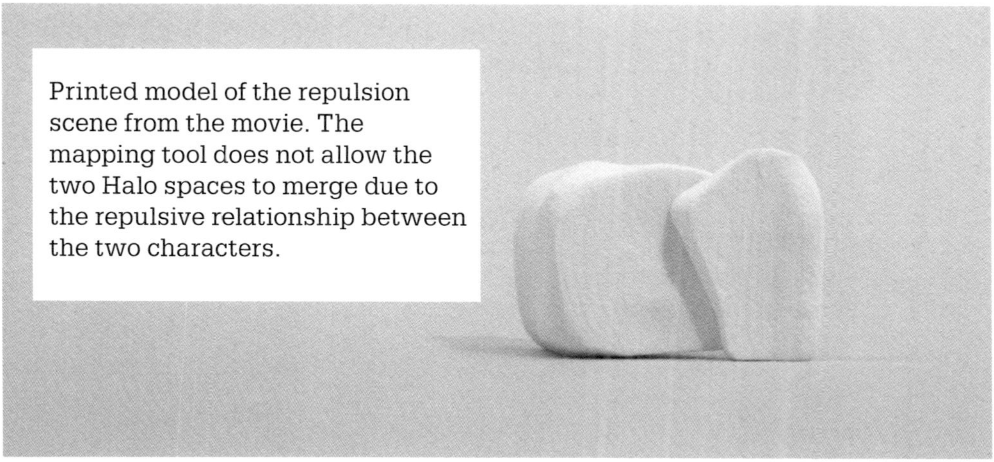

Printed model of the repulsion scene from the movie. The mapping tool does not allow the two Halo spaces to merge due to the repulsive relationship between the two characters.

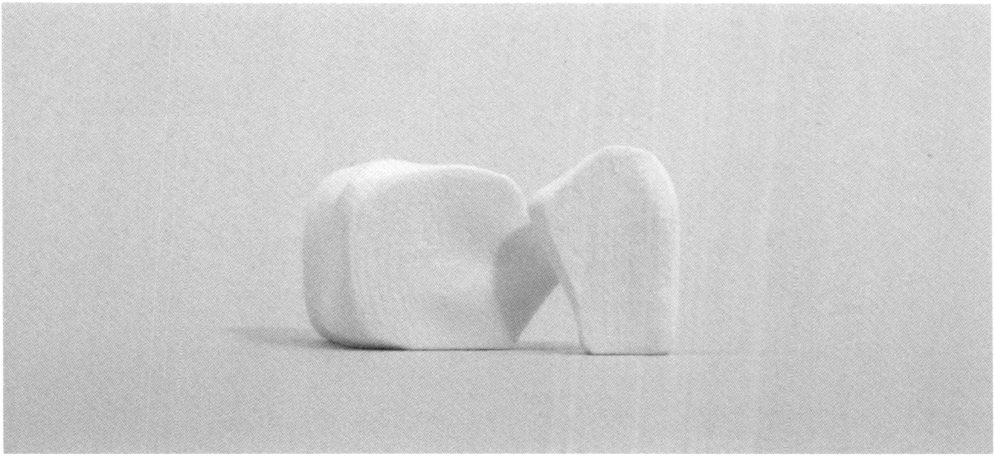

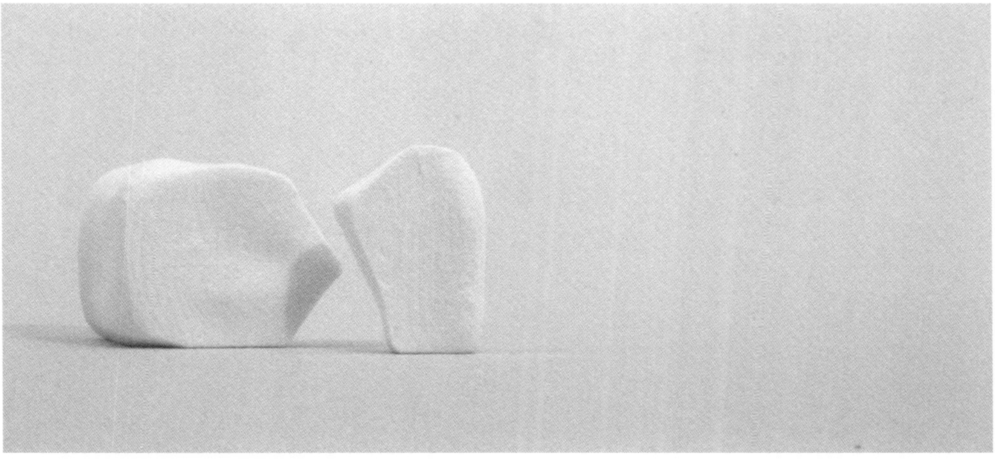

Mapping Different Activities in the Day

The daily activities of a four-person house are simulated and mapped. Twenty-minute time frames are cut from the worm of space usage models and three dimensionally printed. These are models of Zero Wasted Space.

1 Examples of different daily activities simulated and mapped.

[08:00 - 08:20]

[21:00 - 21:20]

2 Zero Wasted Space models generated using the Mapping Tool.

[18:00 - 18:20]

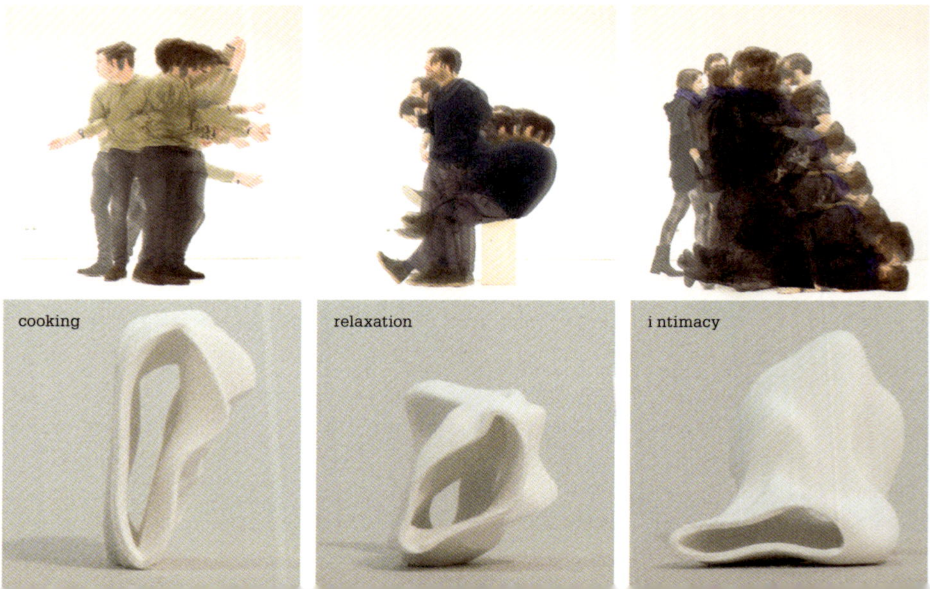

Zero Wasted Space Models for four timeframes in the day a four-person family.

Defining Conditions for Coexistent Daily Activities

What temperature, noise, light and structural conditions need to be fulfilled in order for two people carrying out completely different activities to coexist at close proximity? The role of this chapter is to write a program for a new kind of material. How must this ergonomic boundary behave so that we can comfortably carry out our daily activities?

Studies on the different spatial and environmental conditions necessary for daily activities.

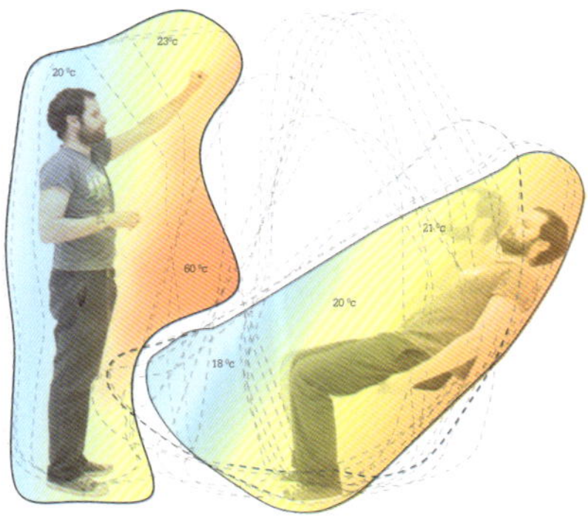

Temperature (0C)
temperature transfer from the cooking to sitting activity must be controlled

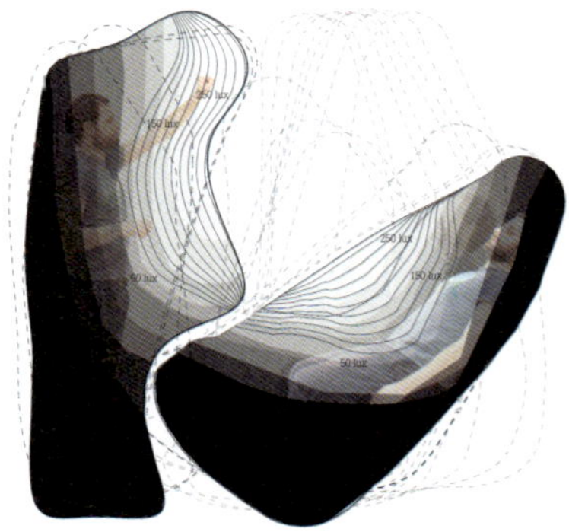

Light (lux)
light requirements for the two overlapping activities

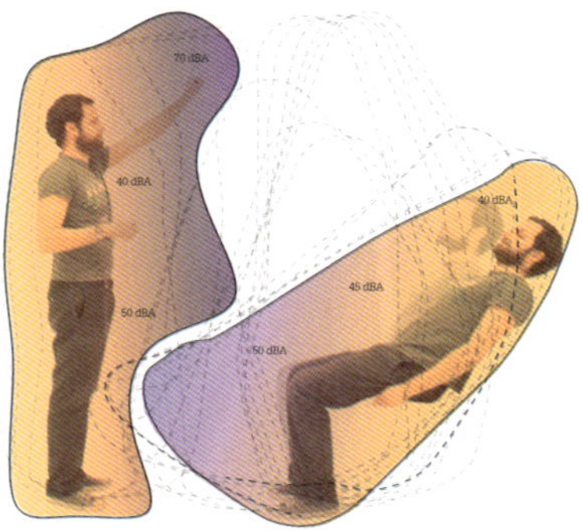

Noise (dBa)
noise transfer should be controlled from the cooking to the sitting activity

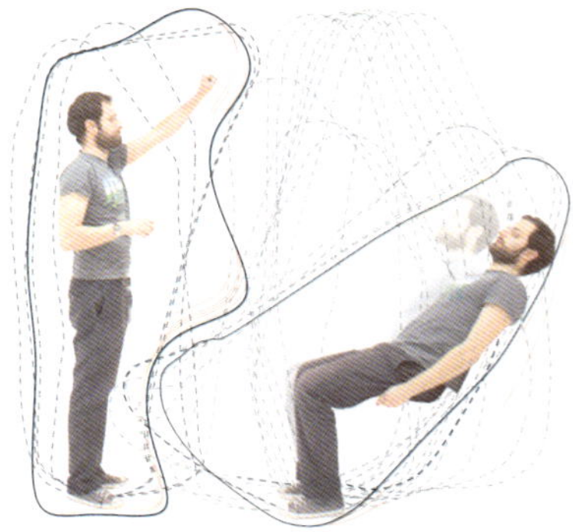

Structural requirements
structure should accomodate for overlapping conditions between the two activities

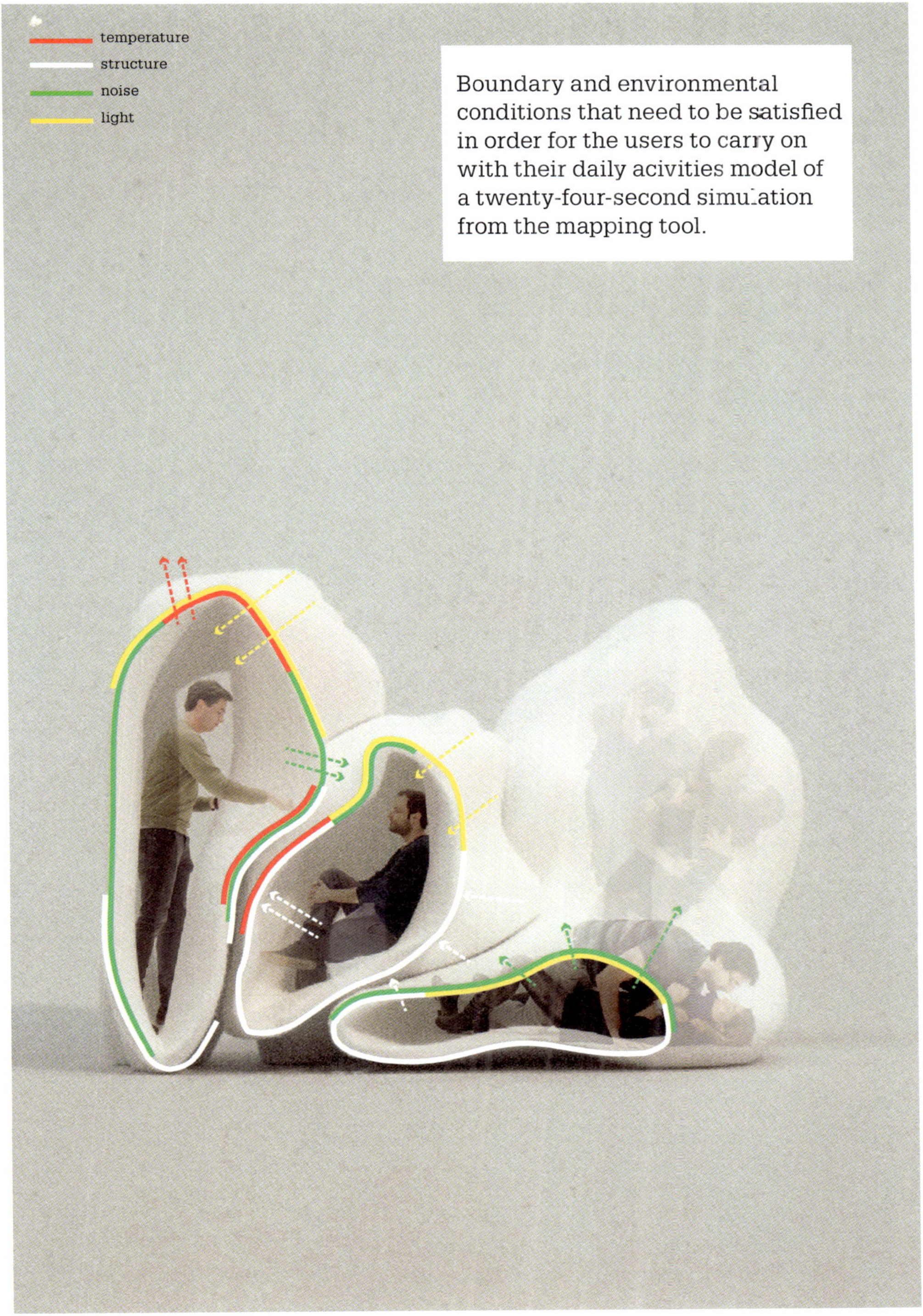

Boundary and environmental conditions that need to be satisfied in order for the users to carry on with their daily acivities model of a twenty-four-second simulation from the mapping tool.

Adding Comfort and Liveability to the Model

This chapter explores the 'ergonomics of comfort' through life-scale tests. Subjects simulated different daily activities while a membrane closed in on them. Their comfort levels were monitored and comfort factors were derived. This allows us to begin to make assumptions about how much space must be added to the model of Zero Wasted Space to prevent it from becoming claustrophobic.

One person studying

One person reclining

One person cooking

Two people talking

Expansion of space after the comfort factors have been added to the model of Zero Wasted Space.

The Interface: Adding Comfort and Liveability

In theory, if a human is capable of interacting with the recorded information he or she can further determine ways to add comfort and liveability to the 'model of full ergonomy'.

The Interface: **mock up**

As comfort is ultimately personal, we have to consider how we might customize Zero Wasted Space. This exploration forms the basis of the interface, the second tool considered in this research.

This work suggested a new one-to-one relationship between the user and the space he inhabits.

The Interface: **prototype**

Toggle menu
User toggles through a menu to chose how to change the space
*right hand on stomach

Opening in room
User selects location and size of opening.
*distance between hands denotes opening size

Furniture maker
A secondary menu allows one to select specific furniture
* right hand on shoulder
– toggle through with left hand

Furniture maker
After specific furniture is selected, the user can chose location and height.
*two hands together – control the furniture height

Mockup of the one-to-one relationship between a user and his space. How Does one add comfort to the Zero Wasted Space model?

From Zero Wasted Space to Inhabited Zero Wasted Space

A new model of increasingly inhabitable spatial efficiency emerges as a result of our own interaction and feedback. The project concludes with a model of the shape created during a twenty-minute time frame in the life of a four-person family.

Mock-up of how each user adds comfort to the Zero Wasted Space Model

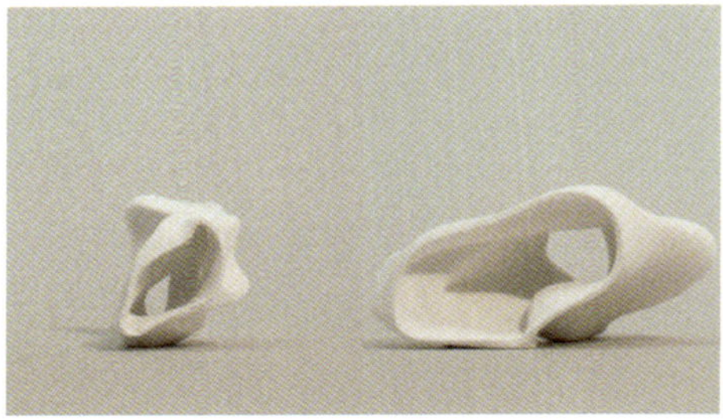

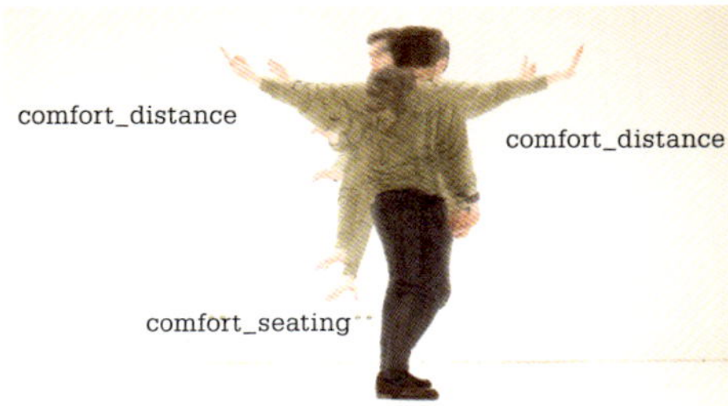

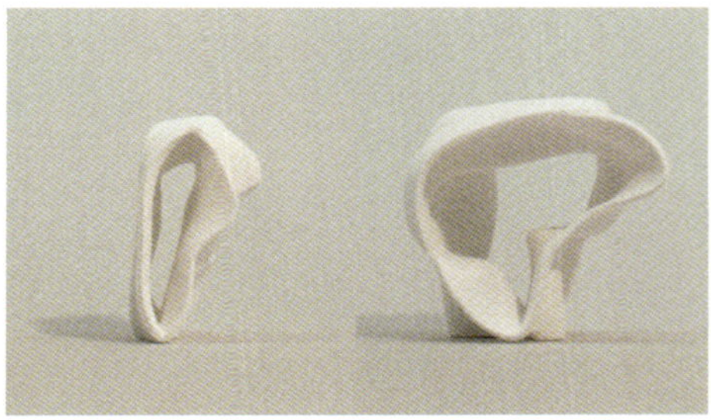

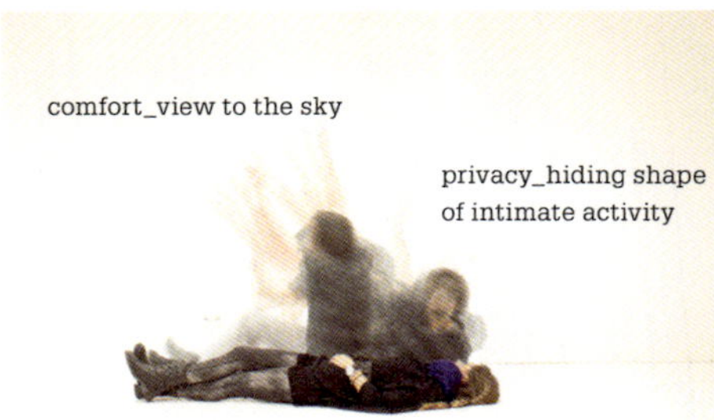

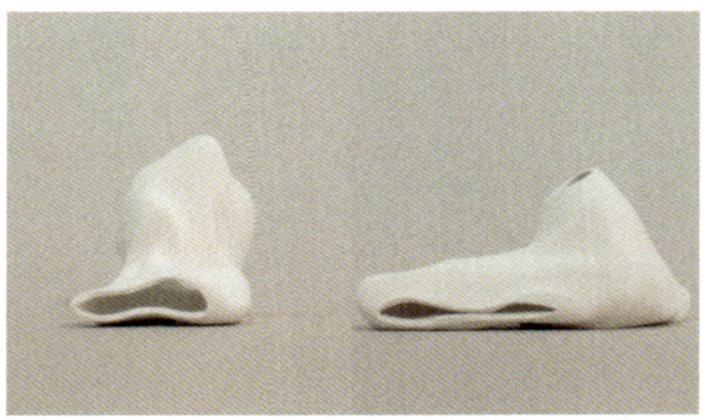

Inhabited Zero Wasted Space
[18:00 - 18:20]

The second project in this group deals with specific desires and dreams related to housing. An image-based analysis provides key indicators with which to design dream homes for a group of real but anonymized clients. These designs are projected onto an urban plot too small to accommodate the extensive homes. In response, the buildings become adjustable and, using a series of adaptive technologies, are able to accommodate different individual 'dreams'.

Aiming to both accommodate individuality and allow for standardization, the project suggests a compelling juxtaposition of the core aspirations of both urbanity and sub-urbanity.

Adaptive Dream Home Tower

The Why Factory: Jayson Johnstone

Adaptive City thesis studio, TU Delft

Teachers: Ulf Hackauf, Adrien Ravon, Huib Plomp with Winy Maas

Sub-Urbanity

City dwellings have been arrays of copied elements, layouts and spaces too long. It's time to give more options to dwellers, to let them be in control of their own homes!

In much of North America and in select parts of Europe, the rapid success of a post-war, suburban revolution led to an unprecedented fulfilment of individual desires. Today, the suburbs seem to represent all that is bad: unsustainability, sprawl, lack of cosmopolitan culture, racial segregation, protectionism, etcetera. [1] While the city, by contrast, can be described as mixed, lively and energetic, its dwellings are becoming more and more generic. Can we find a way to both fulfil our most individual desires and the guarantee the collective benefits of the city?

The 'Adaptive Dream Home Tower' is a means to investigate 'dream home' desires and to configure such dream homes in a city. Dwelling desires and needs are met by physical transformations that reduce compromises.

This one-year thesis research project explores dream homes, desires and adaptations in a series of miniature research undertakings that imagine a utopian means of breaking the city's spatial confines in order to fulfil desires.

This research consists of six parts.

1 Beatriz Colomina, 'Mourning the Suburbs: Learning from Levittown,' in *Public*, ed. Steven Logan, Janine Marchessault, and Michael Prokopow (Spring 2011), 87–97.

The Resident's Profile Study

Can dream homes be designed through social profiles?

We analysed over 1,000 images collected from Pinterest – a social media site where people from all over the world create boards of images that represent their imagined dream homes. These 1,000 images were categorized according to a series of 56 items. This categorization identified the type of room, architectural elements that defined the space, types of views, and types of materials. The image analysis produced a 'Resident Profile Card'. This card holds all the specified resident's valuable design information – from budget and job information to architectural element and view preferences. All the information collected was available publicly, on social media sites from Facebook to Instagram.

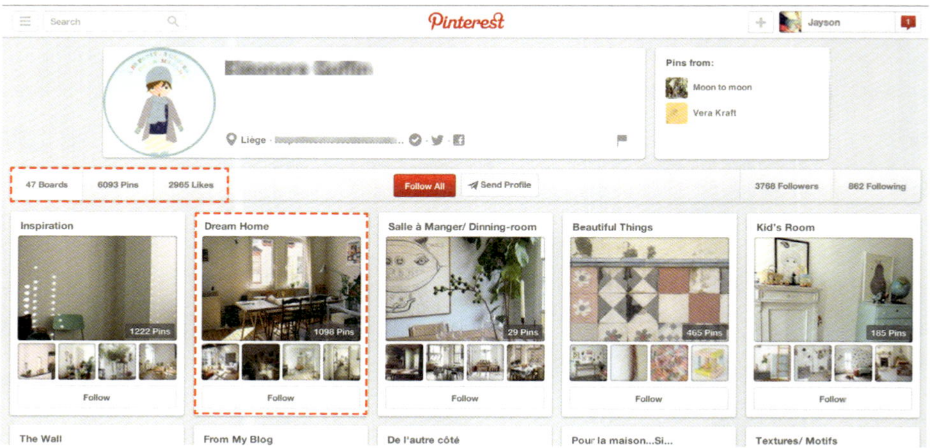

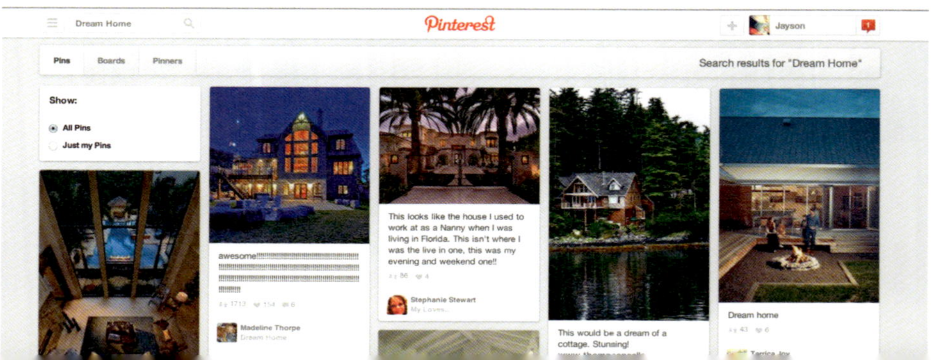

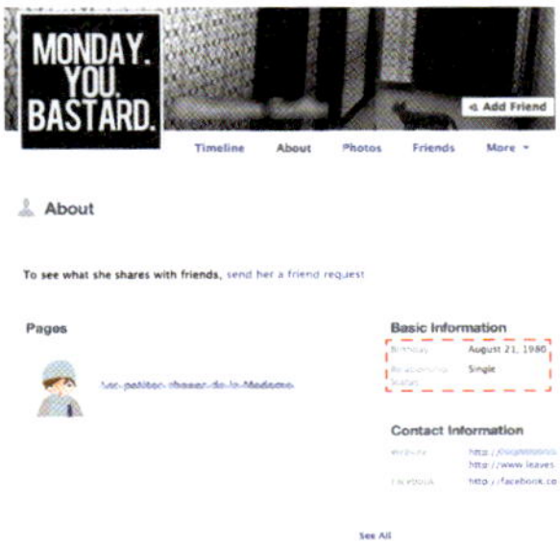

Facebook

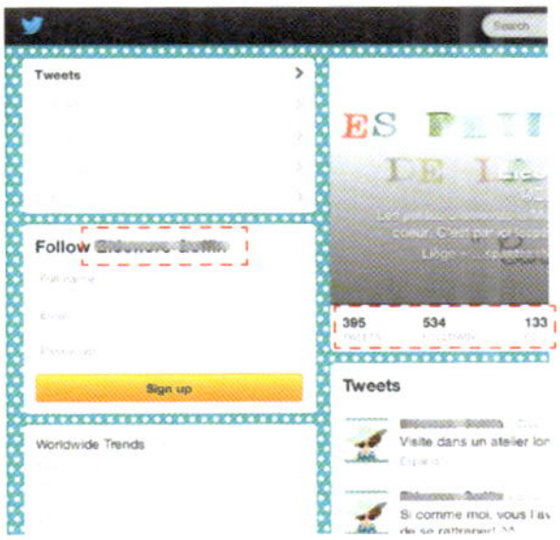

Twitter

Blog

Resident 1
City: Unknown
Gender: Female
Age: 27
Language: Romanian English
Occupation: Freelance writer
Education: Master's Mass Media

Resident 2
City: Liège, Belgium
Gender: Female
Age: 33 (August 20, 1980)
Language: French
Occupation Blogger
Salary: € 35,550/year
Education: Master's Degree, History,
Marit.Status: Single
Children: 2 daughters
Budget: € 142,200 (4 x yearly salary)
Size of House: 95 m^2 (€ 142,000/ € 1,500 /m^2)

Resident 3
City: Toronto, Canada
Gender: Male
Language: English
Occupation: Inv. Director,
Salary: € 72,750/year
Education: Radio & Advertising
Marit.Status: Married

Resident 4
City: San Diego, California
Gender: Female
Age: 39 (1974)
Language: English
Occupation: Ghostwriter, copywriter, content consultant
Salary: $ 56,000/year
Education: BA Fine Arts
Marit.Status: Divorced
Pets: 2 dogs (Maltese and cavalier)
Illness: Ovarian Cancer (2011)

Resident 5
City: Ljubljana,Slovenia
Gender: Male
Language: Slovenian, Croatian, German, English
Occupation: CEO digital media research & development
Education: Economics
Marit.Status: Married
Children: 2 sons

Resident 6
City: Marbella, Spain
Origin: Riga, Latvia
Gender: Female
Age: Unknown
Language: Russian, English, Latvian
Occupation: Singer and Yacht broker
Education: Master's degree Marketing Management
Marit.Status: Married

Resident 7
City: Gangnam-gu , Seoul, South Korea
Gender: Male
Language: Korean, English
Occupation: Co-founder/ Director at XX
Education: Kyung Hee University
Marit.Status: Married
Children: 1 daughter & 1 son

Resident 8
City: Mendenhall, Mississippi
Gender: Female
Age: 60
Language: English
Occupation: Senior systems support analyst
Education: BSBA, Information Technology
Marit.Status: Divorced (2x)
Children: 2 daughters, 1 son
Pets: 2 dogs

Resident 9
City: Sydney, Australia
Gender: Female
Language: English
Occupation: Online retailer
Education: Diploma, Sports Therapy
Marit.Status: Married
Children: 1 daughter,1 son

Resident 10
City: Helsinki, Finland
Gender: Male
Language: Finnish, English
Occupation Information technology and services consultant
Education: Interior Architecture & Furniture Design
Marit.Status: Married
Children: 1 son

The Phantom Image Methodology for Expressing Desire

Can a single image represent the overall character and style of an imagined dream home?

Hundreds of images for each resident were collected and categorized, based on the type of room, the elements in the room, materials, views and size. Then, the data collected was analysed, to draw conclusions of all residents' individual desires. Simply quantifying the images fails to express the ethos of the desire in the first place.
The 'Phantom Image' was created as a means of revealing, in a single image, the data collected through the categorization of Pinterest images. The transparency is used as a direct reference to the distribution of room types that reveal the room the person unconsciously feels is crucial within his or her dream home.

Pinterest images on "Dream Home' Board Categorized:

1. Only count what you see
2. Only count space (No objects)
3. Each images is only counted once
4. Only single images are counted
5. Only real images are counted

Analysing each image to categorize the space, elements and materials

Example: bathroom image

Space:
High ceiling
Vaults
View to forest
Elements:
Large Window
Two sinks
Bathtub
Materials:
Wood

Client H Phantom Image

The Residents' Dream Homes

Can dream homes be designed through social profiles?

A dream home can be designed through distilling and quantifying the desires and needs of each homeowner as made evident in their social media profiles. These profiles can detail the number of rooms needed in the house and the images give insight into the types of materials and style desired. By categorizing and distilling approximately 100 images for each single person online, ten dream homes were designed for ten people from around the world. The goal was to say 'yes' to every desire revealed in the images categorized. No matter how small the request, the imagined detail was included.

Resident 1

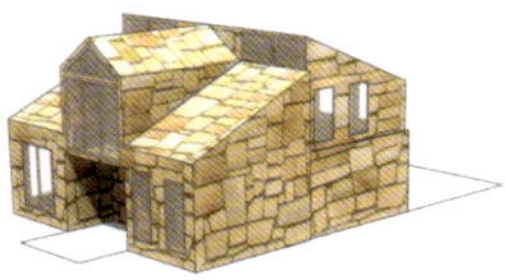

Resident 3

Resident 4

Resident 5

Resident 6

Resident 7

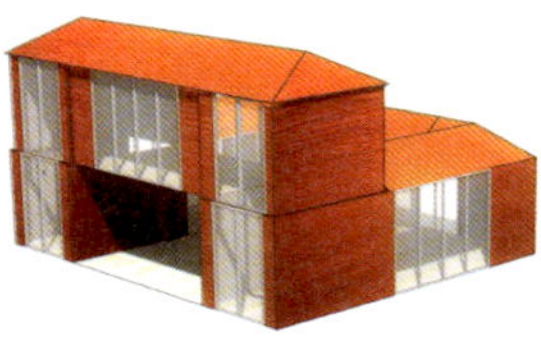

Resident 8

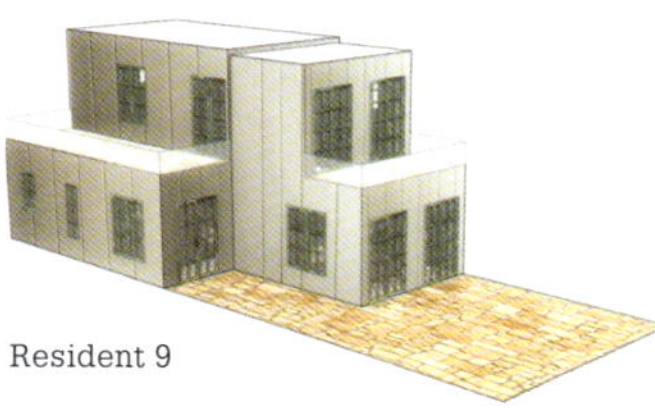

Resident 9

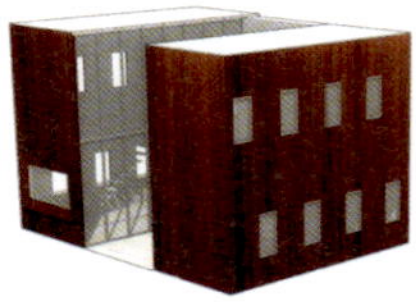

Resident 10

Dining room
large windows (59%)
cathedral ceilings (50%)
open to kitchen (36%)
open to living room (36%)
doors to exterior (18%)
stairs (9%)

Living room
large windows (54%)
high ceiling (33%)
cathedral ceilings (21%)
open to dining (21%)
open to kitchen (21%)
doors to exterior (13%)
fire place (13%)
city view (13%)
stairs (8%)

Bedrooms
garret (58%)
windows (38%)
stairs (8%)
storage (8%)

Master bedroom
cathedral ceilings (58%)
large windows (38%)
skylight (27%)
high ceiling (15%)
open to office (12%)
doors to exterior (8%)

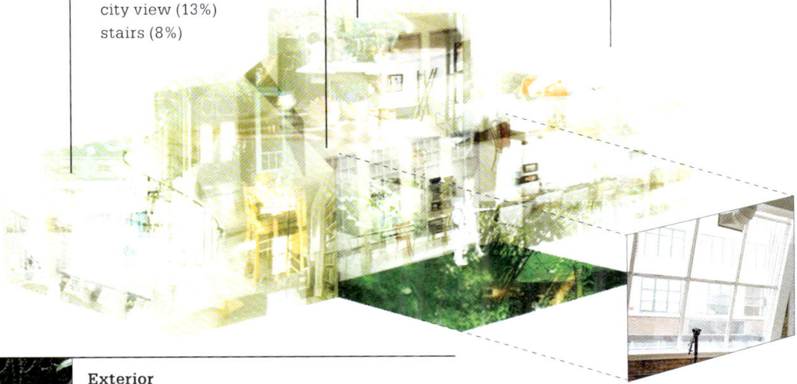

Exterior
courtyard (100%)
doors to exterior (100%)
vegetation (50%)

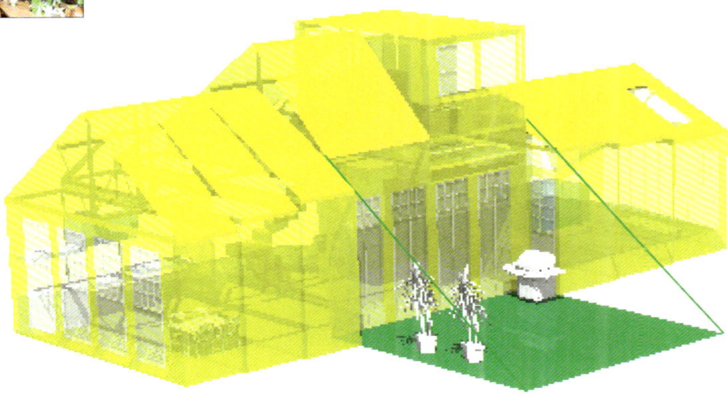

The Dream Home Tower

Can dream homes be reintroduced into the density of the city? Can another model of housing be imagined?

As a test, ten dream homes were stacked on a plot that contained ten 100-m^2 apartments in Rotterdam. The result was a tower that doesn't respect the quality or regulations of city planning and urban life. The alternative is to create ten unique apartments that are tailored to the residents but that represent a compromise of their desires. In essence, this would still be a sort of generic apartment with some customization.

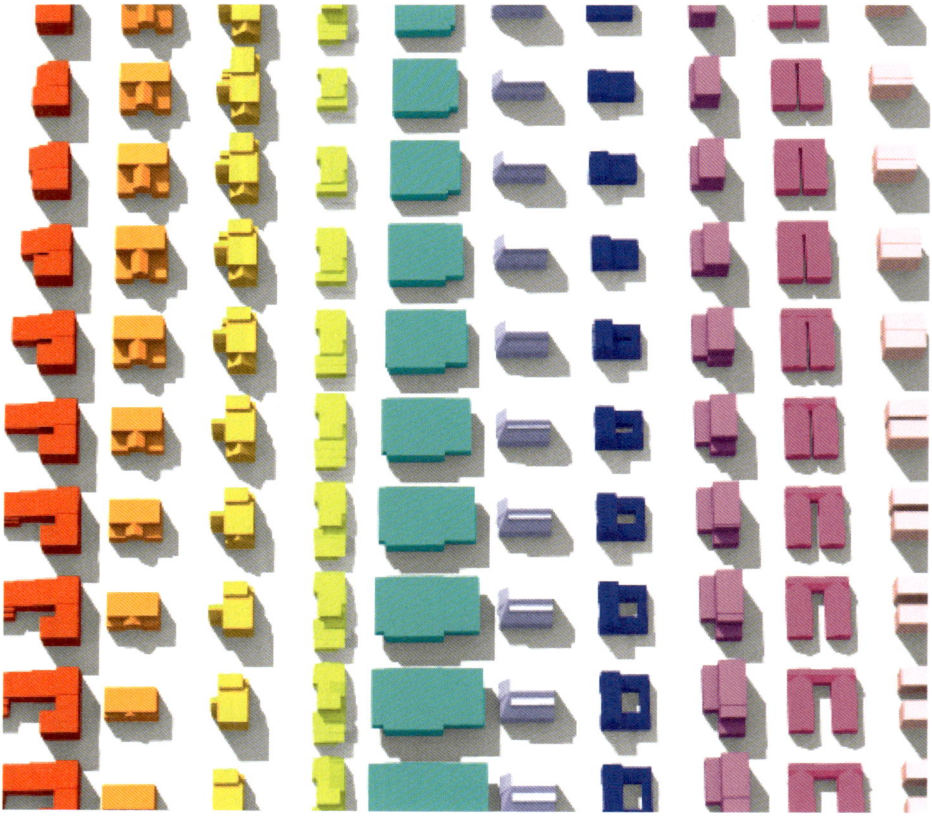

City

Taller than existing building
Housing density is too low
Structure costs are high
Low energetic efficiency
High maintenance costs
Large shadows cast on city
Social separation

Desires

Views & light not 100% satisfied
Accessibility is difficult
Appearance compromised
by structure

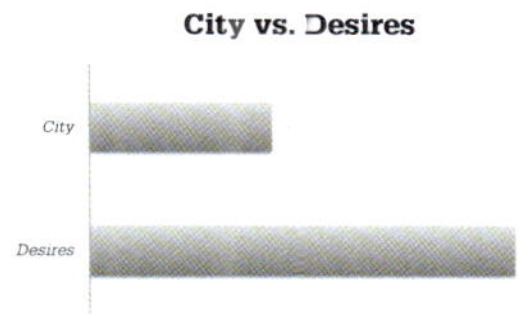

The Dream Home Tower Design - Evaluating maximum desires in the city

City

-

Desires

Room size desires are compromised
Ceilings height not achieved
Ceiling space not customized
Exterior appearance not individual
Exterior space desires are not fulfilled
Doors to exterior are inadequate
Size of windows is inadequate
Views are limited

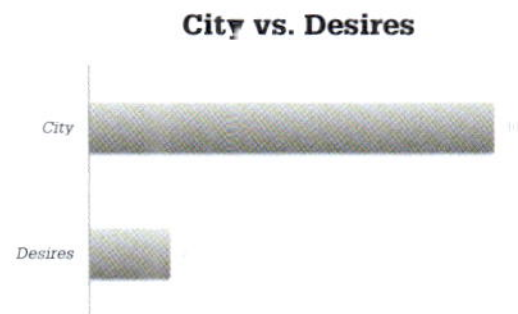

The generic city dwelling Design - Evaluating maximum city compromises

The Adaptive Dream Home Design

How can dream homes be shrunk to fit into the volume of an apartment? Do new technologies need to be developed to adapt the dream homes to scale?

Adaptive, transformable houses were tested as a means to temporarily reduce the volume of the dream home so as to fit into the apartment. The dream homes are 'cut' into pieces to reduce the volumes of the homes by at least 50 per cent. Using quick cardboard models to test methods of compacting the homes guaranteed a certain degree of feasibility. Ultimately, we discovered that we don't need to develop a new technology in order to adapt dream homes. Current technologies – derived for use in shipping or in recreational vehicles, for example – can be reimagined as parts of an adaptive dream home.

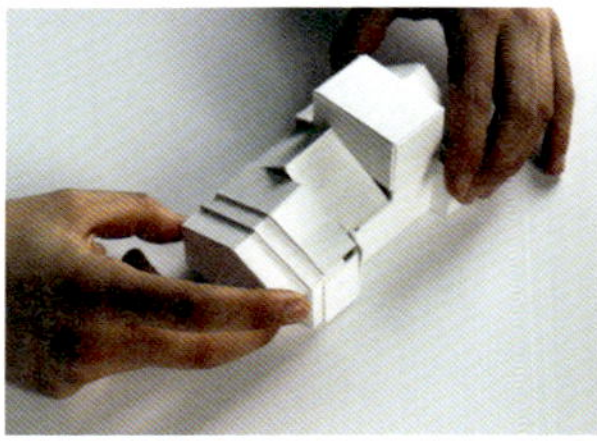
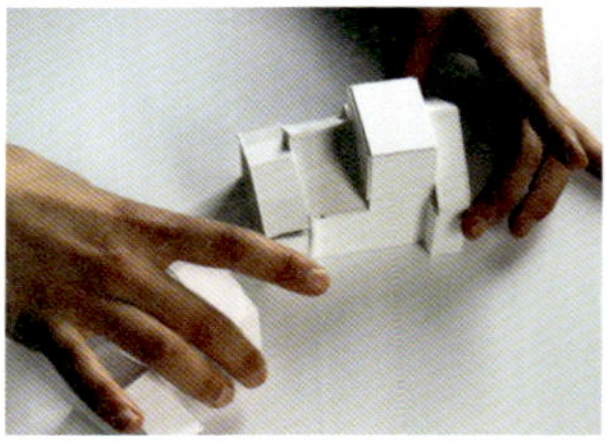

Collapsible second story

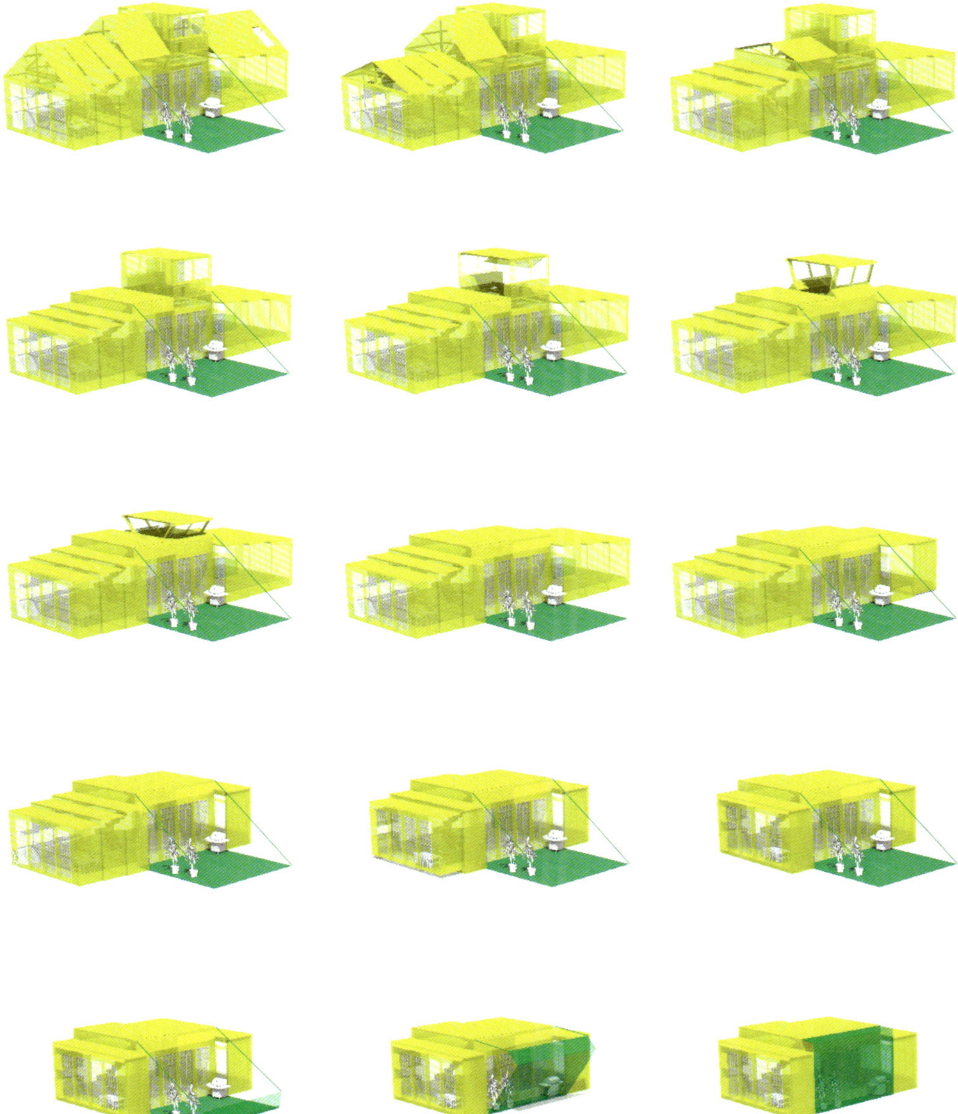

The transformation of Dream Home 2 into its compact state

The Adaptive Dream Home Tower

The culmination of this research was the combination of the ten dream homes and their respective adaptations (ten homes have ten different states). A final model was devised as an interactive tool that allowed the control of each dream home with a smart phone. The smart phone-controlled model demonstrated the state of flux represented by the different graphs of activity of the dream home tower for different times of day.

Dream Home Tower, 8 AM, based on resident's behaviour

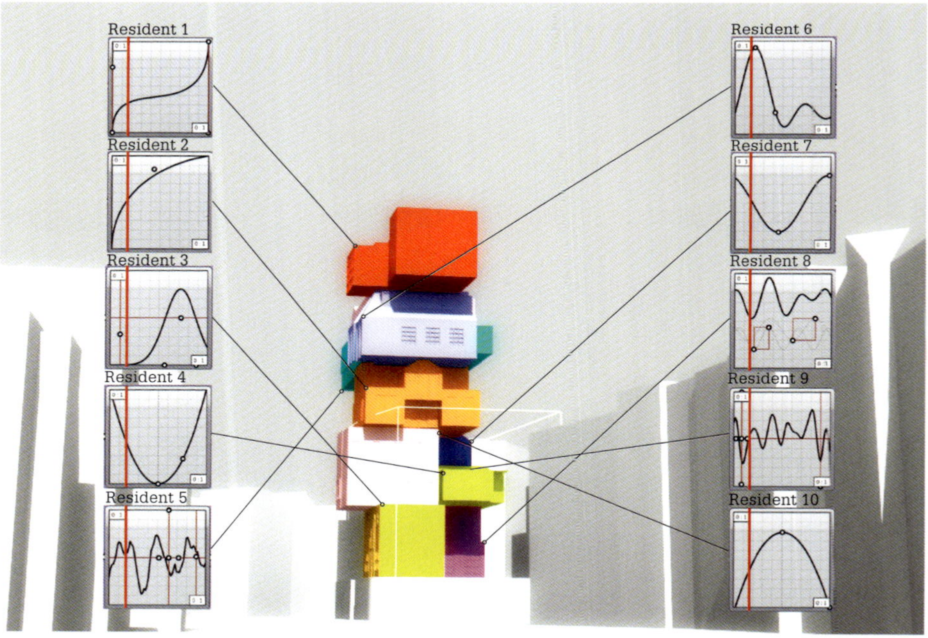

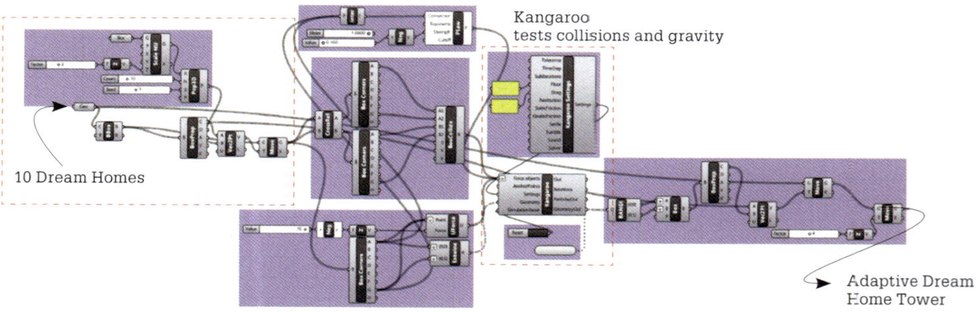

Grasshopper Script of Adaptive Dream Home Tower

Seriously Dreaming

This project takes desire as the starting point for adaptation and proposes a means by which we might accommodate spatial dreams in a limiting context. In the context of our Nano research, the project adds an important aspect, visualizing 'real' dreams expressed by 'real' people. As such, the results do not merely present an imagined and theoretical diversity, as is too often the case in architectural illustrations. What's presented instead attempts to reflect a significant fidelity to average taste. And while the outcome is indeed diverse and rough, it is also fundamentally compelling.

The last project in this series is, arguably, the most realistic one. While the previous projects related adaptation to the room and the house respectively, the focus here is on the urban block. It is a hybrid block, composed of housing mixed with offices and public spaces, and its various functions require different amounts of space over the course of one day. As spaces shrink and expand, programmes can receive more space when needed and less when they'd likely otherwise be left vacant. The project illustrates how adaptation and flexibility indeed allow for higher density and efficiency, while maintaining a high quality of life.

Watch the movie:
www.thewhyfactory.com/barba

Twenty-four Hour Block

The Why Factory: Leo Stuckardt

Adaptive City thesis studio, TU Delft
Teachers: Ulf Hackauf, Adrien Ravon, Huib Plomp with Winy Maas

Basis:

Dissecting the Berlin Block

In order to redesign the Berlin Block as a transforming, adaptive typology, the existing form is abstracted and fragmented into its smallest unit. The apartment, the school classroom or the small, green courtyard can all be reduced to a similar minimum size.
The new block typology is composed of four nested layers that build up from this smallest grain.
Each of these layers acts on a different scale and responds to a different set of rules, ranging from efficiency (volumes) to spatial qualities (light, views) and proximities of different programmes.

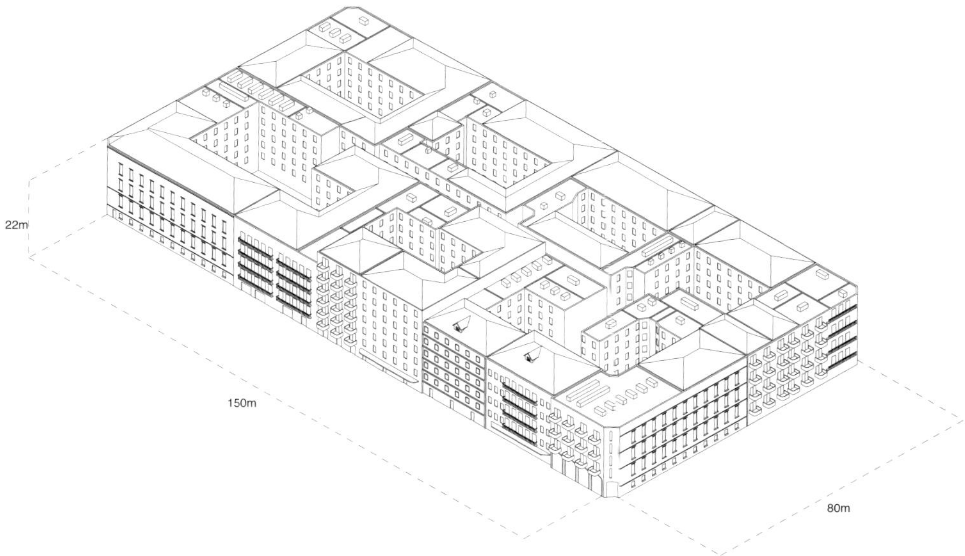

Berlin Block (700 inhabitants)

1 Mass
The block is stripped of its facades, rooftops
and ornament.

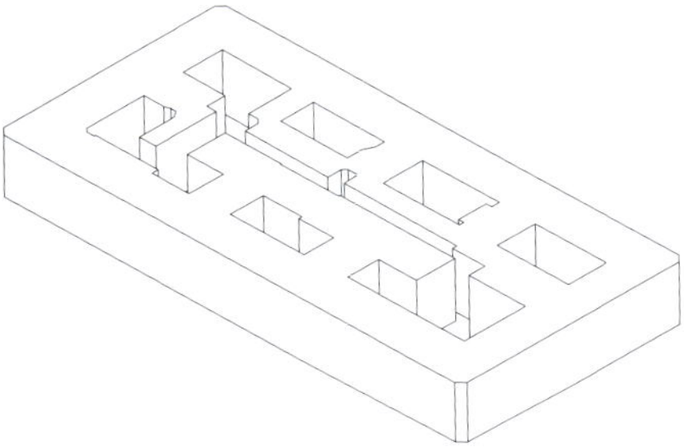

2 Programme Distribution
The public programme is located almost exclusively
on the ground floor.

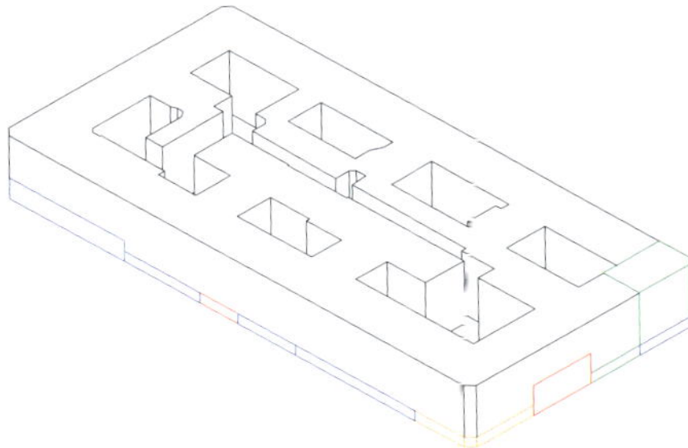

3 Rearrange
The program of the Berlin Block can be sorted and
rearranged to form a barcode of various programmes.

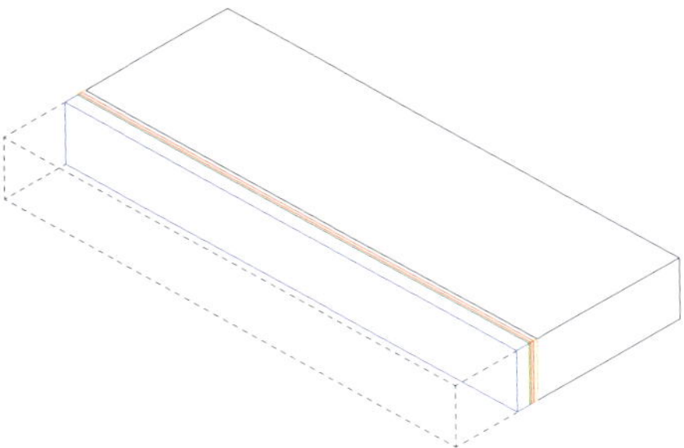

4 Cut 1D
The barcode is cut into several parallel vertical strips.
The programme is thereby fragmented.

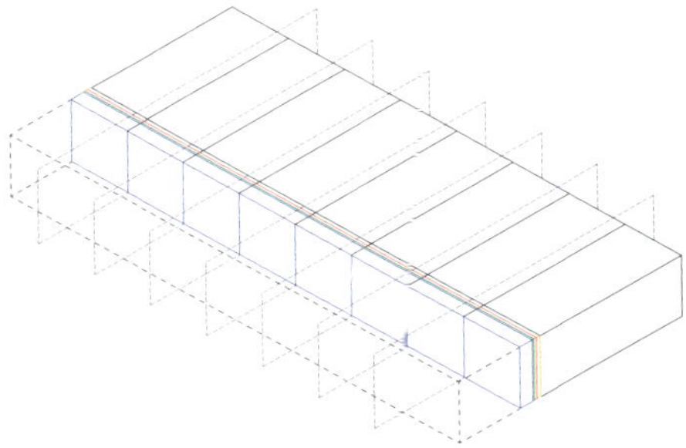

5 Reshuffle and Circulation
The programme is reshuffled into larger clusters
and paired with dwelling.
Fragments of programme have complementary
requirements, proximities and opposing times of use.
Circulation is added between programmes.

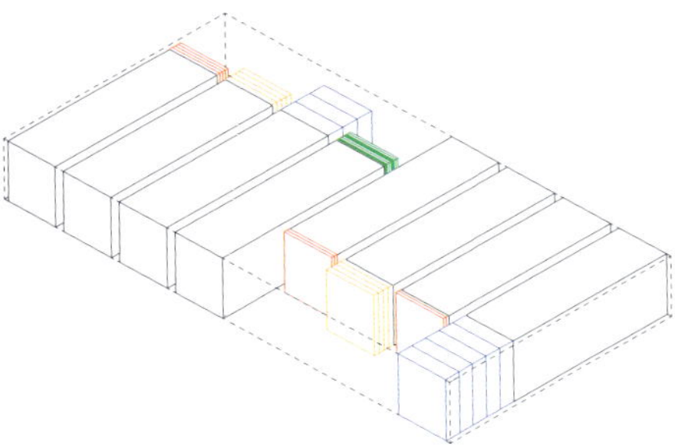

6 Fragmentation
The programme is cut horizontally into equal partitions
according to story heights.
It is cut vertically into equal partitions allowing for
structure, circulation and void in the cross-direction
of the block.

Block Dissection

First Layer:

The Plane [transformations]

The plane is the smallest unit of the block and forms the individual's customizable space. While it needs to follow the rules of the collective framework, it can be freely designed according to individual requirements and technological innovation. An endless matrix of technologies shows different systems for floors (1), ceilings (2), walls and façades (3), furniture (4), services (5) and structure (6). Those elements can be freely combined within one plane. With enough void in front of a plane, it then can mechanically transform itself by extruding, shifting and rotating elements inside of it. Staircases, bathtubs, shelves, beds, screens, desks previously hidden and unused, suddenly appear.

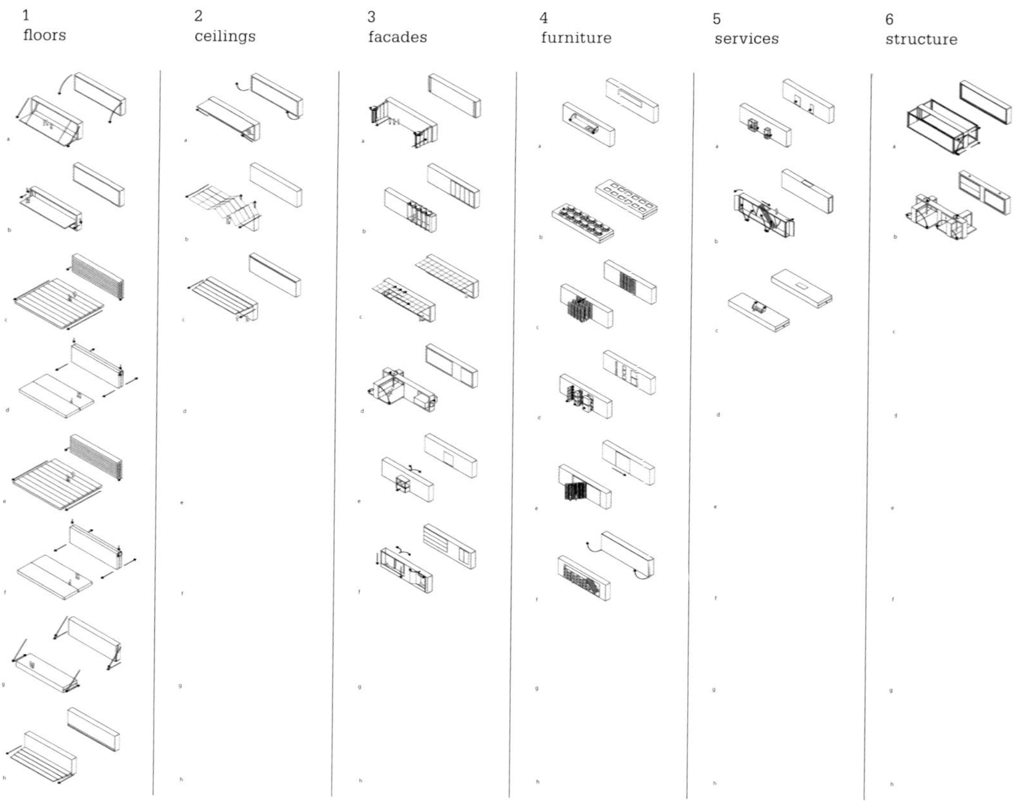

Catalogue of Components

22:28 In the Life of a Cogwheel

81,920 cogwheels were all in motion. Eight wheels per cell, sixty-four per unit, 2,560 per floor, each one endlessly spinning. Some so quiet that one could hear grains of dust crushed under the pressure and others spinning at such a speed that only a hiss could be heard. At rush hours, the mechanical symphony of cogs would conduct the frenetic life of the machine's guests.
Late at night, the wheels rest, cooling down from the friction.

This orchestra follows no recognisable tune, plays an endless improvisation directed by individual desires. But people learned to recognize recurring motives—sounds hinted at the opening of a path, the appearance of a bridge, the closing of a shop.
In the depth of the block, the mainframe, a collective intelligence, was constantly directing this perpetual concert, negotiating between individual desires, recognising patterns and reacting to exceptions.

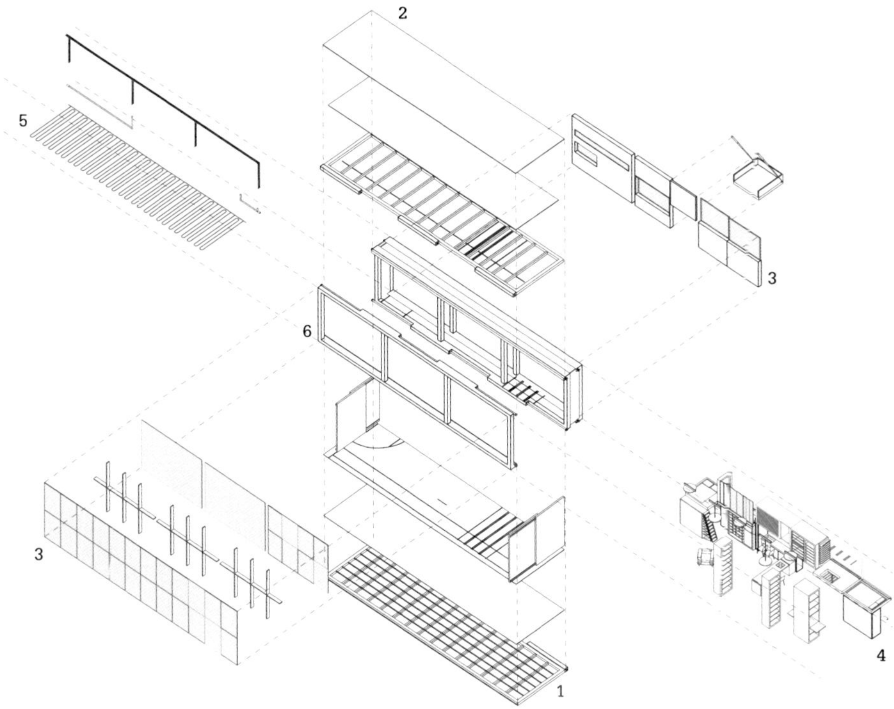

1 Floor:
Flooring, incl. rails, 5 cm
Insulation, 5 cm
Construction, IPE 160, Steel, 20 cm

2 Ceiling:
Panels, 5 cm
Insulation, 5 cm
Construction, 20 cm

3 Facade Type 1:
Rectangular beams, Aluminium, 20 x 4 cm
Translucent PVC Panels with Insulation
Facade Type 2: Cavity Wall Insulation

4 Furniture:
Multiple components, transformable

5 Services:
Top: Water, d = 2 cm
Hot water, d = 2 cm
Electricity/Communication, d = 5 cm
Bottom:
Grey water, d = 10 cm
Floor Heating

6 Structure:
Vierendel Truss (3), Steel, d = 20 cm, t = 3 cm
Engine for movement (2)

Assembly of Components

Second Layer:

The Unit [expansion]

All spaces are created by flexible configurations of vertical planes. The ability of those planes to move allows for changing volumes between them. The thick planes contain storage space and require equipment for their respective functions.

configuration 1
7 people
state 1, unpacked: 170 m^2
state 2, one half packed: 105 m^2
state 3, packed: 50 m^2

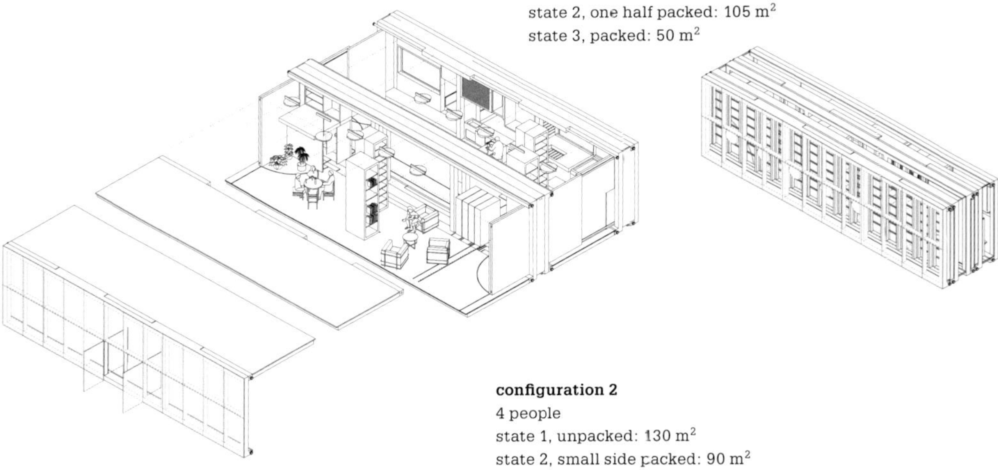

configuration 2
4 people
state 1, unpacked: 130 m^2
state 2, small side packed: 90 m^2
state 3, large side packed: 25 m^2

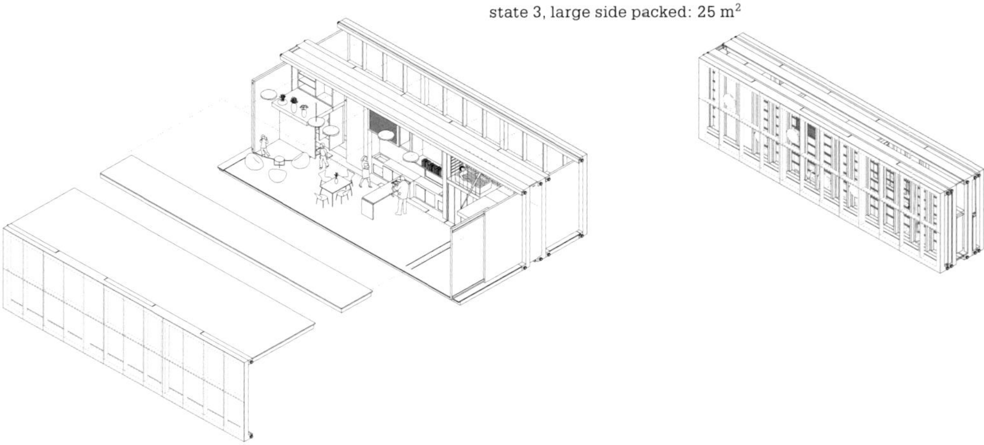

Apartments in expanded and collapsed state

14:24 in the life of an apartment

The motors have almost come to a halt. Only a humming of sensors and idle engines remained. It was not a real sound. And it's only audible because they knew that somewhere in this gigantic block actuators were waiting for instructions on how to match the desires of the inhabitants. 'Last night I dreamt that our apartment closed up while we were asleep', she said. They had already been living in the block for four years. She reassured herself: surely the mainframe could predict their behaviour. After all, if you live in a place for a long enough period you are bound to fall into patterns. The time they woke up, when they left for work, received friends for dinner and went to bed. The views they enjoyed in the morning. The routes they liked to take through the block. Over the years, the machine had learned all their habits and desires.
'Don't be ridiculous', he replied. 'Come over here. The garden on the fourth floor is opening up. The tulips are blossoming'. He moved towards the balcony, which had just then unfolded from their kitchen. The neighbours had left for vacation and their apartment, now collapsed, made room for a view over the city, a view unlike any they had experienced in their previous years living in the machine.

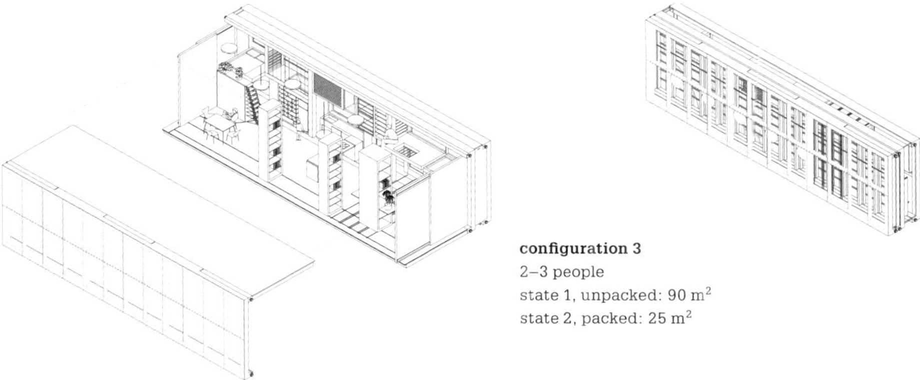

configuration 3
2–3 people
state 1, unpacked: 90 m^2
state 2, packed: 25 m^2

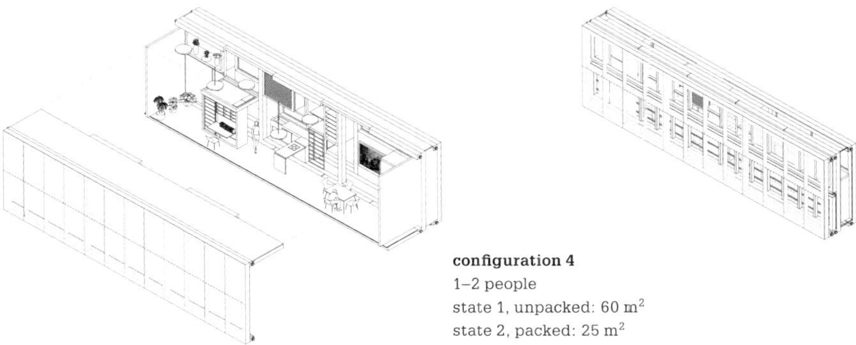

configuration 4
1–2 people
state 1, unpacked: 60 m^2
state 2, packed: 25 m^2

Third Layer:

The Lane [shifts]

Functional planes move together and apart, creating spaces for activities in between when needed. A programme that is needed less often can contract at a moment's notice and make room for one that is used. The drawings show three states (morning, noon and evening) of a lane that contains a school and multiple apartments.

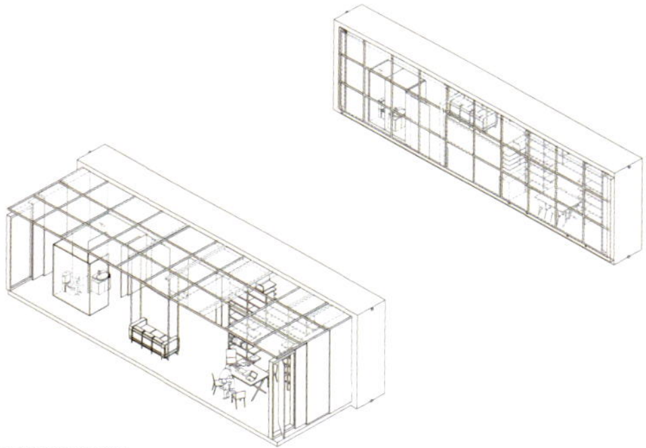

an extruding apartment

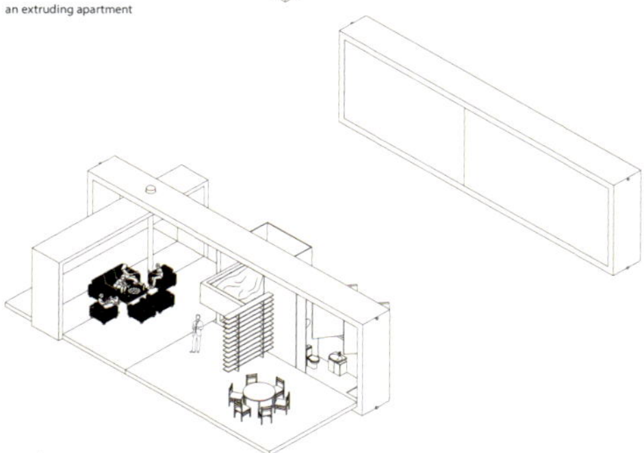

a rotating apartment

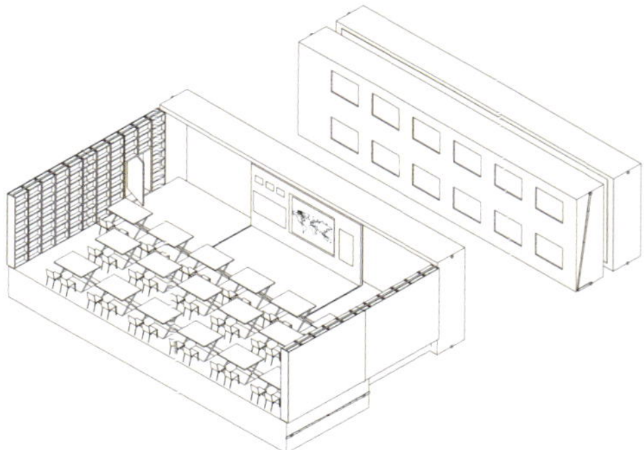

a transformable classroom

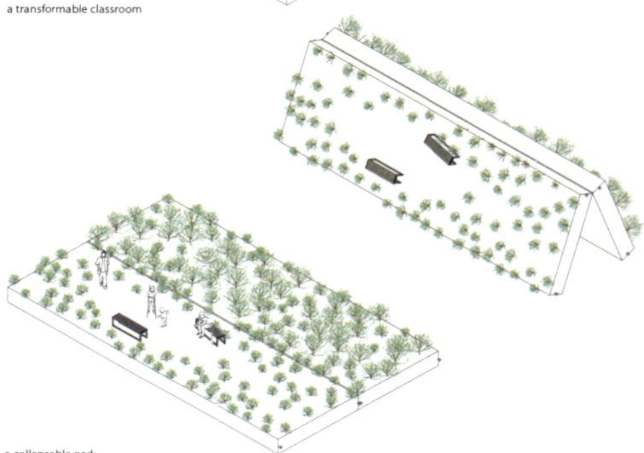

a collapsable park

19:54 in the life of the ground floor

TThe rhythmic sound of sliding metal plates is suddenly disrupted. The engines on the ground floor have begun to accelerate, breaking out of the steady back-and forth with which they've shaped cafes, parks, sport grounds and shops for their visitors. The shops are closing. Slowly they retreat towards the inner depths of the block. The urban life around and within the machine transforms with it. Is the machine changing the citizen's behaviour or is it the dweller who informs the machine through his actions?
Move, look, hear, talk, sleep, work, love, relax, exercise: The Living Machine responds to the shifting desires of its guests.
Only a cascading wall of apartments remains. The overall configuration allows the last glimpses of daylight to reach the ground floor. Inside, inhabitants enjoy the open view of the sunset.

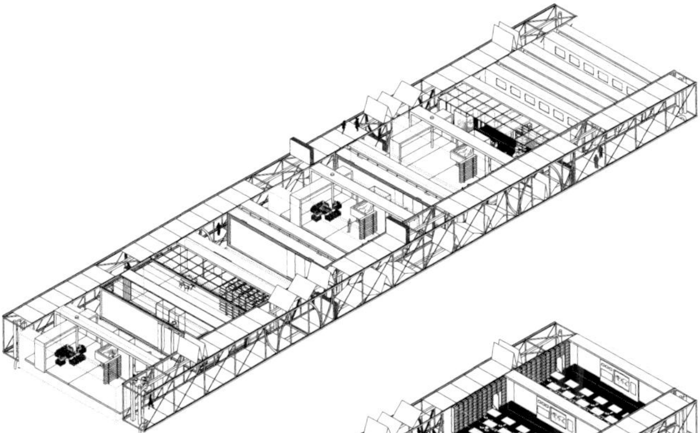

2:30 am
apartments expanded
school collapsed

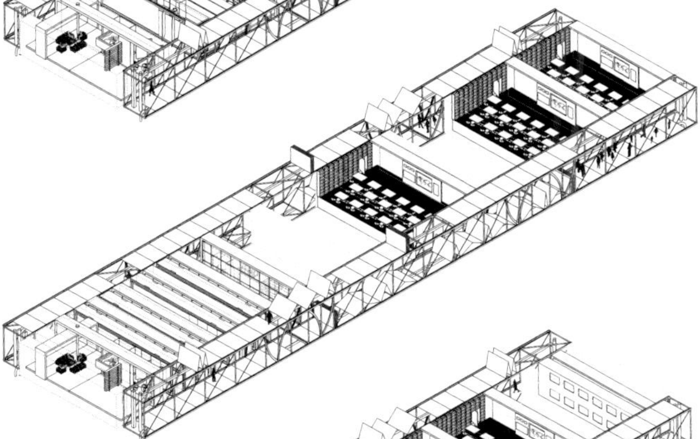

11:00 am
apartments collapsed
school expanded

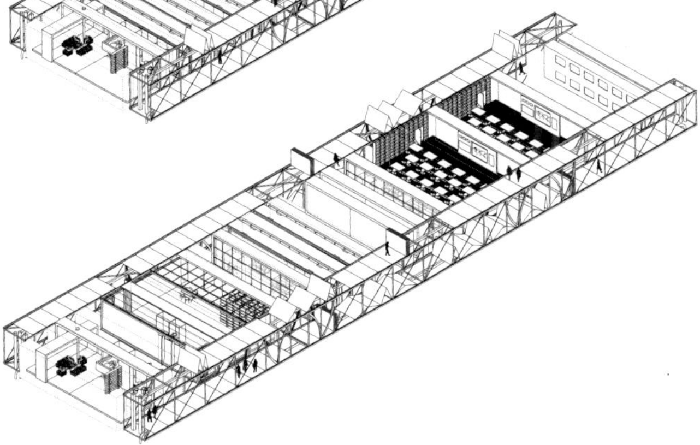

8:00 pm
apartments expanded
school expanded

Fourth Layer:

The Cluster [oscillation]

In the quest to match individual needs, the collective intelligence can take many shapes through configurations of units.
A user who lives on the ground floor might, at some point, a week later, say, find herself on the top of a terraced housing block.

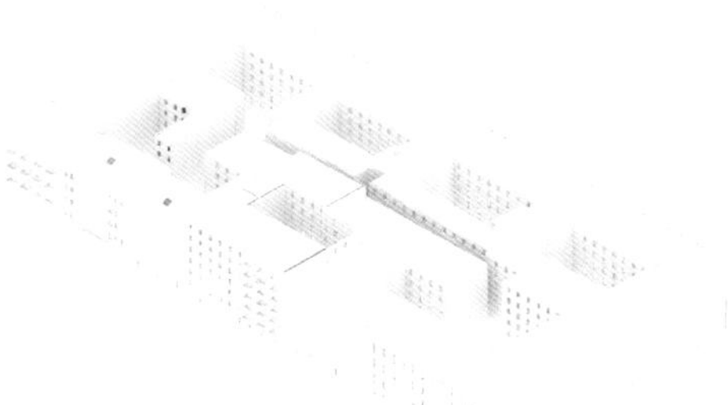

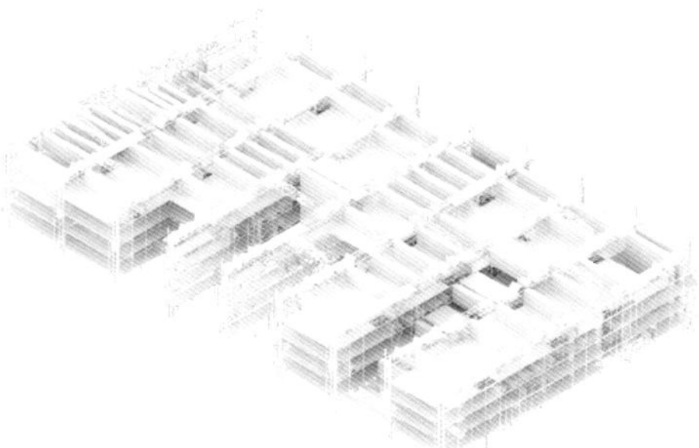

inhabitants / block: 700
number of clusters: 8
people / cluster: 88
people / apartment: 4
apartments / cluster: 22

Quantitative scenarios:

Minimized footprint in use

Minimized footprint vacant

Maximum footprint, maximum quality

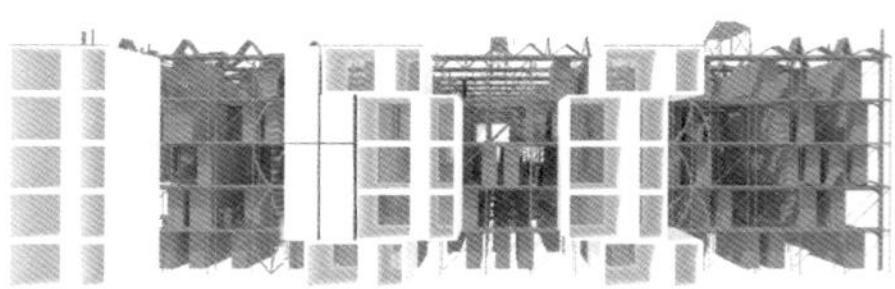

Maximum quality, 25% vacant

Maximum quality, 50% vacant

11:55 in the life of a garden

A hiss flows through the garden, as tA hiss flows through the garden as its bushes move gently in the wind. The frequent change of orientation, which transforms the vertical garden into a horizontal park and back, did not seem to harm them. If anything, they were growing stronger:
The Living Machine cared for them as it did for its human inhabitants. It provided water and fertilizers, exposed them to the sun and moved them from the core of the block to its border. Only this reiterating effort made it possible for the plants to survive in midst of the dense Living Machine. The gardens, surrounded by a condensed complex of apartments, cinemas, classrooms and restaurants, were a calm spot within otherwise endless movement. Sometimes several gardens would merge into a larger park, other times the machine would transform a garden into a greenhouse, surrounding it with other units.
It protected plants and inhabitants from the cold and the rain and created a bridge between corridors and staircases.
No appearance of a garden, however brief, would go unnoticed: the crowd inhabiting the block would gather as if answering the mechanical call of the machine.

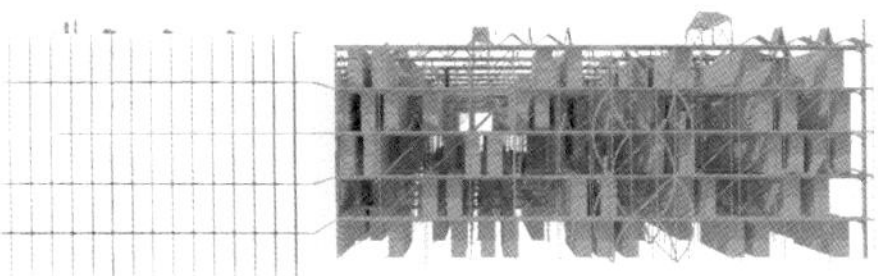

Vacation

Market hall

Everybody at home

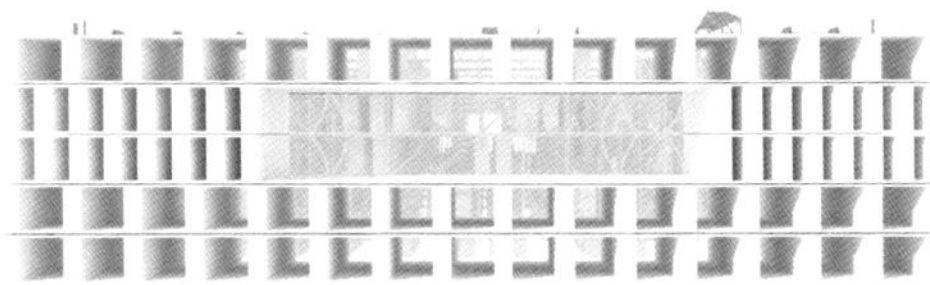

Egoist

Roofed courtyard

Sunlight penetration

Amphitheater

Parasol

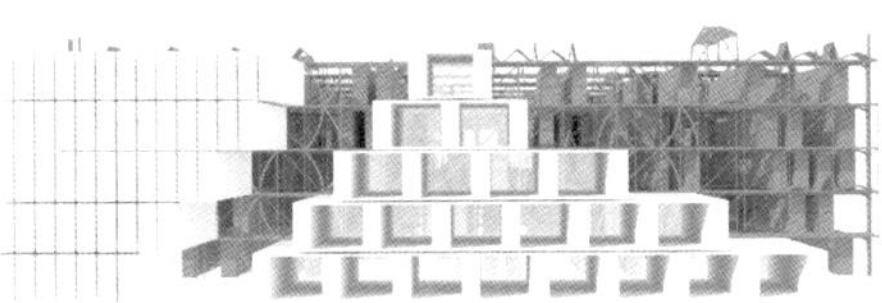

Stairs

Advertising board – NIKE

Fifth Layer:

The Urban Block [meander]

In the quest to match individual needs, the collective intelligence can take many shapes through configurations of units. A user who lives on the ground floor might, at some point, a week later, say, find herself on the top of a terraced housing block.

The 24-Hour Living Block in its context

9 am The block creates terraces towards the sun, open to the city

3 pm People meet on a square that will fade away a few hours later.

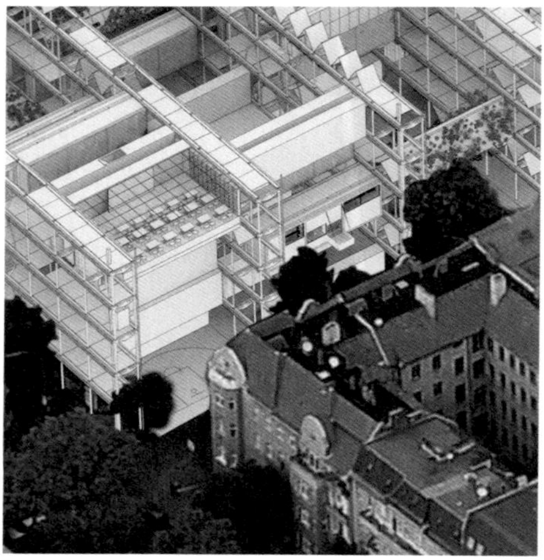

0 am Fully occupied the block comes to a temporary rest.

7 am Planes form a wall, shielding off the block's inside from the city.

8:17 in the life of an urban block

It was all about the noise. The conversations in the corridors, the discussions in the privacy of the apartments, the collective meetings of the tenants; sound was the true sensor of space. It was unlike anything anyone had experienced before, far from the familiar noise of cars or the chatting that could be heard from ground floor flats.
The Twenty-four-hour Living Machine was a transformer, an apparatus that expanded and contracted spaces containing a variety of urban programmes based on their demand. Extreme efficiency of density over twenty-four hours and openness at every single moment could only be achieved through adaptation and spatial transformation.
No individual behaviour was without consequences in the Living Machine. A holiday for one meant a balcony for another; one guest's late night walk would change the crowd's circulation in the morning.
A humming rises from the depths of the block and the machine begins its transformation.

Conclusion:

The Vacant City

Use of space in the city changes constantly. Spaces are dimensioned for the moment of highest demand. In consequence, they are unused most of the time. Periods of vacancy range from entire decades to days or even just hours in which an office building is not in use. The resulting frustrations both create urgencies for the city and open up new possibilities.

If patterns of use are analysed and understood, the city can be intensified, offering an alternative to vertical densification or to urban sprawl. But it's not only programme

and space for activities that could be intensified; the city could also constantly negotiate between spatial efficiency and spatial qualities. The Twenty-four Hour Block is a transformer, a machine that expands and contracts spaces that contain various urban programmes based on demand. Extreme efficiency of density over a period of 24 hours and openness at every moment can be achieved through adaptation and spatial transformation.

Many cities developed their own urban block typologies. How do they compare in scale, density, mix of functions and accessibility? Comparing their patterns of usage over a 24-hour period demonstrates an enormous potential. The Berlin block, for example, could be intensified by almost 100 per cent. This could be achieved by either doubling the amount of programme or the generosity and openness of spaces.

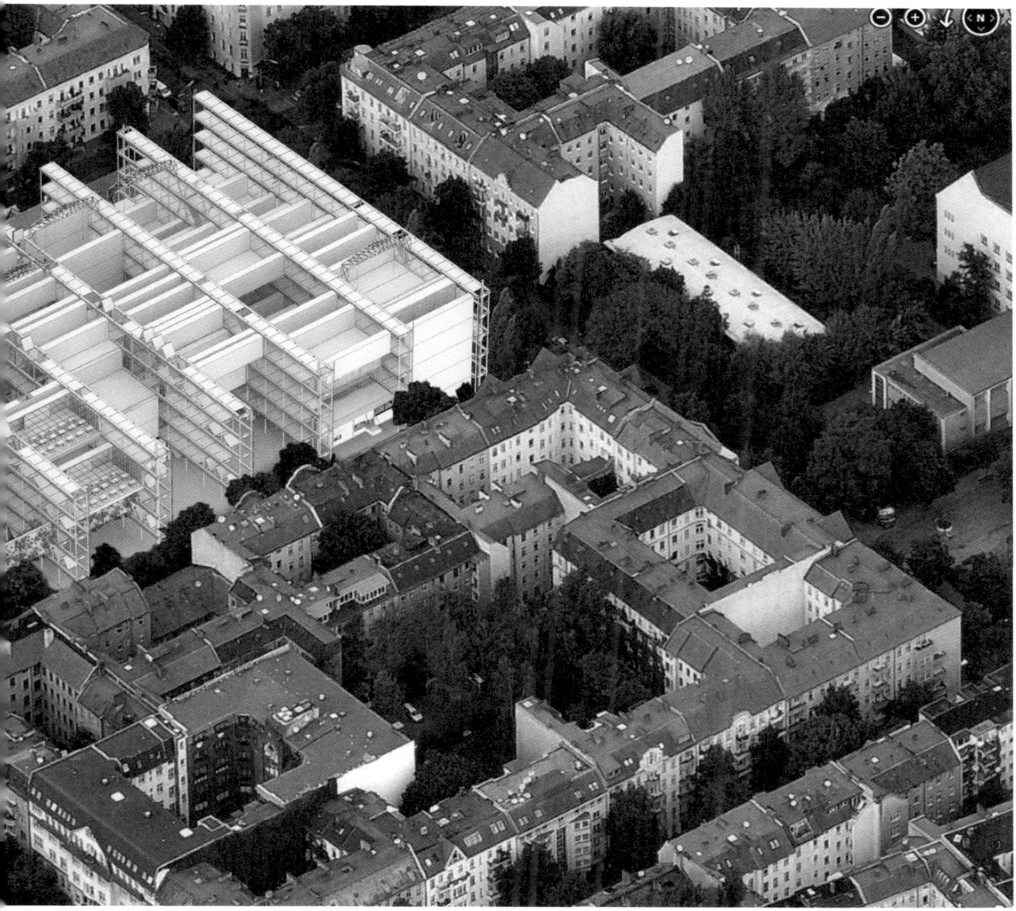

The previous chapter shows how the vision of a fully adaptable environment might be approached long before we are able to master molecules and create the imagined Barba material from scratch. It may be that fully adaptable furniture or products – rather than architecture – represent the most feasible ways to realize the Barba vision. A research project, financed by the European Institute of Innovation and Technology (EIT) and executed in collaboration with the KTH Stockholm and TNO Netherlands, allowed us to develop and build first prototypes of such products.

Prototypes

The studies with felt described in the 'Tests' chapter seemed a very appealing way to develop a fully adaptable home product. We imagine a large panel of felt that could be programmed to bend into different shapes. This flexible felt could become a multifunctional interior object and could take on the functions of many different types of furniture. This flexible felt could even eventually become the only piece of furniture needed. And, if it could bend around you and offer you shelter, it could be the first version of our fully adaptable environment.

The felt panel could be reinforced with a mesh, like the kind of mesh explored in the 'Models' chapter. Proximity sensors could read gestures and allow users to steer the felt's movement. But before we begin to consider actually producing such a magic carpet, we imagine how it might work, how it could change interior space as we know it . . .

Watch the movie:
www.thewhyfactory.com/barba

Magic Carpet

The Why Factory research team

The Why Factory research team, TU Delft
Arslane Benamar, Ulf Hackauf, Winy Maas, Sander Mulders, Lex te Loo, Adrien Ravon and Piet de Reuver

This magic felt panel would look just like any regular felt carpet when not in use.

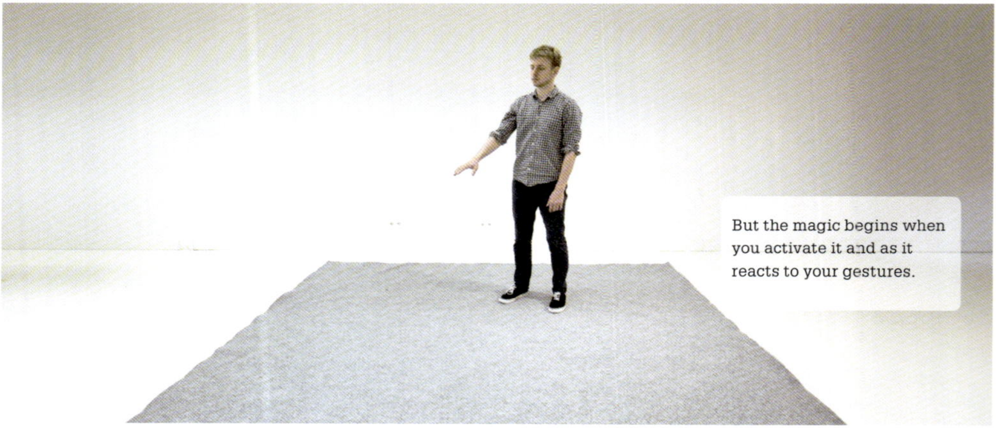
But the magic begins when you activate it and as it reacts to your gestures.

Watch the fabric slowly crunch towards you with a low, humming sound.

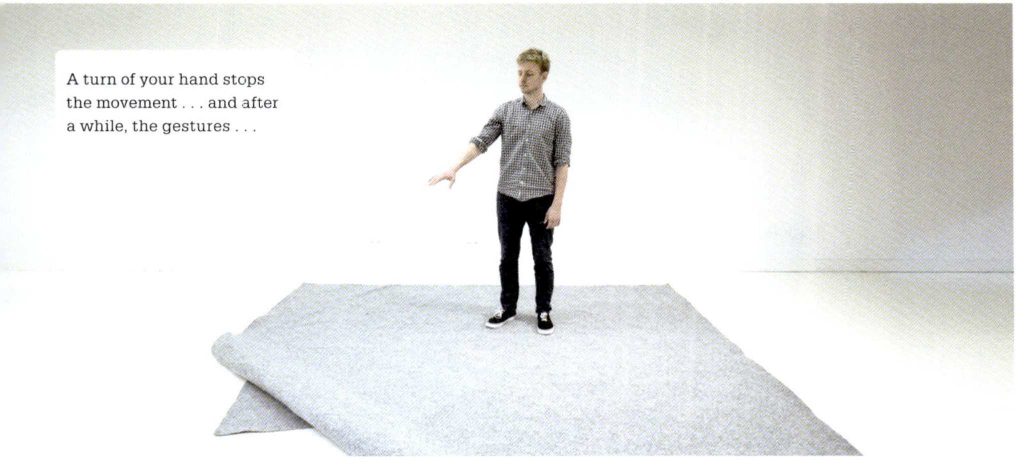
A turn of your hand stops the movement . . . and after a while, the gestures . . .

. . . become intuitive intuitive, feel as easy as swiping and zooming in on your smart phone.

A series of fluent movements that appear not unlike shadow boxing allow you to steer and shape the felt.

The question becomes:
How do you want to use it?

You don't have to stick to
making conventional
furniture.

The carpet can create
new and more ergonomic
shapes . . .

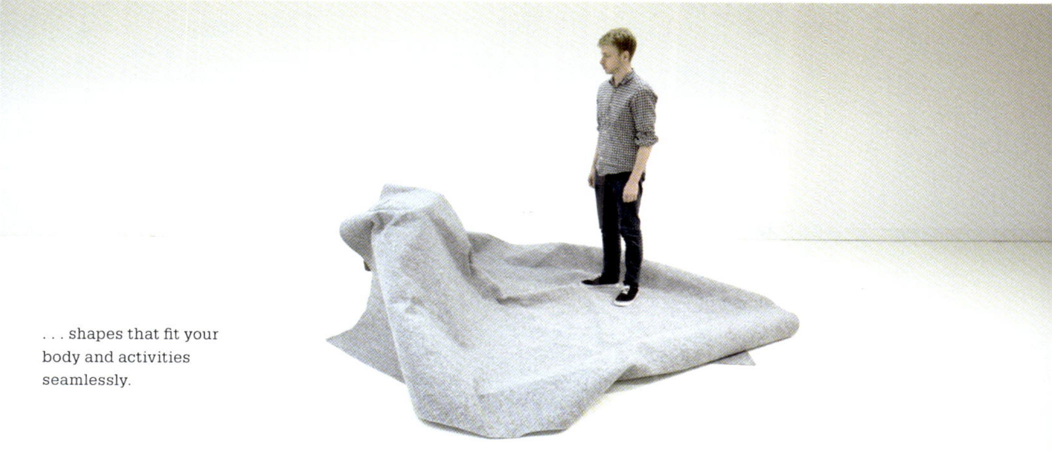

. . . shapes that fit your body and activities seamlessly.

Pleat by pleat, the carpet rises towards you.

It becomes a soft column, a place to rest.

If you want to sit down,
the carpet moves
towards you . . .

. . . and becomes a table.

When you grow tired . . .

. . . the felt is suddenly a soft, recreational landscape.

And, of course, the magic carpet can also become a bed . . .

. . . or even a shelter, a means to protect you from the elements.

Could the flexible felt actually replace walls and ceilings?

Could it become your
new home?

Could you fill it with water,
make a bathtub?

And, ideally, you could take this magic felt with you!

You could wrap it up, stuff
it in your bag.

You could bring it outside
and it could become an
open-air office . . .

Or it could be a bed to
sleep in on a trip . . . or an
apartment to-go . . .

The Magic Carpet requires a soft material, an integrated load-bearing structure, sensors to interact with the user, actuators to change the material's shape and a processor to translate sensation into action. For a first prototype, we've reduced the size and ignored the load-bearing quality for the moment. This led to the idea of a responsive felt panel that changes shape according to different sensors. Rather than a piece of furniture, it can serve as a tool for communication. We've called this the 'Talking Napkin'.

Watch the movie:
www.thewhyfactory.com/barba

Talking Napkin

The Why Factory and EIT ICT Labs: The Why Factory research team with KTH Stockholm and TNO

The Why Factory research team, TU Delft
Ulf Hackauf and Adrien Ravon with Arslane Benamar, Sander Mulders, Lex te Loo and Stavros Gargaretas
EIT ICT Labs
Charlie Gullström (KTH Stockholm) and Joke Kort (TNO)

User Scenarios:

Transformation and Communication

The European Institute of Innovation and Technology (EIT) became involved at this stage. During a presentation at the KTH Stockholm, we showed the preliminary results
of our research on the fully adaptable environment. Researchers there were setting up an EIT ICT Lab project called 'Mediating Presence'. In this project, different tools for communication – in other words, different media meant to simulate presence – were developed as alternatives or complements to existing tools of (tele)communication. The link with our studies was obvious: we had thus far looked into how a user might change the space around him. His behaviour led to reaction by the physical space in his direct proximity. If the action instead causes a reaction further away, in a space with a different user, it can become a tool of communication.
We imagine how this small piece of communicating felt – the Talking Napkin – could become part of a professional as well as a domestic environment. At the same time, we see this study as a step towards realizing the Magic Carpet and, eventually, the adaptive room.

Office scenario:
responsive mousepad

Office scenario:
Privacy table cover

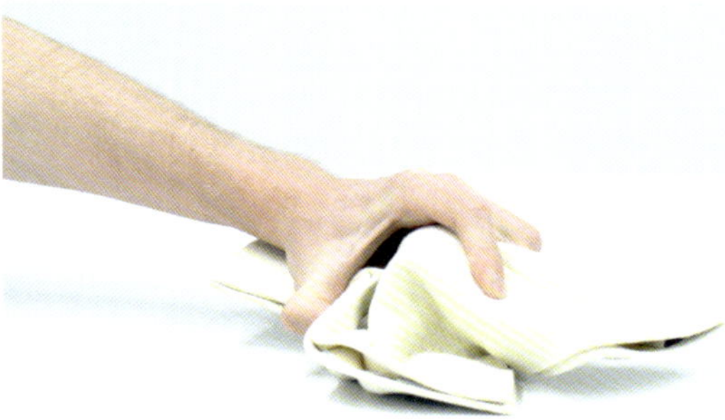

Hospital scenario
emotional napkin

Domestic scenario
talking towel

Prototype:

Concept

We chose a 5-mm-thick wool felt for this project. This material has a number of advantages for our envisioned scenario. In the first place, it is a soft material with nice haptic qualities. Felt is used in clothing and on furniture. We easily associate it with things that are comfortable, personal, maybe even cuddly. As compared to, for example, aluminium or plastic, a felt panel might feel less like an 'intruder' in a domestic environment.

Second, the felt material is stiff enough to have some structural, load-bearing capacities. If you roll up a piece of felt of this thickness, it can bear a reasonable weight without bending. If you hang the felt, you can attach small objects to it without creating pleats or causing sagging.

Thirdly, the material is relatively inexpensive and easily accessible. This makes it practical for the scale and nature of our experiments. But it will be to our even greater advantage if can we apply the prototype to a larger scale, if napkins could become carpets, walls or rooms.

We added a sensor to the felt panel, as well as some servomotors and a processor for computing. Strings are attached to its corners and a light sensor reacts when you put your hand on the felt. The servos pull the strings and the edges of the felt panel bend and turn. These mechanical components are hidden underneath the felt panel. The installation sounds technical but the warm material and its soft, organic pleats make it look and feel as if the felt is almost alive.

A felt panel with a sensor and servomotors becomes a responsive object: if you touch it, it bends and twists. To make this a means of communication, we make two panels and connect their sensors and motors. If you touch the first panel, the second panel bends and twists. Put your hand on panel two, and panel one reacts. This first prototype allows for many variations: we can use different sizes of felt, change the amount of motors or vary the type of sensor. Some variations are tested in the following chapters.

user 1

user 2

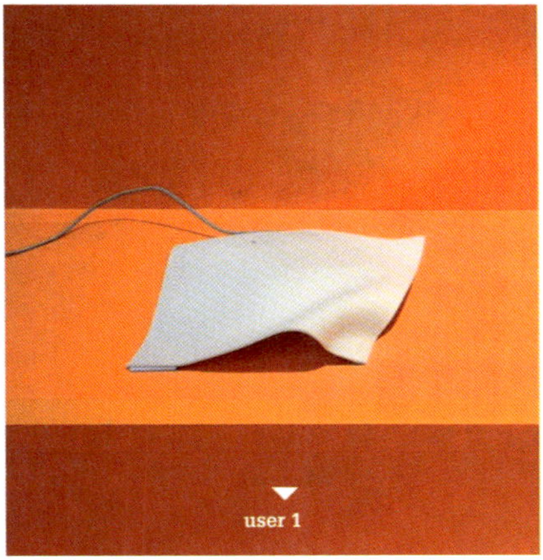
user 1

user 2

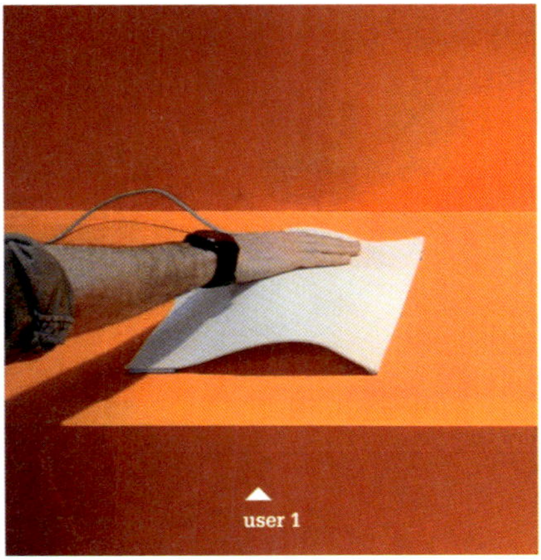
user 1

user 2

Prototype:

Technical Specifications

Base Panel: Felt, 5 mm thick, 30 x 30 cm

We use regular wool felt. The chosen thickness is necessary for the creation of nice, large pleats. Thinner felt panels crunch. The thick felt also stretches back into a flat surface if no external tension is applied. Thinner felt and other textiles do not have this important quality.

Processor: Arduino Uno®

We use a micro controller board manufactured by Arduino. The board works like a small, simple computer. It can be programmed to react to input from various sources through different types of output. It works with a java-based open-source programming language. For our prototypes, we first connected the Arduino to a laptop and used it to control our sensors and servomotors. The script was then programmed in Rhinoceros (three-dimensional modelling software) with Grasshopper (open-source visual scripting plug-in). The second set of prototypes made better use of the possibilities of the Arduino. A small script, written in the Arduino language, was uploaded to the Arduino board. Powered by a nine-volt battery, the board ran as a 'stand alone' without any connection to a computer or power source.

Actuators: Servomotors

To transform the felt panel, we used simple six-volt servomotors typically used for radio-controlled model cars, ships or planes. The model we used provides a 180-degree turning movement, with a power of 10 kg.cm.

Sensors: Light Sensor, Sound Sensor and Kinect®

Different sensors were tested for the prototype.
We used a light sensor to register the touch of a hand on the felt panel.
It was, more specifically, a light-dependent resistor that measures light intensity. The sensor is directly attached to the felt, like a small button might be.
To sense sound, we made use of the laptop's microphone. This of course meant that we did not create a real 'stand alone' felt panel whenever we worked with sound as input, but this could be improved in a next series of prototypes.
To sense a person approaching the felt panel, we used the Kinect® motion tracking device, as previously described in Chapter 4.

Mechanics: Board and Wire

To connect the processor, sensors and actuators, we used a thin MDF plate of 20 x 20 cm. Nylon strings were used to bring the movement of the servomotors to the felt panel. This allowed for a flexible, modular setup. The wires can be stitched to different corners or edges of the felt panel. Each position creates a different possible transformation of the panel.

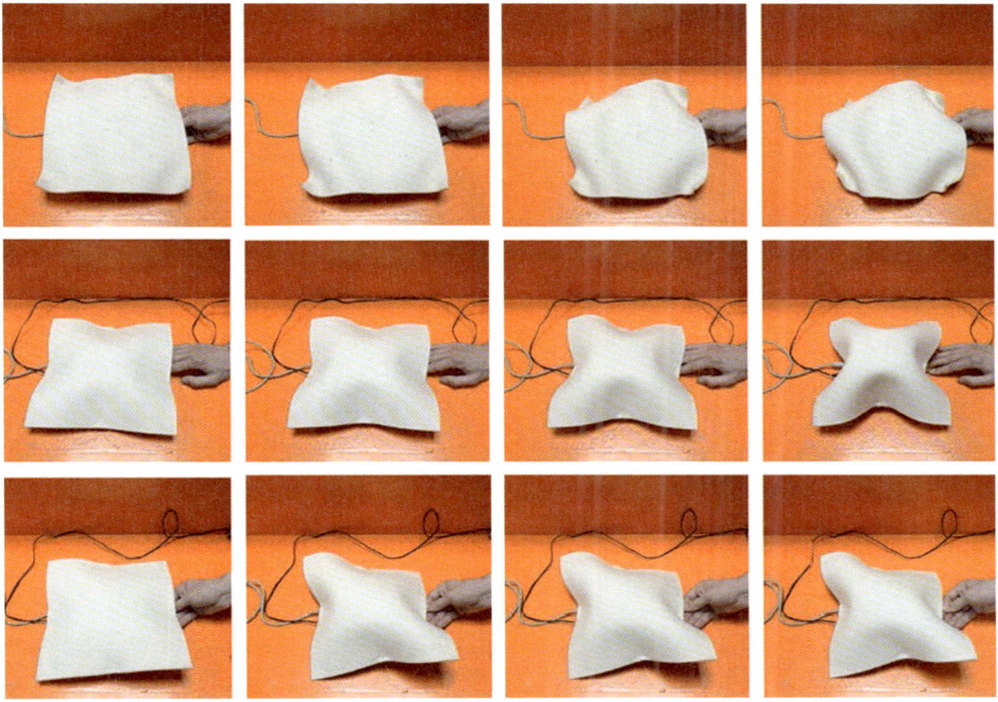

Expressions:

Communication by Shape

We can bend the felt panel in different ways to create different shapes. The modular set up allowed us to play with different complexities of transformation:

Uniform Movement

When there is an input from the sensors, all four servos turn a full 180 degrees in such a way that the panel bends upwards and forms a hill. This simple on-off reaction can be used to create attention. Different shapes can be created depending on where the wires are stitched to the felt panel.

Modulated Movement

We can program the felt panel to react with a stronger or weaker transformation to different types of input. Moving your hand over the panel creates a shadow in such a way that the light sensor receives less light and the servos only turn 90 degrees. Resting your hand on the panel, by contrast, takes away all light, stimulating a 180-degree turn and a full bend. Such a scaled reaction can also be employed in relation to the sound sensors: low and loud noises stimulate proportionally strong transformations.

Differentiated Movement

When we use more than one sensor, it makes sense to control which servomotor reacts to the input. The images on the right show how we applied this in some scenarios:

Look at Me

The panel is connected to a Kinect® motion tracker. When you step within a radius of 3 m from the panel, the Kinect® reads your position in space. It provides input describing your position relative to the panel. The servos close to you remain calm, those further away turn, and the panel bends towards you. Again, we easily interpret this as the reaction of a living being: 'It's looking at me!'
We can imagine different applications of this installation. You could, for example, have a panel that hangs in your kitchen connected to another panel that hangs in the kitchen of a family member in another city. Walking into your kitchen could cause the panel in your family member's place to wave, giving a subtle notion of presence. The information that you emit – I am here, I am walking around, I am calm or frenetic – could be transmitted differently. You could simply e-mail or speak by phone. But using this panel could be an additional, subtle way of adding nonverbal communication to a regular phone or video call.

Who's Talking

The panel is connected to four microphones. The panel bends depending on which microphone receives sound. We tested this in a simulated office meeting. The panel turns to the person talking. It becomes interesting when the microphones provide input to a felt panel on another meeting table, in another room. From afar, I could hear all four people talking and the panel would show me where the speaker was sitting.

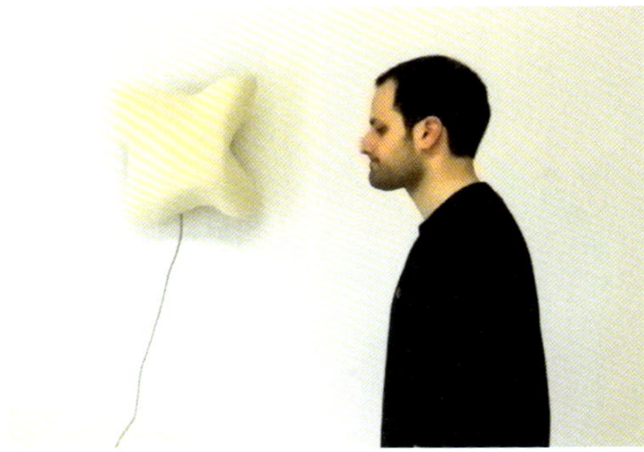

Scenarios:

Improving Communication

The first tests show the potential of the Talking Napkin as a support to other channels of communication. You might use it to supplement a phone or video call. Its main advantage is its 'animated', lively appearance. The felt's organic curves and movements appear generically natural without mimicking any specific life form. And the material's softness adds a positive connotation: it appears approachable and cuddly rather than alien or scary.
The chosen felt quality is both soft and rigid. This makes it possible to make it an integrated material in which surface and structure are one. Together with the use of wires simply stitched into the felt and affordable servomotors and sensors, it becomes a very flexible and robust product. It lends itself to further development through an open source process.
We commonly know felt as a material used in garments, cushions or upholstery. As such, the Talking Napkin easily blends into corporate and private interiors. It holds the potential to become an approachable, friendly 'invader' of domestic space.
We are interested in increasing the size of the Talking Napkin. We want to move on to a larger, architectural scale. Therefore, in the next step, we turn it into a wall panel.

Enlarging the prototypes described in the previous chapter returns the project to an architectural scale. We maintained a focus on communication, which allowed us to explore sensors and interaction models and temporarily postpone considering the structural necessities. Adding a load-bearing substructure to the panel expands its potential from a communication device to an adaptable, multifunctional interior element. Multiplying these elements brings our prototype close to the vision of the adaptable felt house from the 'Models' chapter.

Watch the movie:
www.thewhyfactory.com/barba

Softwall

The Why Factory and EIT ICT Labs: The Why Factory research team with KTH Stockholm and TNO

The Why Factory research team, TU Delft
Ulf Hackauf and Adrien Ravon with Arslane Benamar, Sander Mulders, Lex te Loo and Stavros Gargaretas
EIT ICT Labs
Charlie Gullström (KTH Stockholm) and Joke Kort (TNO)

Mock-up:

Softwall as Portable Device

We temporarily ignore technical limitations and imagine how a larger animated and responsive felt panel could work. We use a 1 x 2 m felt panel and simulate moments of communication between two people: one at a desk and the other moving between different locations in or around an office building. The qualities of felt described in the previous chapter play out nicely in these scenarios. The felt panel can become part a desk. When not in use it is a tablecloth. When you talk to somebody, it can bend upwards. It can mimic the posture of the person you are talking to. And it can lower surrounding noise or your own voice, providing insulation and privacy.
It could be part of a wall, or part of a sofa or bench or a cover for the tables in the canteen. Eventually, it could also be completely mobile, like a large mat or bathing towel, and you could take it outside.
We could surely imagine that such a panel might include a microphone and a speaker. But it would initially be about adding a physical representation of a body to the communication. In this study we felt that the soft material made for a nearly personal experience of communication. The material compelled one to touch the panel; it was surprising to simultaneously see and feel it respond.

Phone booth

Desk privacy

Talking bench

Table cover

At the canteen

Personal talking

Take away

...

Scenario:

Softwall in Office Meetings

In an office setting, the communication around meetings becomes an interesting case for the Softwall panel. There are means today for people at different locations to meet and discuss things without physically being in the same room. And when telecommunication software, webcams and microphones are available and of good quality, such a mediated meeting can be a good alternative to a physical meeting. But, often, the communication before, around and after meetings is more difficult to include in a mediated session. We took a meeting room in our faculty building to illustrate such moments of encounter:

Before Meetings

If you need to talk to somebody, you might choose to walk by that person's desk. You could see from the door if your colleague was working on something and should not be distracted or if it would be okay to have a short talk. If your colleague were on the phone, you might wave and signal that you would like to talk later. Much of this nonverbal communication can take place by making yourself visible, using gestures or suggesting something through your face expressions. And it's the same thing if someone is trying to get your attention: you can choose to wave, nod or turn away.

In Meetings

There may be times during a meeting in which you participate less than do the other people in the room. Think of a situation in which you signal that you'll take notes or answer emails while topics irrelevant to you are being discussed, and in which you actively join the discussion again on a later topic on the agenda. You may even let your colleagues know that they should alert you when they need you and meanwhile continue to work at another desk. The possibility of sitting closer or further away from the meeting table can be an efficient part of the meeting setting.

After Meetings

Eventually, ending a meeting may include nonverbal communication. To sit straight, gather your papers, and cast an open look on your colleagues may make it very clear that you want to round off. As before, standing, waving to your colleagues who continue discussing or standing in the door waiting are often used to signal that you want to switch the mode of communication.

All of these situations come quite naturally in a physical meeting. In a mediated meeting spread over multiple physical locations, such nonverbal communication can easily be lost, leading to miscommunication or unnecessarily long meetings. The next pages describe more hypothetical scenarios and how we imagine that a larger version of the Talking Napkin could work in each case. Each scene shows a part of a nonmediated, traditional meeting in which three participants are involved in different ways. We then imagine how the same situations could play out if one of the three persons were at another location and remotely joined the meeting by making use of a responsive felt panel that supported the communication. These are still illustrations, and they make use of a mock-up of the panel. The question becomes how we can turn this into a working prototype.

Communication before a meeting

Communication during a meeting

Communication after a meeting

"Let me check who is
in the meeting"

"Can you wait
a minute?"

"I am here, whenever
you are ready"

"Welcome to
the meeting"

"Let me show you guys"

"I am losing focus"

"I would like to present the following"

"Let's chat over a break"

Three Prototypes:

Sound, Movement and Proximity

In this part of the study, we developed three working prototypes.
The prototypes are similar to the Talking Napkins from the previous chapter. A felt panel is combined with a wooden board that holds actuators (servomotors), a microcontroller (Arduino board) and sensors. The actuators are connected to strings that pull parts of the felt panel, which in turn bend and twist. The three prototypes in this chapter are larger – approximately the size of a standing person. They do not lie flat but rather hang. This makes it easier to have the felt turn and bend in different directions without a supporting substructure. Such a structure would be necessary to lift a corner of a large panel lying flat on a table, as shown in the mock-up studies on the previous pages. And it would require an even stronger substructure and much stronger motors to create a transforming, load-bearing felt panel that could support the weight of one or more people, as we suggested in the Magic Carpet chapter. One step at a time, we first explore the sensors, then the interaction between the user and the prototype and finally how a user experiences it in a room. We later hope to develop this prototype further, towards the Magic Carpet, so that we might eventually use it to create a completely transformable room.

Large panel prototypes, aligned as continuous surface

Testing smaller panels, continuous surface

Stitching detail: anchor point, force coming from four directions

Stitching detail: anchor point, force coming from two directions

Prototype:

Softwall and Sound

The Sound Wave prototype senses sound and reacts with a wave pattern. With low sound levels, the panel shows a slow, modest wave pattern. When the ambient noise rises, the waves become stronger.

The movements of this prototype are initiated by eight servomotors. Each motor is connected to two points of the felt panel via a nylon cable. There are more connecting points at the top of the panel than at the bottom. This mimics the movement of a human body in which we might argue that the upper body part is more important for communication than is the lower part. As all boards hang from the felt panel, the prototype does not require any additional supporting structure. It can be hung from a beam, from a hook on the wall or from a curtain rail. To be transported it can easily be rolled up, like a beach towel.

Panel:	5-mm wood felt, 750 x 1830 mm
Connections actuator/panel:	nylon cable
Wall mounting:	rubber band
Microcontroller board:	1 Arduino Duemilanove
Electric connections:	2 breadboards
Actuators:	8 Servomotors, FeeTech FS5109M with plastic cross servo horns
Controller/actuator mounting:	3 MDF rectangular panels
Sensors:	1 sound sensor

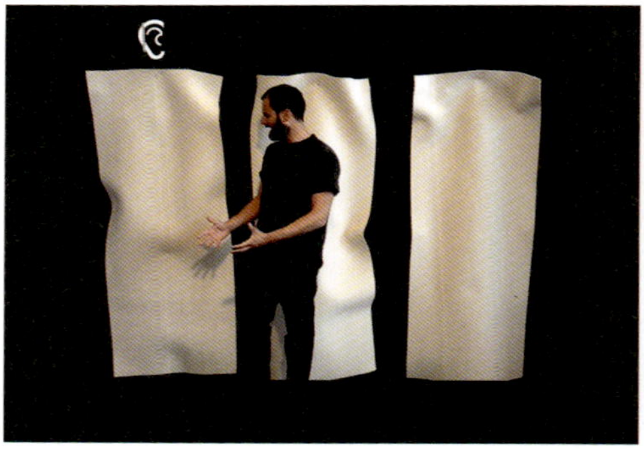

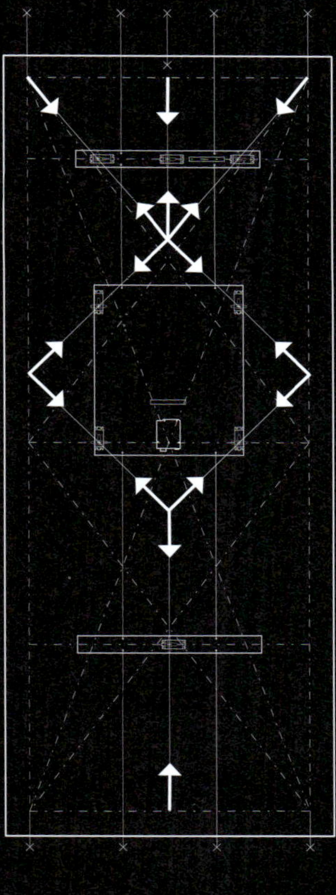

suspension point

board with actuators

board with actuators
and processor

board with actuators

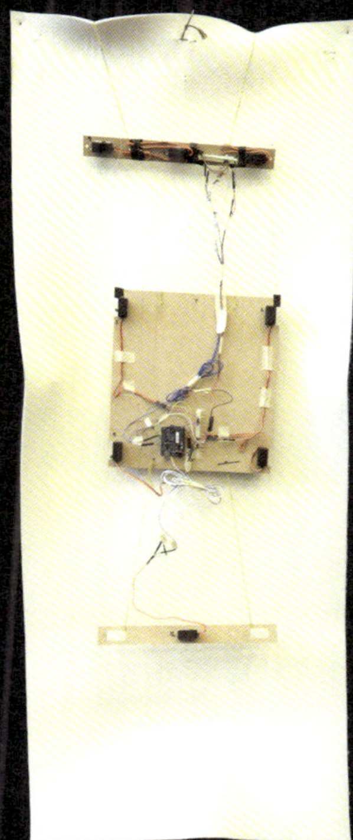

plan with connecting points

panel backside

Prototype:

Softwall Follows the User

The 'Follow Me' prototype senses the position of a person in space and reacts with directional transformation. The panel seems to turn its head, as if in observation of the passing user.
To achieve this effect, this prototype requires a rigid structure that connects the different servomotors. This structure is still light enough to be hung from the felt panel, which in turn can be hung to a wall or from a ceiling with a simple hook. We cut wheels from MDF that connect the strings to the servomotors and thus increase the effect of a single turn of the latter.
For the input, we use a Kinect®, which makes it necessary to have a computer between the sensor and the Softwall panel. In a next step, the Kinect can be replaced by two proximity sensors, which together calculate the position of a person in relation to the panel. This would allow running this prototype independently.

Panel:	5-mm wood felt, 750 x 1830 mm
Connections actuator/panel:	nylon cable
Wall mounting:	rubber band
Microcontroller board:	1 Arduino Mega 2560
Electric connections:	2 breadboards
Actuators:	8 Servomotors, FeeTech FS5109M with 8 MDF circular rotating discs
Controller/actuator mounting:	4 MDF rectangular panel with pvc tubes
Sensors:	1 Microsoft Kinect® sensor, alternatively two proximity sensors

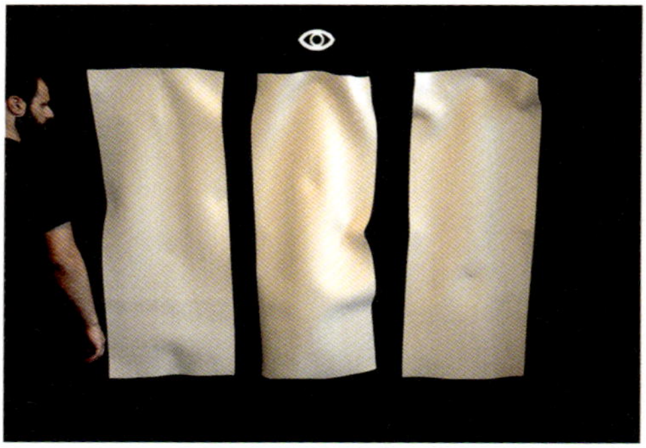

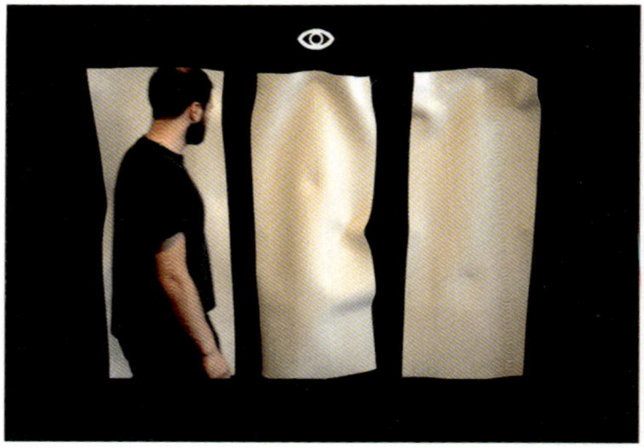

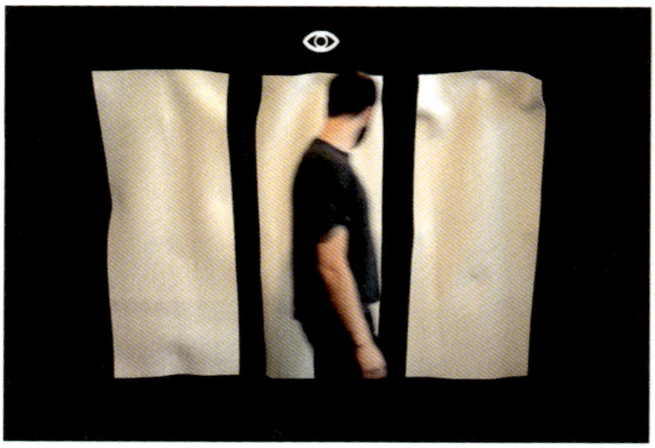

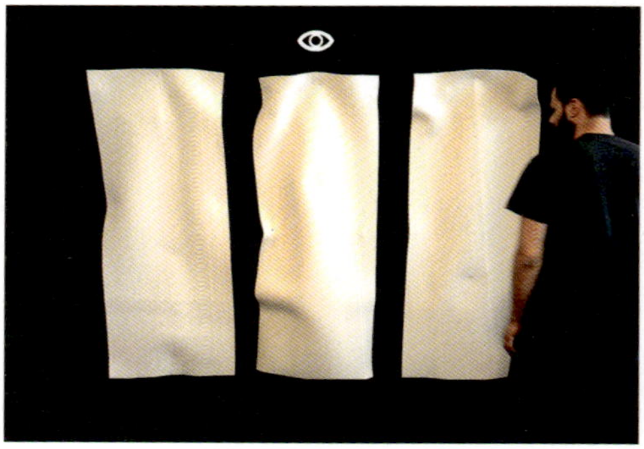

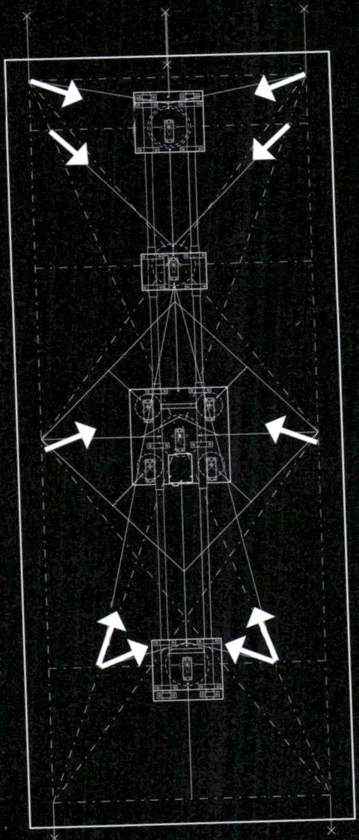

suspension point

board with actuators

board with actuators

board with actuators
and processor

board with actuators

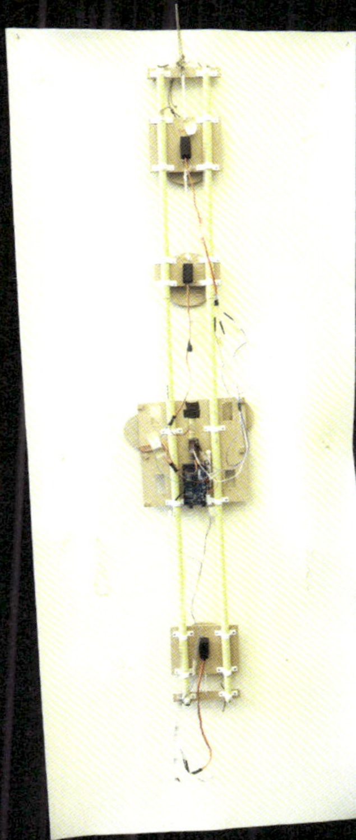

plan with connecting points

panel backside

Prototype:

Softwall Reacts to Proximity

The 'Proximity' prototype senses a user's relative proximity and reacts by bending, creating the impression of hugging the user. The bent panel not only provides visual and acoustic privacy but the hugging gesture also has an emotional effect on the user. This third panel has the most optimized backside. The wooden boards, which hold the Arduino board and the servomotors in place, are minimized. Attached to the felt panel, they make optimal use of the felt's load-bearing capacity. This panel can be easily rolled up for storage or transportation. It is light and materially efficient. This optimization is not specific for the 'hugging' movement of this panel – other transformation patterns can be achieved through this combination of customized wood boards and wheels.

Panel:	5-mm wood felt, 750 x 1830 mm
Connections actuator/panel:	nylon cable
Wall mounting:	rubber band
Microcontroller board:	1 Arduino Uno
Electric connections:	1 breadboard
Actuators:	3 Servomotors, FeeTech FS5109M with 3 MDF circular rotating discs
Controller/actuator mounting:	4 MDF rectangular panels
Sensors:	3 proximity sensors

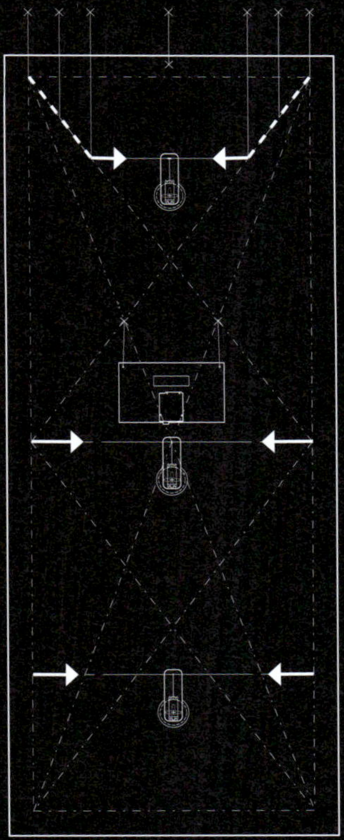

suspension point

board with actuators

board with actuators
and processor

board with actuators

board with actuators

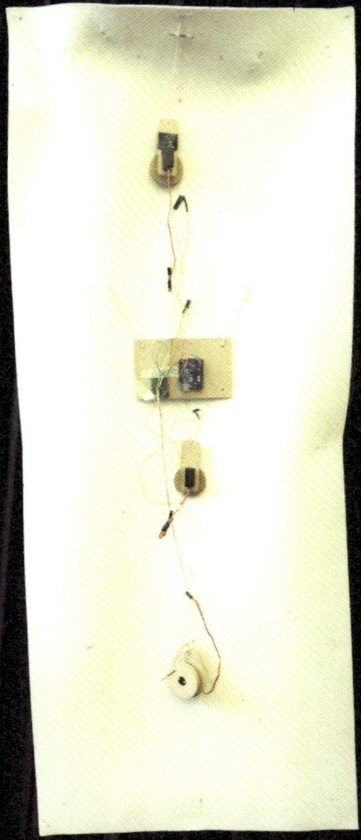

plan with connecting points

panel backside

Mock-up:

Applied Scenarios

The previous chapters show the prototypes in single settings in which the Softwall panel reacts directly to the immediately proximate user. But, as pointed out before, the desired intention is to connect the sensors of one panel with the actuators of another panel at another location, and thereby make it a tool of communication. How might this play out in an office scenario?

Office Scenario

Imagine you collaborate on a project with somebody working in a different location. You could use the Softwall panel to signal availability. When you're sitting at your desk, the panel in your collaborator's office would bend in a discreet wave pattern. When the panel in your room starts moving, you know that he is moving around. Another wave pattern could indicate whether there is noise in the room, possibly a sign that the other person is talking to someone and should not be disturbed. When it's quiet again, you'd contact your collaborator to discuss some issues. During the conversation, the panel shows you where the person you are talking to is in the room. Is your collaborator sitting or moving around? Eventually, you could step up to the panel for more privacy. This time, the panel in your room reacts to you directly. Your collaborator sees this and knows that it might be better to keep the discussion short now.

Domestic Scenario

The use of the Softwall panel would not have to be limited to an office environment. Imagine how you could use the panel to keep in touch with a family member: a parent or grandparent living alone, a child that recently moved out or a partner staying abroad for work. Besides phone or video calls, you may want to share your presence with each other, letting each other know that you are in the kitchen, for example, or in the family room. Seeing something of the other person's movements might be very comforting.

Hospital Scenario

Another specific scenario that arose in our study occurs in hospitals. Mediating presence and alternative ways of communication play important roles in helping children who are hospitalized regularly or for longer periods. Easy, low threshold channels of communication – such as that provided by the Softwall panel – could help these children and their families to stay in touch with each other and feel a kind of physical presence outside of visits and phone calls. The child could see and experience something of the activity at home, even if too tired to call and talk. And the parents on the other side could see if their child is awake or how well he or she is sleeping.

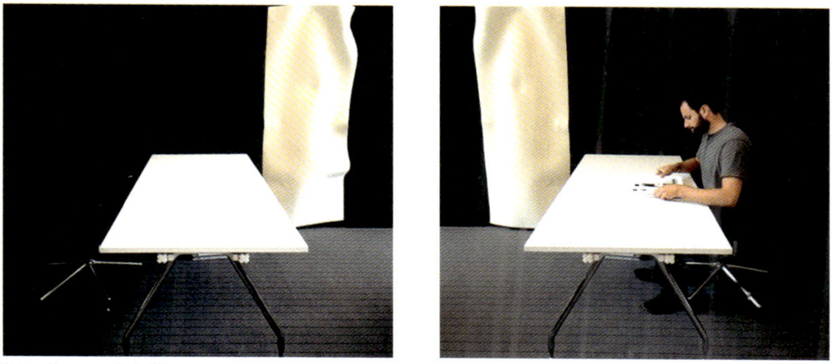

Signaling availability

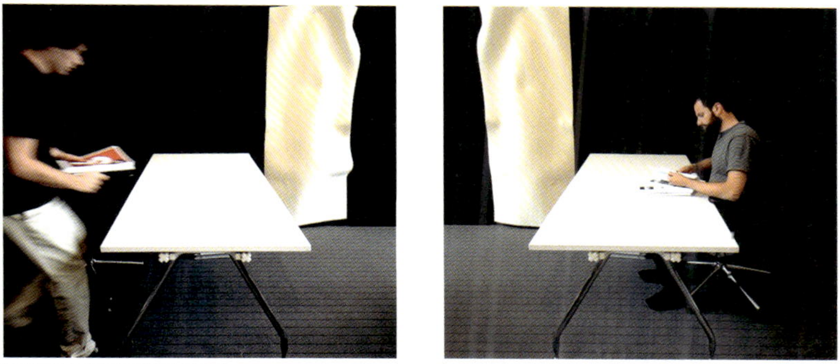

Sound becomes movement

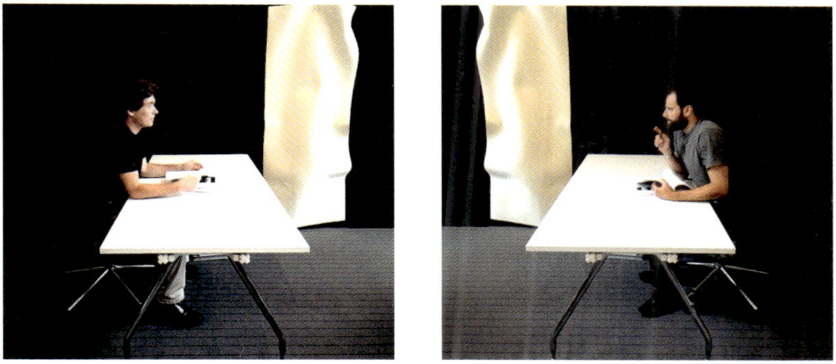

Position in space

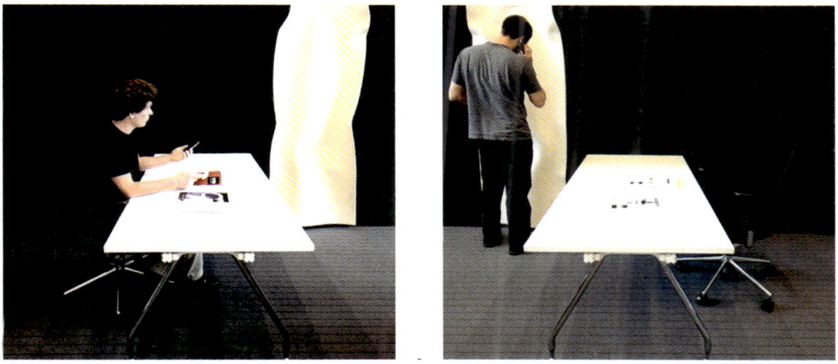

Private talk

Next Steps:

Office Space Integration, Furniture and Magic Carpet

Thus far, prototypes have not led to a necessarily saleable product. We plan to continue developing the technologies further and in different directions.

Office Space Integration

Together with our research partners from the EIT ICT Lab project 'Mediating Presence', we plan to test several Softwall panels in an office environment. The Swedish company Ericsson is involved in the project and could be part of the testing. In order to use the prototype in a permanent installation, we need to make it more robust and easier to maintain. The open structure of the actuators is advantageous, as it allows for the easy adjustment and repair of the mechanics. At the same time, this accessibility is also a potential weakness, as the cables can easily get stuck. It will be interesting to see how the basic principle can be adjusted to integrate it into a space. The material's colour can, of course, be chosen to match a room, but the size and depth of the panel and sound of the motors should be further developed. Ideally this would become a product that interior designers could modify and seamlessly integrate in their designs.

Furniture Design

Turning the Softwall into a piece of furniture is another interesting possibility. Think of a movable, free-standing Softwall that could be used as space divider that allows you to communicate with your colleagues. It could also become part of tables or seating, as the mock-up explorations in the beginning of this chapter illustrate. This would require a different substructure – hanging the panel from a wall would not be enough. The next step here requires developing this mechanical substructure.

Magic Carpet and Transformable Room

Our main focus for the next step, however, is to work towards the Magic Carpet: the responsive, programmable, load-bearing felt panel that can be turned into different pieces of furniture. This product will fill the gap between interior object and interior space . . .

Prototyping in progress

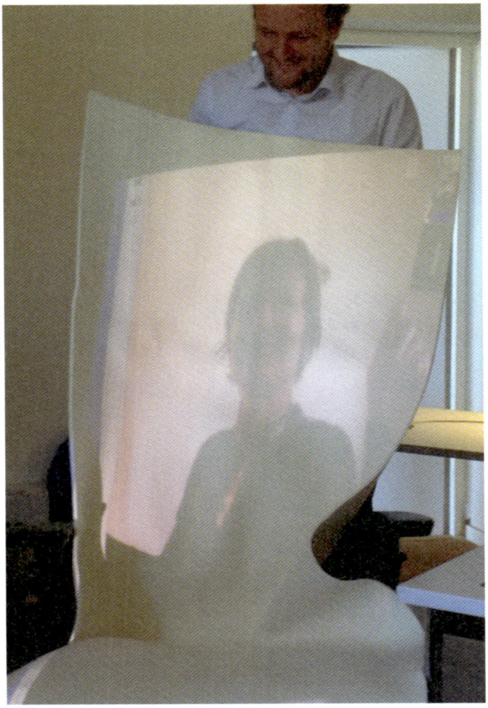

Testing projections

Maintenance

Panels forming a room

Our vision of Barba and the fully adaptable city is meant to generate discussion. But is it ultimately a proposal, a prediction or a warning? Decide for yourself!

Implications

Today's cities are primarily composed of stable buildings. These buildings provide orientation, borders, property – they define space as we experience in. In Barba and the fully adaptable world it could create, this material stability would disappear. What would be the consequences for us, for our daily lives, for living together and even for political systems?

Map of Questions

The Why Factory and ETH: Transformer at ETH Zürich studio

Transformer at ETH Zürich studio:
Paolo Birrer, Tania Carl, Riccardo Caruso, Melina Cerfeda, Elena Chestnova, Nicholas Ganz, Marcel Hodel, Léone Kündig, Marisa Muscionico, Sebastian Oswald, Kai Peter, Pascal Ruckstuhl, Enrique Ruiz, Paula Schilliger, Vera Schmidt, Henry Stehli, Ying Yi Tan, Yangzom Wujohktsang
Teachers: Winy Maas, Ulf Hackauf, Adrien Ravon, Stefan Wülser
Guest critics: Knut Brunier, Benjamin Dillenburger, Joris Fach, Manuel Kretzer, Felix Madrazo

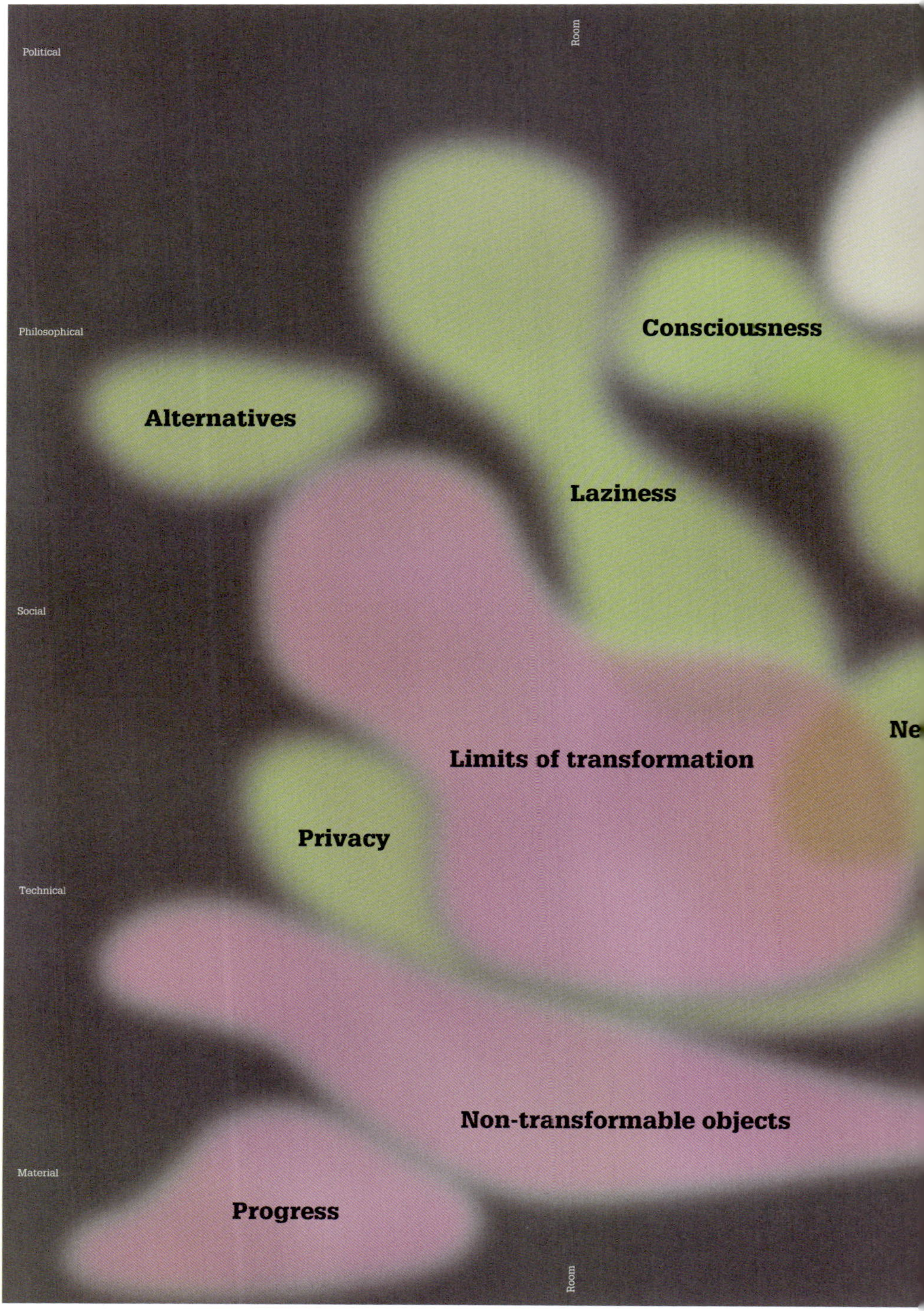
Political
Room
Philosophical
Social
Technical
Material
Room
Consciousness
Alternatives
Laziness
Limits of transformation
Privacy
Non-transformable objects
Progress

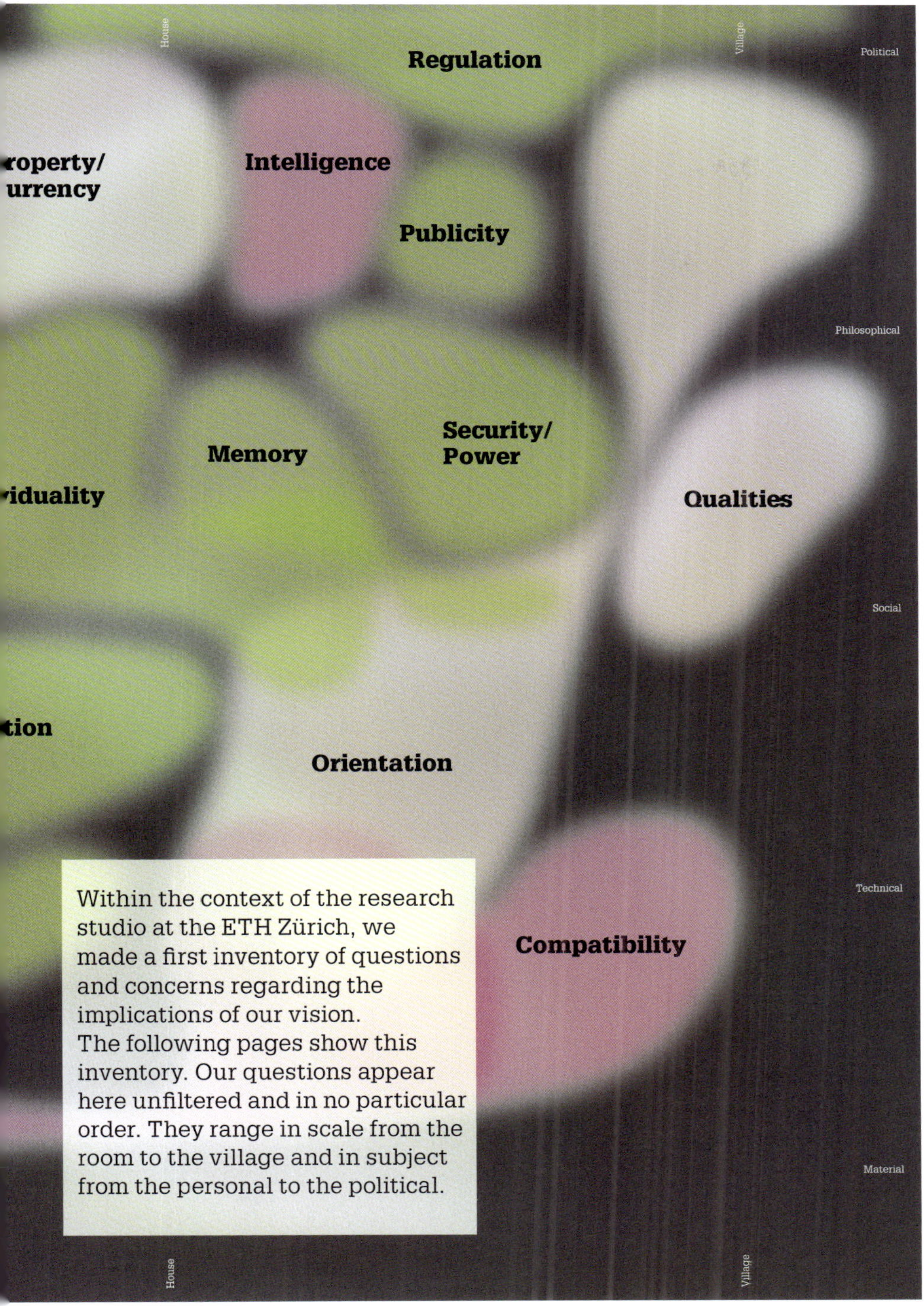

Within the context of the research studio at the ETH Zürich, we made a first inventory of questions and concerns regarding the implications of our vision.
The following pages show this inventory. Our questions appear here unfiltered and in no particular order. They range in scale from the room to the village and in subject from the personal to the political.

Political

Philosophical

Do we really want the world to be changing around us all the time?

Alternatives

If anything can be synthesized at any time, does the concept of 'reality' still have any meaning? Or does the distinction between the virtual and the 'real' become obsolete?

Why should the environment transform? Why don't we transform instead?

What if we no longer have needs because the room or the city or the environment provides everything for us?

Social

How do we communicate with the room?

How can one distinguish his or her property from

Is space the new currency?

If space is, inc
new currency,
need new reg
to prevent the
being squash
the rich?

Is wealth measured by the amount of space you have and the kinds of transformations it can undergo?

Propert
Currenc

If private pr
longer a nec
live our who
fluid city, w
use for mon

Will there still be something like a house? Or will there simply be one big homogenous village?

Would this scenario mean more or less freedom for each user?

Consciousness

How powerful are my thoughts?

What if having all we need within an adaptive space isolates us?

Will such a transformable environment lead to laziness? How will we be able to stay fit?

Will there still be the need for individuality? Or will people start to celebrate the community?

Will individuality and creativity increase or vanish completely?

Laziness

Will the system work equally for everybody?

Individuali

Do comfort and freedom of choice lead to isolation?meaning? Or does the distinction between the virtual and the 'real' become obsolete?

How could the space differentiate between what we need and what we want?

What would our personal leisure activities imply?

Will a flexible system cultivate and enforce the culture of individualism?

Would stepping out of t
adaptable environment
enjoyable?

Will microclimates be made modifiable within one's own space?

Will there be minimum and maximum limits? What happens, for example, when people meet up in

How does the cat interact with the transformable rooms? What about the other pets?

How would you negotia
between your desires a

Will houses require different kinds of regulations and building codes?

If space is, indeed, the new currency, will we need new regulations to prevent the poor from being squashed by the rich?

Is space the new currency?

Can cities think?

:s wealth measured by :he amount of space you ıave and the kinds of :ransformations it :an undergo?

Property/ Currency

Intelligence

Will the material be able learn?

Will there still be something like a house? Or will there simply be one big homogenous village?

If private property is no longer a necessity, and we live our whole lives in a fluid city, will there be any use for money?

Will the conclusions drawn from one user's experience be stored 'in the cells' and so help the other users with their own experiences? Would the material then function as some kind of shared brain?

Is there such a thing as collective intelligence?

Can there be monuments if buildings keep changing?

re still be the need ·iduality? Or will ·tart to celebrate munity?

Will individuality and creativity increase or vanish completely?

Memory

Individuality

tem work everybody?

What will I be able to remember, when the buildings move and everything around me is constantly changing?

How can we go back to the place where we met last week if that place no longer exists?

What would our personal leisure activities imply?

a flexible system vate and enforce the re of individualism?

If everything changes, can there even be history at all?

Would stepping out of the adaptable environment be enjoyable?

ow does the cat interact ith the transformable ·oms? What about the :her pets?

How would you negotiate between your desires and those of your neighbour?

Do we need any kind of stability or nontransformable elements? For example: sites of collective memory

What would be the function of politics? How would police interact with the flexible city?

Village

Political

egulation

Who maintains, controls or understands the system?

If there are regulations, how can we do better than today's world, in which money dictates what is built and what remains impossible, a utopia?

Who would be responsible for public space?

blicity

How would this flexible scenario change the relation between public and private space?

Would there still be such a thing as public space? Would there still be a need for public space?

Philosophical

How will relationships between people change?

Would corporations gain control?

Security/ Power

Would power still materialize?

How to achieve a spatial sustainability within this transformable world?

Are people in rural regions going to be left out by this new urbanization?

What about criminality? How can we achieve security?

Qualities

What is the border between utopia and dystopia?

One of the most wonderful qualities of our contemporary cities is the informality and tension provided by unplanned circumstances. Can this quality be maintained?

What economy would we have? And wha kind of jobs?

Social

How can orientation be provided in a potentially endless space? How can you orient yourself in a city that keeps changing?

Social

How do we communicate with the room?

How can one distinguish his or her property from that of others? Is private property still needed? And, if so, how would it express itself?

Limi

Who is allowed to control the transformations of a room?

Can children transform spaces, too?

Privacy

What happens to dead people?

Will interchangeable spaces result in a lack of privacy?

Technical

What would we keep? How do we select the things we want to keep?

What are the properties of this material?

Nor

How far along is material research for transformable material today?

Material

Can this become real?

Progress

Can things make themselves?

Are there things we can learn from this concept that might be useful for today's design tasks?

Will microclimates be made modifiable within one's own space?

What happens to my room if I'm not there?

Will there be minimum and maximum limits? What happens, for example, when people meet up in one room and it grows and grows?

of transformation

How far can a room adapt to a human being? Does it have any limits? Will there be just 'usual' changes or will it even adapt to our mood, our dreams, our feelings?

Do we have a forward and backward button?

Where are the limits of the Halo? What can and what cannot be done with the Halo?

How does the cat interact with the transformable rooms? What about the other pets?

How would you negotiat between your desires an those of your neighbour?

Negotiation

What will happen to existing structures that become useless?

Can materials be attache to other existing buildin to allow the building to grow?

Will there still be objects with emotional value? Would they survive in an ever-changing city?

How would we deal with different kinds of nontransformable objects?

ansformable objects

Will the material have the ability to absorb waste and reuse it?

What happens with objects that cannot be transformed in a room?

oes the cat interact
ıe transformable
? What about the
ɔets?

How would you negotiate between your desires and those of your neighbour?

Negotiation

ill happen to
structures that
useless?

Can materials be attached to other existing buildings to allow the building to grow?

Do we need any kind of stability or nontransformable elements? For example: sites of collective memory and monuments, or streets and static houses meant to provide orientation.

Is there any fixed structure?

Orient

Will there be city neighbourhoods?

If all the space around us is constantly changing, what will happen to nature? Will there still be animals?

Social

How can orientation be provided in a potentially endless space? How can you orient yourself in a city that keeps changing?

What will carry the load of such dense agglomerations?

n

Is there a spatial hierarchy? Are there rules of negotiation?

Technical

Would different cities be compatible with one another?

Can materials be attached to other existing buildings to allow the building to grow?

Compatibility

What will happen to nature? Can trees grow on the material?

What will happen to existing structures that become useless?

Material

Village

**In previous chapters, we described current research that could, eventually, allow the development of programmable and transformable materials. But such a material hardly exists today and, if it did come to exist, it might look very different from how we've imagined it or might take a very long time to become available. We're forced to acknowledge that, ultimately, an omnipotent and flexible material might be impossible to make.
Would this mean that the story presented thus far is pointless? I don't think so. What we've presented is not a prediction, it's a speculation.**

What If Not?

The Why Factory: Ulf Hackauf

A speculation that is useful exactly because of its speculative nature. We aim to suggest questions that carry weight, to suggest things we'll have to confront in the future, to present and propose new moral and spatial dilemmas. Our story is open to interpretation and its aim is to inspire. And then, we can also see the story as a visualization of several ongoing developments that already have an effect on how we live. Our project is relevant even if it never literally comes to be true as we describe it. This makes it well worth taking the story of the fully adaptable environment a step further and looking at its possible consequences – consequences with which we may be confronted even before the material is available. Where would a fully adaptable environment fail? What is the dark side of this story?

Space Currency

The wall that divides two apartments usually forms a clear border. You can work, eat and sleep in your apartment without discussing any of these things with your neighbour. If anything, it's primarily noise – the sounds of daily life, the television's murmur – that can sometimes be noticed despite such a border, and as such is a typical source for disagreement between neighbours. If you hold a party, for example, or practice playing the drums, your noise might directly influence a neighbour's well being.

But in a fully adaptable environment, there would be no fixed dividing walls. If you hold a party, it would not only be noise that penetrated your neighbour's apartment, but it might also be your whole living room. You would have to develop some sort of agreement, rules on when you would be allowed to extend your rooms and thereby limit those of your neighbour, and for how long. In an ideal world, neighbours would easily come to mutual agreements about these kinds of things. You could have guests one night, and your neighbour would give up some space for you. Tomorrow, if your neighbour hoped to swim in his bathroom, you would plan to shrink your living room to make more space. In a not-so-ideal world, we'd likely need rules to govern how we negotiated this dynamic use of space. We could, for example, define a time-related budget for space, kept account of with a kind of space currency. If you rent an apartment in today's world, you have the right to use it – in its fullest capacity – at any moment of the day. In the fully adaptable environment, you would instead rent space related to time. And at the end of the month you'd only pay for the square metres that you actually used. Or you could pay for a budget in advance and then use it as you need. There could be different allotments and distributions of this space-time currency. It'd be rationed, almost. And if, say, you had a lot of visitors at the beginning of the month, you might only have a small apartment left to use at the end of the month. Such temporary uses of space already exist today in the form of 'flexible workspaces' where office workers might sit at a different desk each day of the week, depending on the overall occupancy of the building. This concept allows for a more efficient use of space

and can support a dynamic, inspiring and flexible working environment. That said, such an arrangement also has the possible drawback of generating unrest and stress.
This could also be a problem in the fully adaptable environment: flexibility could be accompanied by a kind of restlessness. The constant adaption and fluctuation of space could actually create a stressful environment. You might even decide to sacrifice the luxury of a large space and keep your room smaller if it meant that it wouldn't have to suffer through change for an evening or for the whole day. Or you might program your room to change size very slowly over the course of the day – it would expand like a flower blossom unnoticeably following the sun. The balance between the efficiency of real-time adaptation and the liveability of a more stable environment plays a key role in the calibration of a fully adaptable environment.

Protect Me from What I Want

Ultimate ergonomic comfort is a fundamental part of our vision for the fully adaptable environment. A space would fit smoothly to your body and you would control all parts of the room so they'd be just as you wanted them. Whether you would like your room to have an organic or rectangular shape or even feature a gothic vault – anything would be possible. You could live in a different place every day without the pain and hassle of moving house. And all of this, the possibilities actually available to you, would ultimately depend on your imagination. Your ideas feed the space. Without inspiration, you could end up with a purely ergonomic and functional surrounding. But what happens if you run out of ideas? Where do you get new inspiration if you are only ever surrounded by an exact reflection of your own ideas?
The programmable material is a simple, neutral and, in a way, 'tasteless' ingredient from which you can create all kinds of shapes and spaces. It would be without any distinct qualities of its own. If you compared it to food, it would be like wheat, water and yeast but would not even hint at how to actually make bread. The material would be like letters of the alphabet, and it would depend entirely upon you to form words or stories. Without any creative inspiration, the fully adaptable environment would eventually become a boring and repetitive mix of ergonomic comfort and fake nostalgic expression (something, we could argue, that can already be seen in many cities and suburbs).
We'd need surprise and we'd need inspiration. The exchange of ideas would be essential. Actual interior designs (and not only their photographs) could be shared as music and videos already are. Architecture, design and fashion would merge and gain speed. At the same time, we might hesitate to replace all 'real' materials with programmable, omnipotent matter. Wood, steel or ceramic – non-transformable materials of all kinds – could gain value, because of their tactility, smell and even the simple idea of material authenticity. The adaptable environment might require quite a lot of surprise and authenticity to avoid becoming dull.

Lethal Laziness

The comfort provided by the fully adaptable environment could not only lead to boredom, but could also actually be unhealthy. Instead of walking from one room to the other, you'd simply allow the room around you to change. You might still have to move to get to a collective workspace, but it might be closer to you and you could, potentially, be moved up and down and between buildings without actually walking or cycling. The consequences would be obvious: too little physical exercise would likely make us sick. We have already seen the effects of a hyper-comfortable, sedentary lifestyle in Western societies. If the fully adaptable environment served to increase this trend it could become a dangerous or even lethal environment.

On the other hand, the flexible material could be programmed to challenge us. It could stimulate us to move and, at the same time, provide a permanent resistance to our bodies. Our dining table could actually move away from the kitchen, so that we'd have to walk back and forth. Corridors could be slightly inclined or have a soft floor, so that walking around the house would become exercise. Streets could be sloped to allow for exciting bicycle rides and roofs could become hills for hiking and climbing. Whether unhealthy comfort or healthy challenge prevails depends on how the surroundings are programmed. The question is who would decide – would you ask the street to incline or would the street decide to do so by itself, based on your condition and on the dinner you'd had last night? In such a responsive environment, could you still decide not to be healthy? Would you have that freedom?

Soft Borders

We initially imagined that a fully adaptable environment – a world of spaces that could extend and shrink spaces depending on their use – would lead to the more efficient use of space. This, in turn, would allow for much denser cities. To make full use of this newfound efficiency would require the evaporation of fixed borders between different homes or city quarters. If I were not at home, my apartment could shrink, and my neighbour's living room could take over that same space.

But this new flexibility could also be seen as a loss of stability. A soft border between two homes provides the freedom for negotiation but also takes away our freedom to not negotiate.

A very dense version of the fully adaptable environment could feel like a permanently crowded subway train. Avoiding your neighbour's elbow in your back while trying, simultaneously, not to step on somebody else's foot – such a situation can feel very stressful, effortful. A home that constantly adapts to your and your neighbours' activities could become equally as stressful.

This stress could actually inspire quite the opposite of a dense environment. To avoid constant negotiation, inhabitants might choose to move away, to actually create more distance between each other. The fully adaptable environment could turn into a vast suburb in which inhabitants disconnect from one another and a sense of community diminishes.

With such a 'threat' of the suburb in mind, there could actually be reason to desire increased density, resulting from inhabitants' desire to live together. Here, the new flexibility could provide new, more nuanced ways of cohabitation, a communal space beyond the common units of room, apartment, building and village or district. One could decide to live in a more separated way at one moment and to seek the company of others at another. It would be interesting to see to what extent density would eventually triumph, as this balance between privacy and community are negotiated.

The Death of History

The fully adaptable environment would offer complete flexibility of shape and form. Every inhabitant could himself become a designer of objects, as well as of interiors or buildings. At the same time, inhabitants' different activities would shape the environment. Social relations would become visible. As you walked through a fully adaptable city, you'd be able read a lot about the lives lived in it. Conscious design and unconscious expression would mix and form an ever-changing, complex sculpture.

But this incredible richness of shapes would always be temporary. Like clouds in windy weather, shapes would constantly appear, change, disappear and appear again as something new. You would never be able to go back to the place of your childhood or to the place where you first met your lover. The essential meaning of 'place' might actually be lost, replaced by what resembled, instead, a constant event. The layers of history revealed by streets, buildings and urban spaces could be lost completely in a constantly changing city. There would be no monuments; you'd feel no nostalgia and have no memories when returning to places of the past.

But this wouldn't be inevitable. New, adaptive architecture could, actually, further increase the desire for historical monumentalization and memorialization. The instability of some parts of a new adaptable city would be balanced by historical parts, islands of urban space dedicated to the preservation of the past and the maintenance of memory. But even this would be very different from the constant, physical manifestation of a history of the everyday we experience in our contemporary world.

Accumulated Power

In the fully adaptable environment that we've described, the computation of the responsive environment leads to a physical reaction. As such, we would not be able to control every adjustment of the environment. If we say 'kitchen', for example, a complete kitchen appears. In the mornings, this might even happen without us having to say anything, simply because the environment 'knows' our habits: that we have breakfast every morning. Such an attentive responsiveness requires the constant observation of our activities and the complex computation of our needs and desires. We would need to cede control to the adaptive structure surrounding us. We would need to forfeit some of our self-determination to a powerful structure that, if it turned against us, could be very harmful.

A potential danger lies in the very number of inhabitants using the adaptable environment simultaneously. The permanently high amount of user input and complex spatial reactions would not work in concert with a central computing 'brain'. Only a distributed system, one in which reaction is computed close to the input, could cope with the complexity of such a system. Some decisions would be made on the scale of the city, others on the scale of a block, building, home, room or piece of furniture. It would be your apartment, and not the city, that would change your shower into a breakfast table each morning and it would be your armchair and not the apartment that would softly recline when you fell asleep while watching television.

It would obviously be important to keep this system safe. On the level of the apartment, the local 'brain' would ensure that no harm befell you. It could, for example, automatically stop moving walls when they come too close to you and prevent you from being squeezed. It could also negotiate with the 'brain' of your neighbour's apartment. If his walls move into your living space, it could 'tell' it to not touch you or simply resist its force and push it back. But what would happen if more and more neighbours around you extended their apartments at the same time? Could the slight extensions of some hundred rooms add up to a tsunami-like wave of matter, a force too powerful for your own room to resist? Would it be possible to tame the ocean of changeable matter at all times, to make sure that nobody was ever squeezed to death by the power of the masses? The safety of the individual would meet a new scale of challenge if walls started to move . . .

Controlling Space

A related, yet different challenge lies in the possibility of consciously overriding the fully adaptable environment's adjustments. In my own home, I would like to be able to create openings wherever I want to, to move from one space to the other and to create windows for light and views. At the same time, I want to control who can enter my home and who has to stay outside. For a family, the possibility of opening and closing spaces would be more complex. Each family member could be allowed to walk in and out of the home and maybe to bring guests. Young children could have limited 'power to override', to avoid them crawling away from home. Access for strangers could be denied.

In any collective space inside or outside of a building, this 'power to override' might come into conflict. Changing public spaces or moving buildings might require complex rules and computing. If different inhabitants want different spatial changes, the system would need to negotiate between conflicting interests. Sometimes, the majority could win and other times the urgency of an individual's needs could overrule the comfort of the majority. But how desirable is it really to make the public domain changeable and thus

negotiable? If the desire of the majority would manifest directly in physical space, minority groups might be spatially disconnected, in a very literal way. The fully adaptable city could easily turn into a segregated city.
And there might be an even greater danger. The open and distributed system could be difficult to protect against intruders. If the city were hacked by some external and hostile force, the collective adaptable mass could become a powerful weapon in the hands of a few. But the biggest threat might even come from within the system itself. A system that aims to predict the diverse desires of a large number of inhabitants would need to be able to learn and evolve. Conflicting desires and inputs would constantly need to be weighed, considered and, ultimately, judged. Such a capacity to form judgment would, in a way, suggest a form of artificial intelligence. What if the whole environment were to gain consciousness? What if the system had to weigh the needs of inhabitants against its own desires? What then? Can we really be sure that we could trust the powerful environment?

Serious Speculation

As I began: to create a fully adaptable environment could, quite possibly, be impossible. Such a world might never materialize, or at least might never exist in the ways we've described it. But we could nevertheless be faced with questions and consequences and we can already see many of these things beginning to stir and to become reality today. Here lies the value of speculation as a tool for both research and development.
When we founded the Why Factory some years ago, our first publication centred on the need for vision and visionary thinking in the fields of architecture and urbanism. We wrote then that a 'vision is . . . a dream for the city and for its spatial translation that offers a long-term, cohesive, seductive, and strong perspective for future societies'. [1] As our projects became ever more specialized and detailed, I came to find the term 'speculation' – rather than 'vision' – a more appropriate one, as it allows us to work from a more singular hypothesis towards an imagined scenario. Unlike a 'vision', I feel it is a humbler proposal and that it doesn't implicitly claim, for example, to provide a complete model for a future society. Whichever term is used – speculation or vision – this speculative approach allows us to give answers to questions that have not yet really been asked; it allows us to leap forward, to imagine possible futures beyond the strict extrapolation of ongoing trends. This kind of 'serious speculation' requires transparent and consequent argumentation. And, if it approaches a relevant topic, it might inspire further research, give insight into the implications of ongoing development or provide new perspectives on current trends.
Such a brand of 'serious speculation' is familiar from literature and film. Thomas More's *Utopia*, Aldous Huxley's *Brave New World* or George Orwell's *1984*, for example. But also Ridley Scott's *Blade Runner*, or Steven Spielberg's *Minority Report*. These books and films, and countless others

of their kind, provide a view of the future and simultaneously and implicitly comment on the present. To produce such hypothetical narratives at universities, as part of otherwise 'concrete' academic research, seems a less obvious mode. But it is here that serious speculation can be tested without the pressures of critical acclaim, of profit or box office success. And architects, urbanists and other spatial designers have the tools to turn such speculations into spatial constructs, constructs that have a link to everyday routines and existing spaces and, thereby, become tangible and understandable. I think that serious speculation deserves recognition, that it belongs to academia and that it can actually be an essential research tool, something fundamentally of use in the fields of architecture, urbanism and beyond.
'What if' has a value even 'if not'.

1 Winy Maas, 'Why Be Visionary', in: *Visionary Cities* (NAi Publishers, Rotterdam, 2009), 212.

Through the Glass Darkly

Patrick Healy

Reading, electronically, a book allegedly the first to be made entirely of straw left me gasping at virtual reality. I need to experience paper, need to be able to touch it. I am immensely grateful to be able to consult this book – written by Mathias Koops, entitled *Historical Account of Substances used to Describe Events* and held by the La Biblioteca Nazionale in Florence – thanks to the library's Internet archive, but I still feel a little moody, have a sense of deprivation and loss. The book was not only a pioneer in its use of material, but was also a history of paper making, a study of the diverse materials upon which writing has taken place, from snake skin to tortoise shell to stone. And Mathias Koops may have been the first to take out a patent for recycling materials, as early as 1800.
Continuing with my research the following day, I found a reference to a work about paper models in architecture, a book that claimed to be the first of its kind in English. It is entitled The Art of Architectural Modelling in Paper, and was written in 1859 by A.T. Richardson, an architect, and John Wayne, in London. On page 20 I read: 'The paper I use, and have always found best for all purposes, has a surface similar to that of Whatman's double-elephant drawing paper, and is I believe sold under the name of crayon paper: a specimen is bound with this book, forming the next page.' I click. No. A blank page. Instead of the cream texture, a shining tabula rasa. Here I am, already at metaphysics, required to say that there are material and immaterial objects.

As I mull over this loss, consider the degree of loss entailed by this experience, a friend who works as a computer programmer tells me he has a new Kindle reader, and that the screen is the closest to 'paper' he has ever seen. I should get one, he says. I do not mention the fata morgana of my earlier paper chase.
The clash, or collision, between experience and 'experience' – which is surely what is at stake here – leaves me thinking about the nature of our reading and research, thinking about how the technological means available necessarily alter and add to knowledge. The university, according to ministerial and management mantras, has become a workplace for the knowledge economy, a kind of marketplace in which the ideas of research are relevantly proposed to complement both the overarching phenomenon of technology itself and broader social needs. It is easy to object to such a drive towards pragmatic and measurable realization based on broad and largely incommensurable claims. Nevertheless, reality forms, even as enabling fiction. The rule of measure quickly takes over working lives.
It is often easy, in such a constructed account of education, to confuse the instruments of research with the goal of research, and to leave aside the concrete and specific ways in which human creative learning is achieved outside of any metric system.
Conflating means and ends, we lose the perspective and richness of search and are restricted to a kind of 're-search' that blocks the open and the new. Something truly new is often not

recognizable; its value awaits a longer future than instant gratification demands.
Invention and novelty are imperatives in the constant progression of technology.
In some way, the idea of experience is also motivated by a response to technology: it is truly not contractual but experimental. It is in this collusion, if you like, of early modern science and the intensification of the subjective that a fatal overlap occurs. Descartes makes the removal of doubt, immunity to scepticism, depend upon clarity of a mathematical kind. From there, it is a simple step to the proposition that such knowledge and truth are equal. Truth belongs, then, to science. Further, if the world were viewed like a mechanism, like a great clock, the modern age was essentially tasked with understanding this machine. Angels were not welcome in the laboratory and the fantasy of the animation of material and things, the redolent atmosphere of the world of appearance needed to be placed under anathema.
Human experience is constantly vulnerable to illusion; even the senses remain suspicious in themselves, until clarity is reached. Rationality – clear and distinct ideas guided by logic – does not allow for the value of rumination, spontaneous fantasy, the imagination. Or it treats them as fundamentally secondary. It eliminates everything that is messy, fuzzy and it ultimately eliminates the open itself. But sometimes one has to be 'away', outside of oneself, to realize that the cogito tells a very small part of the tale of human knowing.
One could say that this is the essential modernity of the clash of experience and 'experience'. However one looks at the role of *techne* and research, the intimacy of the relation can never be sundered. Think of how we are only now coming to terms with the result of early modern science. But if one abandons the so-called 'progress of knowledge' thesis, then Newton's ability to research physics and alchemy, for example, points to a richer constellation of study than the academy now permits.
t?f, Now called The Why Factory. The question mark in the hybrid acronym and logo pointed to the newness of what was introduced to the Faculty of Architecture at Delft University of Technology with the arrival of the brilliant and vivid Winy Maas, and through the research aims he hoped to introduce.
The question mark was a 'why' and not just a 'what' or 'when' or 'where'. There are no secrets here. Maas, in a lecture presented upon becoming a Professor in the Faculty, presented a mission statement and a vision. He was guided by a commitment to create a platform meant to guide both understanding of and prescriptive intervention in the new urban development.
He announced a version of what fundamental research 'by design or imagery' required, and indicated that through models and visualization it was intended to produce a series of critical alternatives 'through images'. The first agenda of The Why Factory was proclaimed the 'future cities' programme.

With new media, installations, films, games and virtual realities, the need for what Maas called 'argumentation' increased in relation to complexity.
In a time of crisis, the model of research for architecture was neither to retreat into a specialist autonomous discourse nor to simply pursue individual projects. There was and is a need for a wider agenda in which architecture takes responsibility in the public debate on space and in which the architect uses his or her skills to visualize the process of the development of future social, economic and political realities as emergent processes.
The architect is a spatial specialist who derives lessons from traditional discussions on space, light and form. A new communicative expansion was needed in light: not only of rapid urban expansion but also in the way in which the 'information' age had to negotiate both material and immaterial values. This was not a matter of the tradition of the past being handed on; there were rather new actions that required a 'speed' of response that media changes enabled. Maas was sensitive to the fear of the overload of the image, and saw the need for the argumentative and theoretical to guide research against the incessant Blitzbild that seemed to overwhelm the everyday. Maas's was also a frank acknowledgment that the search for the iconic in architecture needed to be critically set against the social.
A key modernist issue was how to understand the collective and the nonhuman world as paramount, even as existential requirements insisted on being given priority. Another tension that had to be understood was a commitment to where, as a 'device', architecture would broaden research into the critical, the cultural and the social without becoming bogged down in discussions about the formal requirements of an autonomous body of knowledge.
Architecture is not a sign whose meaning is already known and therefore legible, nor is it dependent on other signs to signal its intention. Instead it is both messenger and message. This was not a solution involving a linguistic turn in architectural discourse, but rather an extension of the fundamental communicative requirement of the art. Maas was calling for an expanded concept of architecture. Further, and the point needs to be developed in a fuller study, he turned away from the 'scenario'-inspired research that had become current: the scripting of the architectural that moved design towards symbolic isomorphisms. Good examples of this can be seen in proposals for museums imagined as cultural packages and social transformers in the United Arab Emirates. There may be a tilt here against Remment Lucas Koolhaas, whose early work in film, *De Blanke Slavin* (The White Slave-Girl, with Renée Daalder) was one of the most expensive productions of the 1960s in the Netherlands and preceded his move to full-time architectural practise. Bart Lootsma has written on the difference between Constant and Koolhaas in terms of architecture's view of the megacity and projection towards

the future. Following Lootsma's schema, I imagine Winy Maas as much closer to Constant than to Koolhaas.
This search for explanation and meaning was accompanied by a robust pursuit of information gathering and analysis. Ultimately, there could be no form without content and no content without form. The group accepted but downplayed the issue of autonomy and understood, it seems, that tradition is maintained only by constant de-struction – that is: a de-structuring of previous assumptions, historical narratives and restraining theories of development.
In a sense, Maas wanted the key element of a workplace that was committed to production: busy with modelling the plethora of current thoughts and anti-thoughts and challenging the 'what-if' approach of scenario hypothesis with the more trenchant and critical 'yes-but' emphatic response.
His was a call for a wider agenda, a new constellation of the university and the workplace, a deeper remit for architecture, the creation of a research model in which a productive laboratory allowed angels to return from their seventeenth- century exile.
Exhilarating, confusing and, above all, challenging. The relation between the research group and the university, the relation of the architecture office and public life, a new criticality, a commitment to image research, and yet something that remained familiar, the issue of production via vision itself.
It is worth dwelling on the term 'factory'. In architectural discourse it first arose in the 1740s in Switzerland, when the model for chalets was being developed and the whole was assembled in a *fabrique*, later *Fabrik*. The word is synonymous with the Industrial Revolution and, by the end of the nineteenth century, someone like Edison, the great inventor, will have what he called an 'Invention Factory', much later to be inhabited in mysterious circumstance by Andy Warhol, then again just referred to as simply 'the Factory'. There is in the later use an idyll of creativity and work, a place to hang out, search, talk, make things, and all of this far from the monastic strictures of the Bauhaus, say, and indeed a tilt at the void, which factories had become.
Maas was aware of the fine line between the university as a production line and the requirements of resistance, freedom of untrammelled research, the deadening allure of efficient causality in which only questions like 'how much' and 'does it work' stifle creative and imaginative life. The old pedagogical nexus of the teacher as the one who has learned and teaches of the need to learn in order that students learn how to learn – the transmission of bodies of information, and the creative and existential exigencies of a life – would not be easily contained within any programme. The initial days of work in March 2008 were in collaboration with the DSD, a joint venture in colloquia, workshops, discussion and even, 'hanging out' – a proposal that lingering together in free speech might be the greatest and only gift university life offers.
Six publications and many realized projects later, such an assessment

requires little elaboration. Here, again, the paths of communication are quite intrepid, from paper architecture to image, from imagined to built, the dream is allowed to breathe and circulate in open air. Fantasy is not strictly forbidden. At the heart of the group's belief was the importance ascribed to information and communication.

Behind this trenchant dedication to discovery lay, I think, the broader problematic of the modern, certainly the aspect first announced in Riegl's 1905 study on the problem of the monument, namely the rejection of the venerability of the past for its own sake and the search for the new as the main requirement and demand of the emergence of mass culture.

A spectacular development in Berlin of 1909 confirms Riegl's thesis, namely the development of Luna Park, which features so wonderfully in Elske Lasker-Schüler's *Mein Herz*.

One entered Luna Park from Kurfürstendam 124. It was, at the time, the largest entertainment park in Europe. The guesthouse Wirtshaus am Halensee had been acquired by the head chef of Kempinskis Aschinger who, with Bernd Hoffmann, transformed the property into an amusement park based on Coney Island in New York. It included swimming pools, wave machines and a *Shimmy-Treppe* that lifted skirts, boxing matches, beauty contests. The park provided daily popular entertainment to 50,000 visitors. No one wanted to talk about pop-architecture, and yet it was not so far away from the demands of the everyday glorified in other discourses.

It is the problem of the role of mass entertainment, something that inspired a moralizing and negative assessment in Siegfried Giedion. According to Walter Benjamin, Giedion showed that the principle 'welcome the crowd, seduce it to stay' leads to corrupt architectural practices in the construction of the department store. For Benjamin, Dada and surrealism had as mother and father the Parisian arcades. Alfred Gotthold Meyer, in his text on iron construction of 1907, accepted engineering as a new and dynamic base for architecture, and saw a new beauty and 'steely sharpness' of the Eiffel Tower. In Benjamin's account, architecture gathered as dreams around the technological innovations of iron and glass. A future unconscious could be read from the kernel of material things, preparing already new developments. The development of glass and iron construction during the Industrial Revolution would dream a future that reaches forward to today, to the new Markthal in Rotterdam. However contemporary the research, the past comes to meet you in an explosion, an image at a standstill. You do not need to go searching for it, to locate some authentic 'how it really was' of time since long gone.

But no study that claimed to deal with the changing reality of and context for architecture could ignore urban modernity and the attendant masses. The question of experience as repetition and a reconfiguring of an awakening is central to Benjamin's study in the *Das Passagen-Werk*.

I grasp something of what he means when, standing in front of the

Parliament in Berlin, I see Charles Fowler's glass conservatory at Syon House, Middlesex (1827) and wonder about the crystal dreams that stretch all the way back to Geneva in the sixteenth century. Or I think of Meyer, cited earlier, who held that the origin of all modern day architecture built in iron and glass was the green house (this is cited by Benjamin in *Das Passagen-Werk*). All of this and of course I begin to understand that the kaleidoscope can be shaken again to produce other constellations.

It is also possible to read modernist shock as a question of speed, transparency, a generational reaction against the 'sins of the fathers', the embracing of the demotic, overturning of elitist conglomerations, a will to power and an imperative to treat force as finite.

What has proved much more intractable is grasping the atavism and avatars that haunt every fresh start, dealing essentially with the issues of limit and edge that also guide, through questioning the path to the open, the truly cleared space of the future.

In light of Benjamin's research, it seems hardly accidental that Koolhaas went from co-writing film scenarios to producing work, as an architect, in which the notion of the metropolitan was already central. Winy Maas rejected this approach as adequate in itself in his inaugural lecture.

It is difficult to avoid a kind of jumble given that the diachronic interpretation of the tradition of Dutch modernism has itself such complex crosscurrents and roots. I mean here that modernism took on different identities in Vienna, Paris, Berlin, London. It was marked by its cities, and in very different ways, not to speak of its notable impact in California. While it is possible to show a long arc of such concerns, it is nevertheless significant that the problem of study for this group has always been about transformation, transfers and open transmission.

As I struggled with the loss of tactility in reading, and with the endless distraction characteristic of information gathering on the net, I received an e-mail from Ulf Hackauf, enquiring as to my availability for an interview to be transcribed for a forthcoming publication. Having agreed to meet and discuss this suggestion, I visit with Ulf at The Why Factory space within the Faculty of Architecture and the Built Environment. The space is in the east of the building, in which a vast monumental interior and street-like circulation seem to flip all inner/outer, public/private dimensions.

But being there and talking, I realized it was not just a matter of reversals. The steps are as deep as at the approach to the Pergamon Altar in the museum in Berlin. I was struck by the ancient measure and the emphatic colour, of which Kandinsky once said it denoted enormous confidence. The design is aware of its own resources; the space has been given warmth through orange. It is a work station and a theatre both. The relation to the monumental scale of the tribune, and its various activities, create a 'public interior'. There is something uncanny in this; it also reaches to the Markthal, where hybrid functions are a dream space for future possibilities.

The 'Tribune' was housed in a glass house. My mind wandered to this feature of such significance, to the crystal dream that has powered utopian feeling from the time of alchemist's operations to the work of Scheerbart in his Glasarchitektur of 1914: as if the scrying in a glass allowed the spiritual transformations sought in the quest for the philosopher's stone. It is the search for a prismatic paradise.

The public interior invites dreams like the openness of a bird's nest; it somehow releases a sense of movement and of possibility. I enjoy the idea of the Roman rostrum, on which the architect must stand and give a speech to the actors who are wearing the family masks. Attending the funeral of the honourable ancestor, or, facing north-east as the menacing skies outside, leaden and pewter grey, shadowless, imply a drama about to happen.

As Proust said of the railway of St Lazare, he had to enter the cavernous place to fulfil the experience of the intensity of the difference between arrival and leave taking. Like certain skies painted, by Mantegna or Veronese, with an almost Parisian modernity. Skies beneath which could be accomplished only some solemn and tremendous act, such as a departure by train or the Elevation of the Cross.

I was wandering in thought, thinking about the way the railway stations in the Netherlands function as 'public interiors', thinking about the weather and the oldest function of the arcades, protection from the weather. And I was awoken from my reverie by Ulf, who is very affable. He introduces me to the wall-works, and guides me around the room, which has all the accoutrements of a work in progress, Styrofoam models, pages laid out for publication, the assembled thesis work of students piled on desks, a photocopy machine with a paper basket full of discarded A3 pages.

Not really feeling up to an interview, I offered a written essay in response to a guided tour of the current research project. I realized later that I could only offer stray and wandering thoughts, some suggestions. The work at the Why Factory is so immense in scope and so fundamentally engages of body, movement, zero waste of space, transformation of materials, haptic and rhythmic studies. I mutter something about Bruno Taut and say something like 'different ways of living produce different habitats'. We discuss the study of movement in the photography of Muybridge, the relevance to Duchamp, the strange ways of pyrites and how their capacity to form cubes points to the late work of Gehry, how small the entrance to the Vuitton museum is, whether Koolhaas's new work in Rotterdam was a torsioned classicism in the make. The recent and sad passing of Michael Graves. A new shape-shifting metal in China is announced, I cannot remember the name, gallium. Something about origami and biodegradable robots, and still in all of this there are bodies and things, there is matter and stuff, there is often a shining in the object.

The turn in the research seems to me cartographic; there is extensive mapping in which even the frames

are moving, and in the movement there are emergent eidetic properties.
I think it should be said that the factory stakes its claims in no ambiguous terms. Winy Maas articulated it in short, punchy prose, prose that is markedly 'on message' and is fired constantly by an unrelenting vision. The social commitment, the reinvigoration of what it means to be modern, the absolute and resolute and ferocious commitment to questioning and moving towards the future. It is of course a high-risk strategy and yet.
Yes. But.
The current research, guided by resilient and hard-working teachers, expands upon the notion of an ongoing research unit, which is not directly and uniquely tied to graduation projects. The vision has endured. Research by design is taken to involve long-term commitments, and the The Why Factory functions as a collaborative research platform and as a think-tank . It is rare to have a close view of such work evolving, and to see that what was at stake in earlier research has not been shunted aside and rather has become increasingly evident.
I wish them all well there, and must now return to the magical mystery tour of my online reading.

Patrick Healy, Amsterdam
19 April 2015

Barba is the product of a series of education and research collaborations. Over the course of five years, numerous students, teachers and researchers were part of the project, and each of these contributors informed the result you see in this book.

Evolution

The Why Factory, TU Delft, ETH Zürich, Buckylab, EIT ICT Labs

Barba was made by:

Callum Andrews, Louisa Au, Brendan Bakker, Roxane Belot, Arslane Benamar , Marcel Bilow, Paolo Birrer, Laura Ospina Botero, Nicola Campri, Tania Carl, Riccardo Caruso, Melina Cerfeda, Elena Chestnova, Gloria Chen, Huihua Chen , Yuxiao Chen, Sam Chia, Phillipa Elliot, Rachel Julia Engler, Nicholas Ganz, Stavros Gargaretas, Maosen Geng, Khalid Ghobashi, Federico Gobbato, Marino Gouwens, Alice Guarisco, Ulf Hackauf, Nadia El Hakim, Pirjo Haikola, Ching He, Marcel Hodel, Ruth Hoogenraad, Alise Jeksabone, Aleksander Jensen, Jayson Johnstone, Mark de Klijn , Léone Kündig, Niek van Laere, Maria Laisi, Cathrina Lee, Eun Yu Lee, Chen Li, Christopher Loh, Tomas de Loo, Lex te Loo, Winy Maas, Marine Manchon, Vincent Marchetto, Jeronimo Mejia, Alejandra Martinez Micolta, Andrea Migotto, Alberto Minero, Julia Ting Yan Mok, Rob Moors, Lino Moser, Sander Mulders, Marisa Muscionico, Kai Peter , Huib Plomp, Daniel Pulitin, Marie d'Oncieu de la Batie, Vladimir Ondejcik, Sebastian Oswald, Adrien Ravon, Piet de Reuver, Guendalina Rocchi, Pablo Roquero, Pascal Ruckstuhl, Enrique Ruiz, Dominik Saitl, Virginia Scapinelli, Paula Schilliger, Vera Schmidt, Ansis Sinke, Izabela Slodka, Henry Stehli, Leo Stuckardt, Ugur Sütcü, Ying Yi Tan, Yi Tang, Robbert Verheij, Igor de Vetyemy, Hans Vlaskamp, Kaegan Walsh, Eline Wieland, Stefanie Winter, Fei Wu, Stefan Wülser, Yangzom Wujohktsang, Rui Xuan, Jordan Yerbury, Saba Zahedi Asl, Barbara Zajac

Robotic City: Transform
Graduation Studio, TU Delft, Faculty of Architecture

September 2008 – April 2010
Khalid Ghobashi, Edoardo Felici, Maarten Haspels, Ahmet Korfali
Teachers: Ulf Hackauf, Pirjo Haikola, Winy Maas

The Robotic City thesis studio developed visions of a future in which all parts of the city – including transport, food production or logistics – would be automated. Khalid Ghobashi proposed a rubber-like building structure that would allow individual rooms to contract and extend depending on their use. His project inspired us to continue and deepen the idea of fully adaptable spaces.

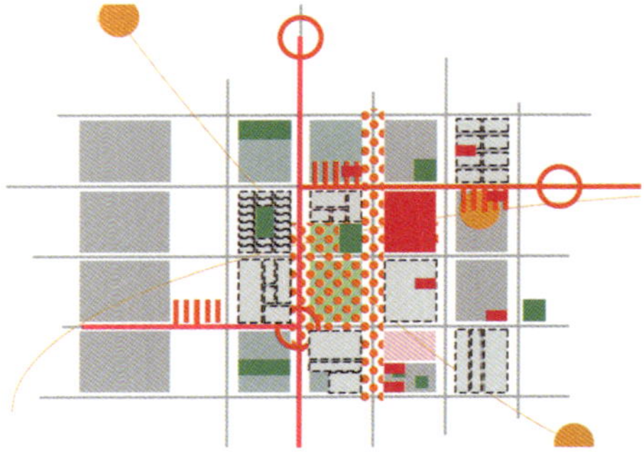

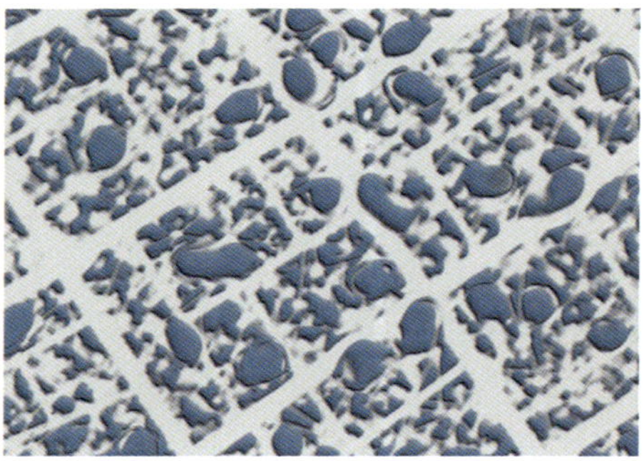

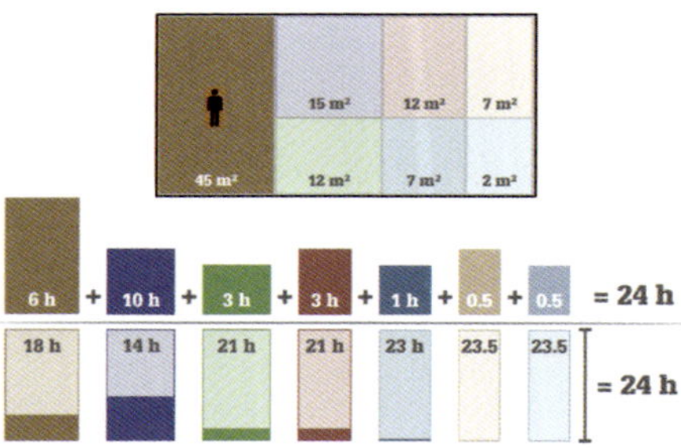

Barbapapa and T-1000

Reference

Throughout the different phases of our studies, two icons from popular culture served as references for us: 'Barbapapa', who inspired the title of the whole project, and his evil counterpart 'T-1000', who appeared in the 1991 movie *Terminator 2: Judgement Day.* Both Barbapapa and T-1000 are *shape shifters*, able to transform their bodies to mimic different objects or people. Such a fluent and effortless transformation from one shape to another became our goal for the Barba project.

Transformer Studio

Research and design studio, TU Delft, Faculty of Architecture

September 2010 – January 2011
Huihua, Chen, Alice Guarisco, Maria Laisi, Eun Yu Lee, Tomas de Loo, Alberto Minero, Julia Ting Yan Mok, Lino Moser, Virginia Scapinelli, Igor de Vetyemy, Stefanie Winter
Teachers: Winy Maas, Ulf Hackauf, Jeronimo Mejia

This studio assumed that the fully adaptable environment would be made from a soft but load-bearing and programmable building material. Using only line drawings, it sketched a complete first scenario for the environment. Research began with a description of the smallest grain: the hypothetical Transformer molecule. From there, walls, rooms and buildings are described. The ideas of 'cavities and spaces', 'peristaltic transportation' and 'ripple movement' are developed in order to make full use of the material. Eventually, this fragmentary dream continued to urban spaces and extended to the whole world as a kind of Transformer environment. The results of this studio formed the base for all following steps of this project.

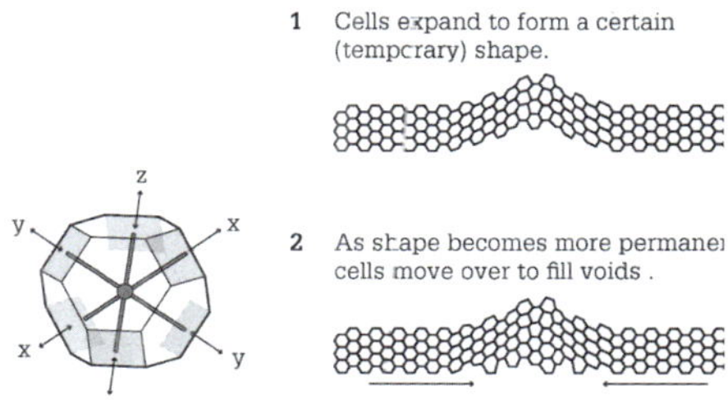

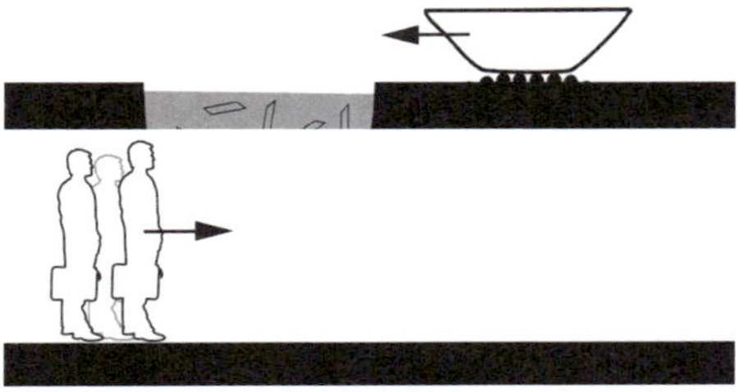

Household items stored in between material of the space

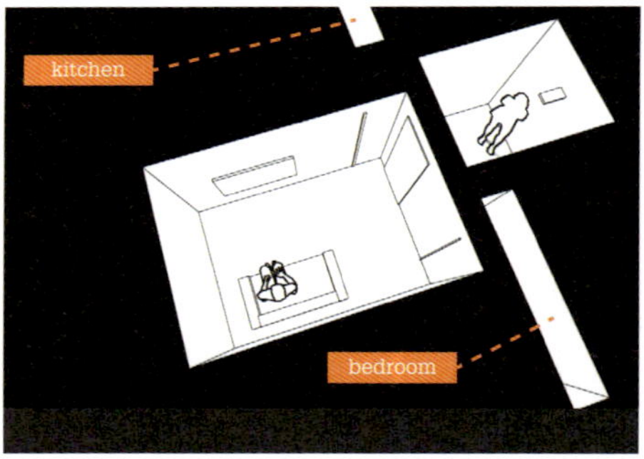

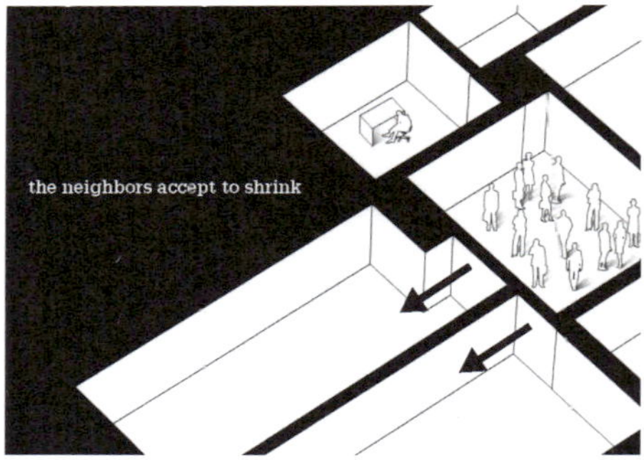

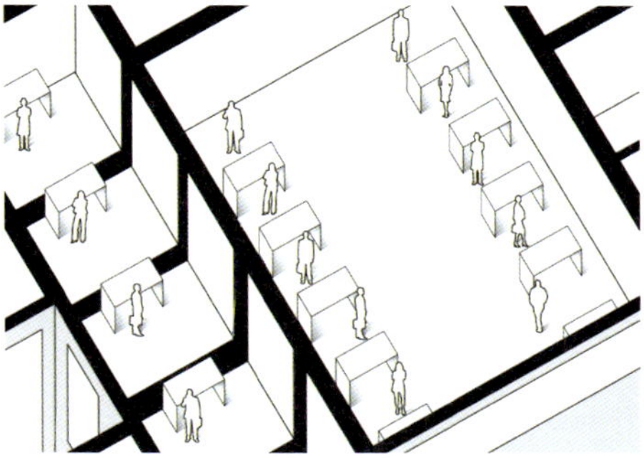

WEEKDAY 3.00 am
WEEKDAY 2.00 pm
WEEKEND 3.00 am
WEEKEND 2.00 pm

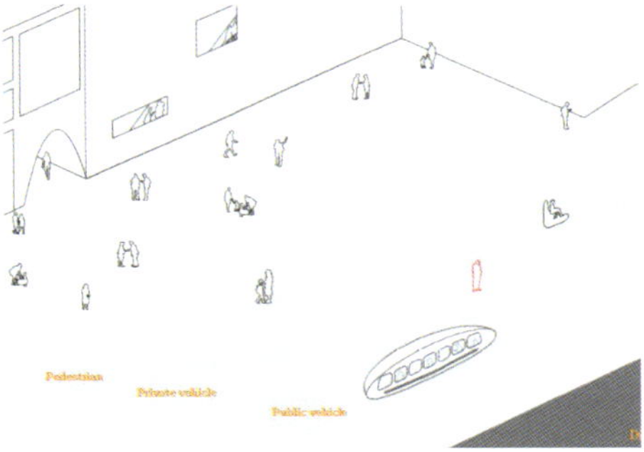
Pedestrian
Private vehicle
Public vehicle

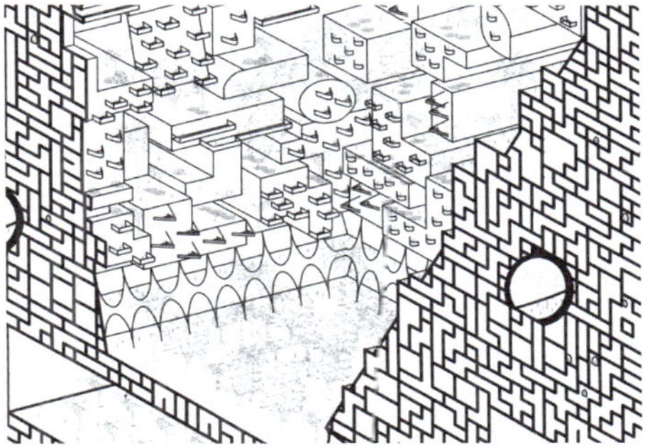

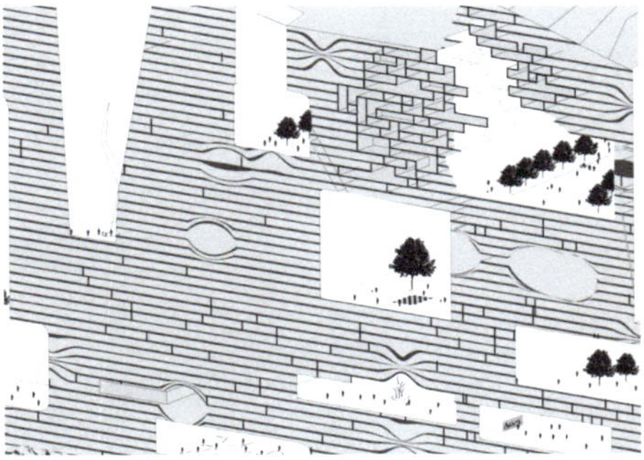

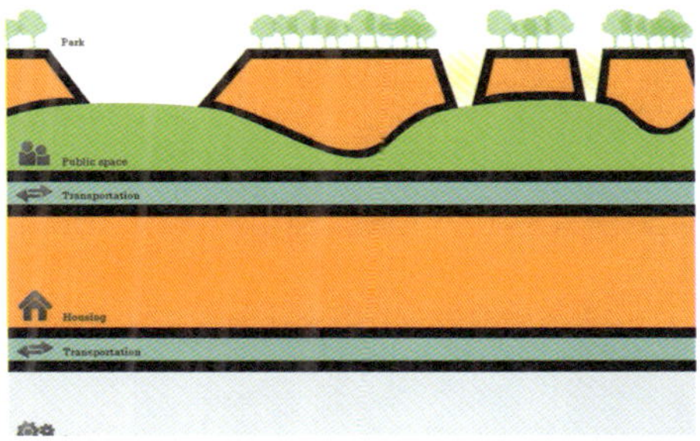
Park
Public space
Transportation
Housing
Transportation

The possibility of covering the entire world would bring us
a need to deal with a problem related to our
architectural/cultural heritage

Transformer Workshop
Research and design workshop,
TU Delft, Faculty of Architecture

March 2011
Louisa Au, Brendan Bakker, Yuxiao Chen, Phillipa Elliott, Christopher Loh, Alejandra Martinez Micolta, Vladimir Ondejcik, Daniel Pulitin, Izabela Slodka, Yi Tang, Rui Xuan, Saba Zahedi Asl, Barbara Zajac
Teachers: Ulf Hackauf, Pablo Roquero
Guest critics: Winy Maas, Marino Gouwens

This one-week workshop filled gaps in previous studio work. It focused on public space and activities, such as shopping, health care and leisure, but also considered farming and the relation between the environment and its surroundings.

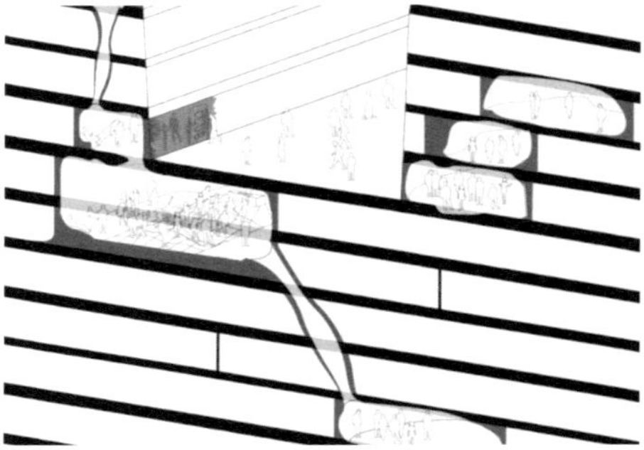

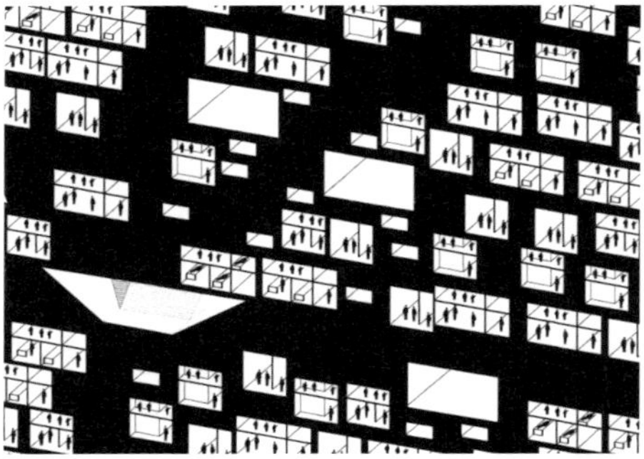

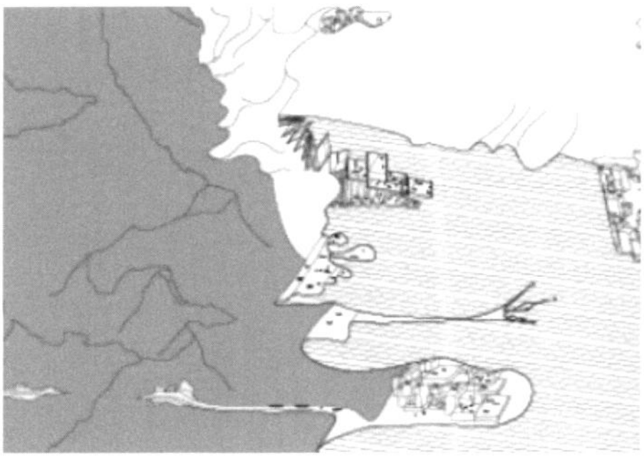

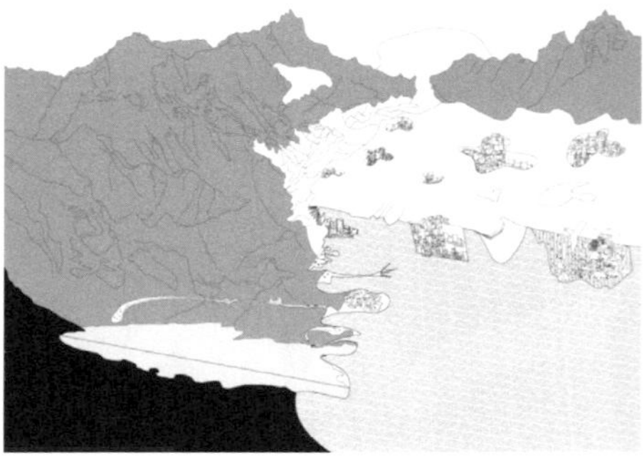

Sensor City and Dancing Towers

Design and research studio and workshop, TU Delft, Faculty of Architecture

February – June 2012
Xue (Snow) FANG, Ying (Mendy) MENG, Kelwin Palmer, Benjamin PHUA, Iraklis Romanopoulos, Leo Stuckardt, Nate Weems, Kar Yuen (Melissa) WONG
Teachers: Adrien Ravon, Ulf Hackauf, Winy Maas with Sander Mulders and Bas Kalmeijer

How can sensors trigger physical responses in the city? While this studio was not directly related to the previous studies, it provided interesting insight into the possibilities of interaction between inhabitants and buildings.

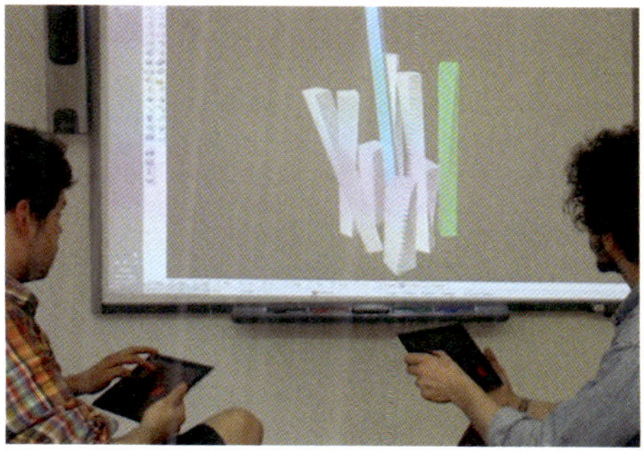

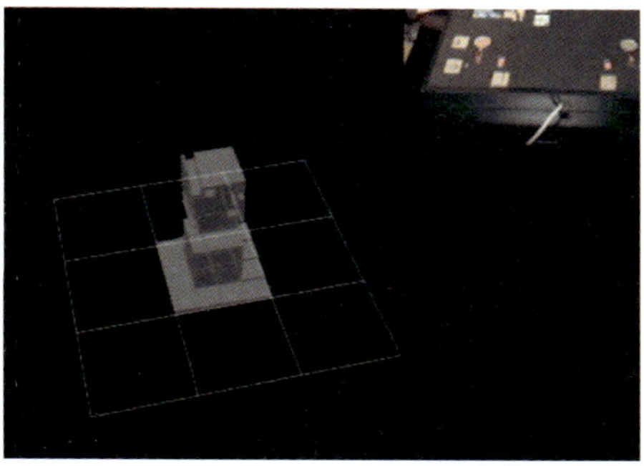

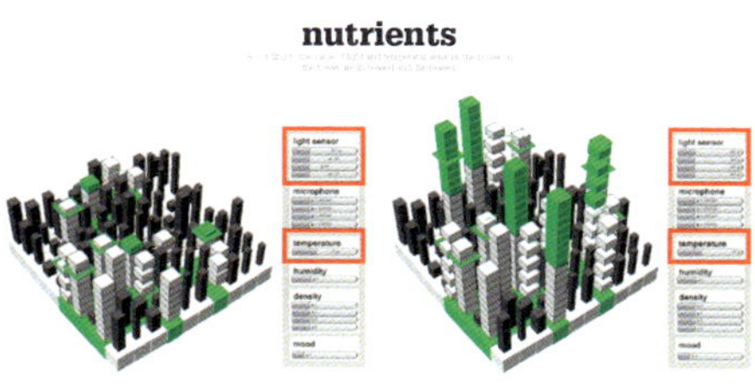

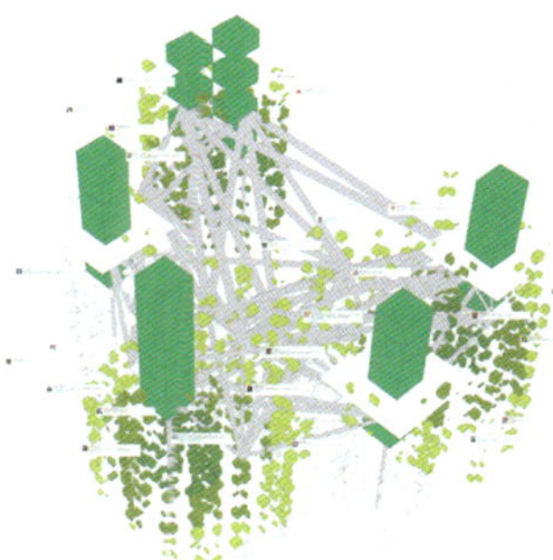

Transformer at ETH Zürich Studio

Guest studio at ETH Zürich, Faculty of Architecture

September – December 2012
Paolo Birrer, Tania Carl, Riccardo Caruso, Melina Cerfeda, Elena Chestnova, Nicholas Ganz, Marcel Hodel, Léone Kündig, Marisa Muscionico, Sebastian Oswald, Kai Peter, Pascal Ruckstuhl, Enrique Ruiz, Paula Schilliger, Vera Schmidt, Henry Stehli, Ying Yi Tan, Yangzom Wujohktsang
Teachers: Winy Maas, Ulf Hackauf, Adrien Ravon, Stefan Wülser
Guest critics: Knut Brunier, Benjamin Dillenburger, Joris Fach, Manuel Kretzer, Felix Madrazo

The Barba concept was greatly advanced through further research undertaken in collaboration with the students at the ETH in Zurich. This time, the fully adaptable environment was not only sketched but also modeled and scripted. In this manifestation, the vision received inherent rules and regulations. The vision was consistently modeled and described on the scales of a room, a house and a village. And the idea of interactivity entered into the project: between different spaces and between the user and the fully adaptable environment. Concepts such as 'spatial inertia', 'zones of activation' and 'spatial negotiation' were developed. The studio resulted in diagrams, interactive models, physical models and an amazing animation. While the results of this work are not presented as a separate chapter in this book, they greatly informed all parts of the research.

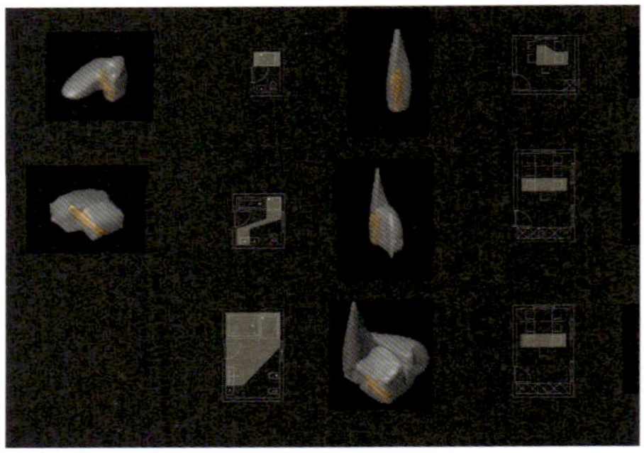

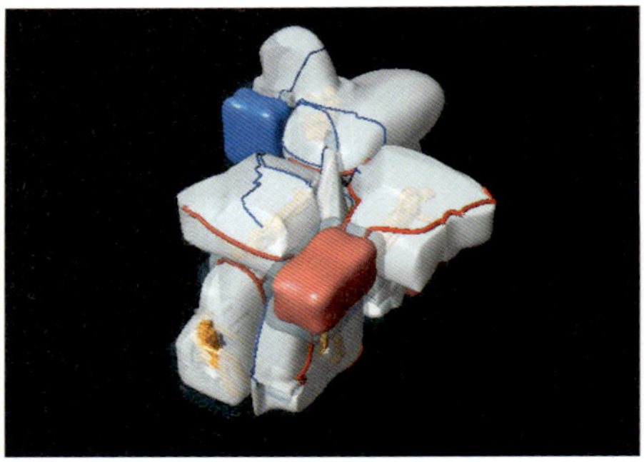

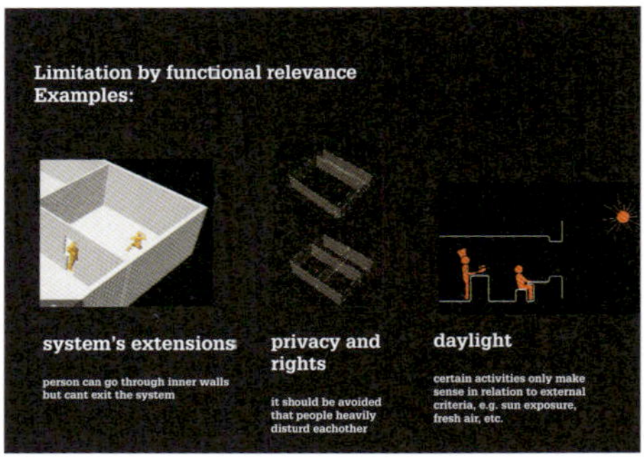

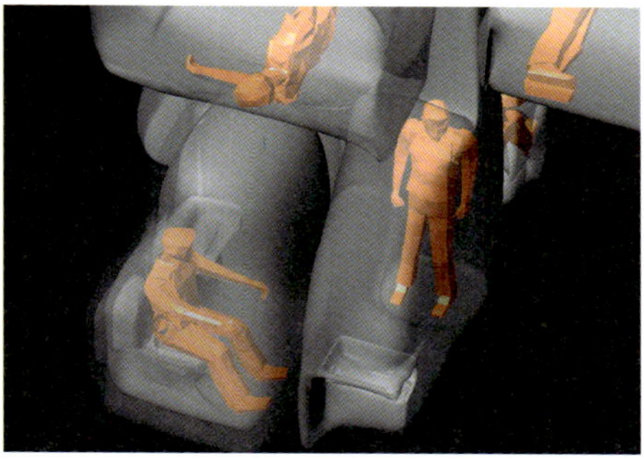
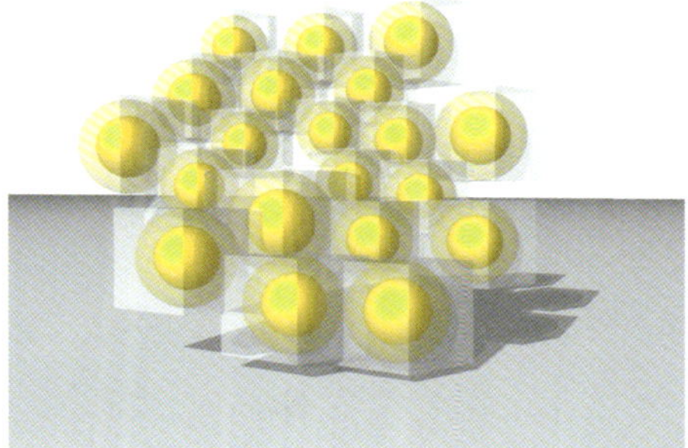
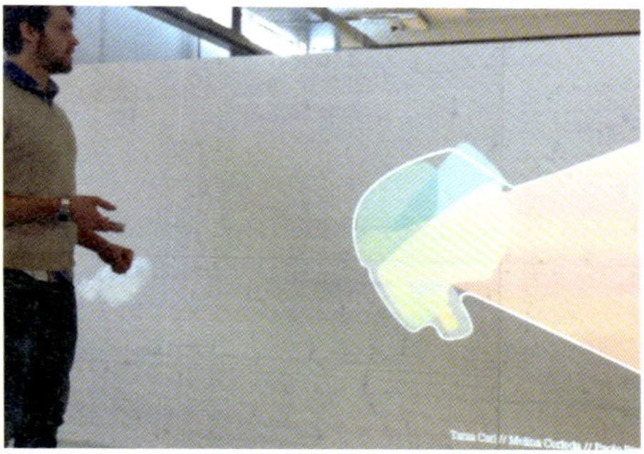

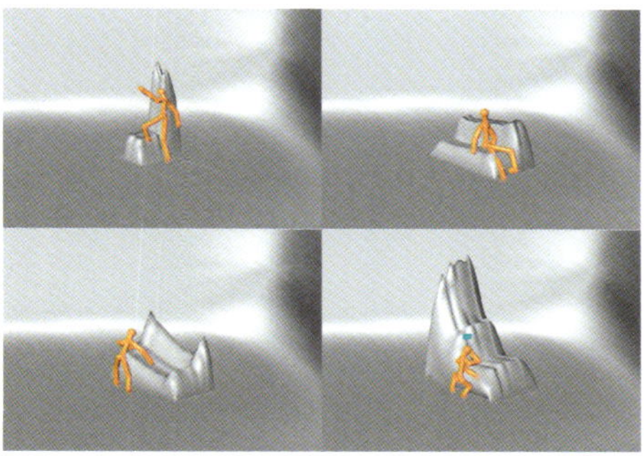
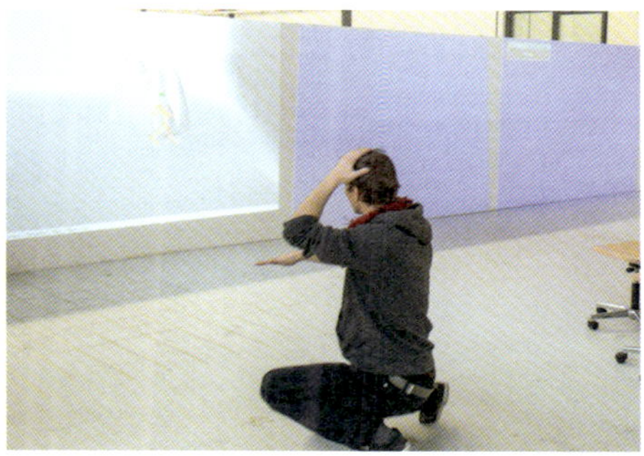

The Structuremaker

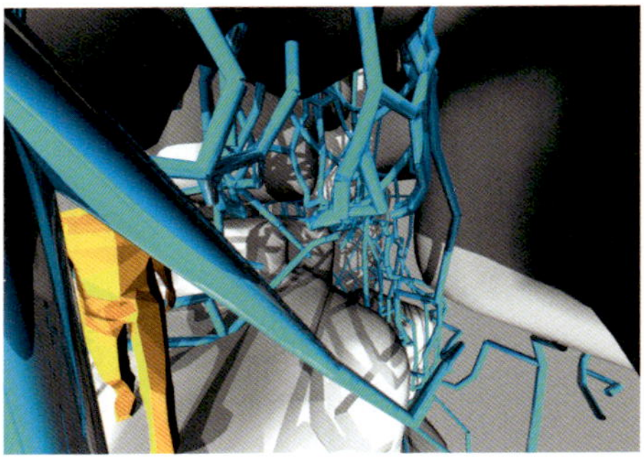

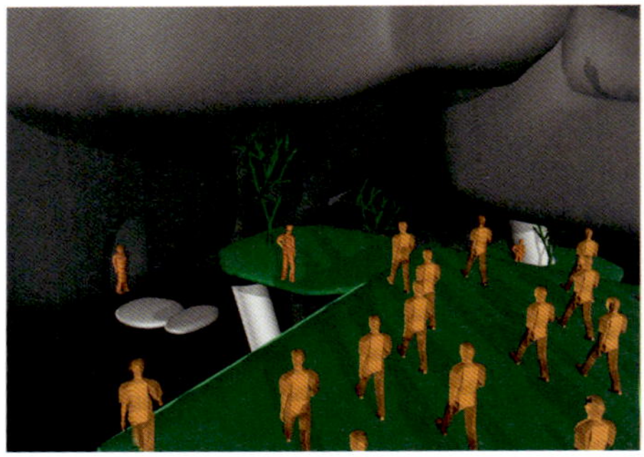

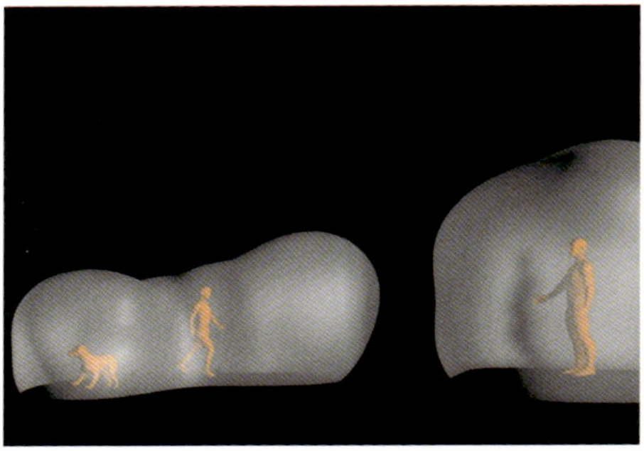

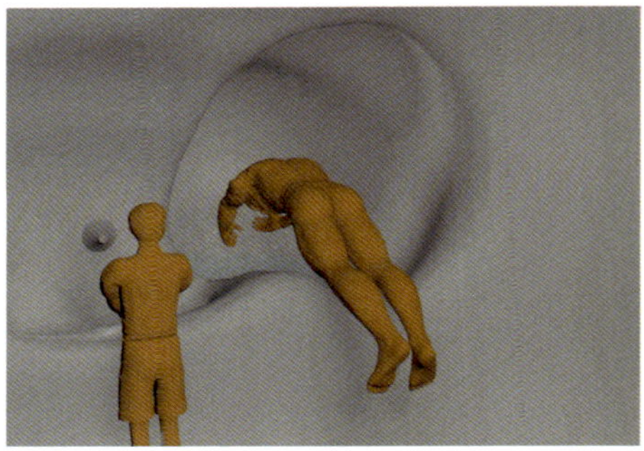

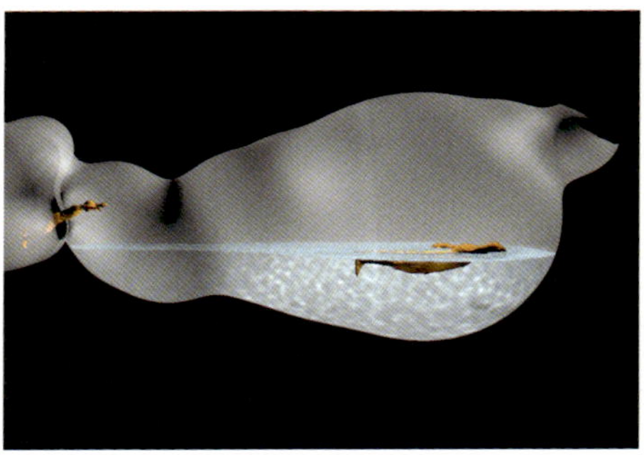

Watch the movie:
www.thewhyfactory.com/barba

Building Barbapapa

Workshop in collaboration with Buckylab, TU Delft, Faculty of Architecture

May 2013
Callum Andrews, Laura Ospina Botero, Gloria Chen, Maosen Geng, Ching He, Aleksander Jensen, Mark de Klijn, Niek van Laere, Chen Li, Marie d'Oncieu de la Batie, Guendalina Rocchi, Ansis Sinke, Ugur Sütcü, Fei Wu
Teachers: Marcel Bilow (Buckylab, TU Delft), Ulf Hackauf, Adrien Ravon
Guest critics: Ulrich Knaack (Façade Research Group, TU Delft), Winy Maas

While Barba became interactive in the previous studio, it still remained virtual. The material is based on custom made material-grains – something that at this moment it is still impossible to fabricate. But we still wanted to get an idea about how the material would feel, and so we got our hands dirty and used different materials to make the vision tangible. This not only gave us a better impression of the environment but also gave us first hints at how the concept could be approached with different existing materials, things already available at the moment.

Transformer Installations

Design and research studio, TU Delft, Faculty of Architecture

September 2013 – January 2014
Roxane Belot, Arslane Benamar, Nicola Campri, Sam Chia, Federico Gobbato, Nadia El Hakim, Ruth Hoogenraad, Cathrina Lee, Lex te Loo, Marine Manchon, Vincent Marchetto, Andrea Migotto, Rob Moors, Piet de Reuver, Dominik Saitl, Robbert Verheij, Hans Vlaskamp, Jordan Yerbury
Teachers: Ulf Hackauf, Adrien Ravon, Sander Mulders, Huib Plomp
Guest critics: Winy Maas, Marcel Bilow, Agata Jaworska

The 'Building Barbapapa' studio necessitated a more structured follow up. Different materials were used in a scale model to actually create spaces capable of transforming in real time. The proposals were then enlarged, providing an idea of how the material could be made (we termed this 'experiment') and how it would feel to live in it (we call this 'experience'). From an initial vision of a distant and speculative future, the vision became something much closer to our world today . . .

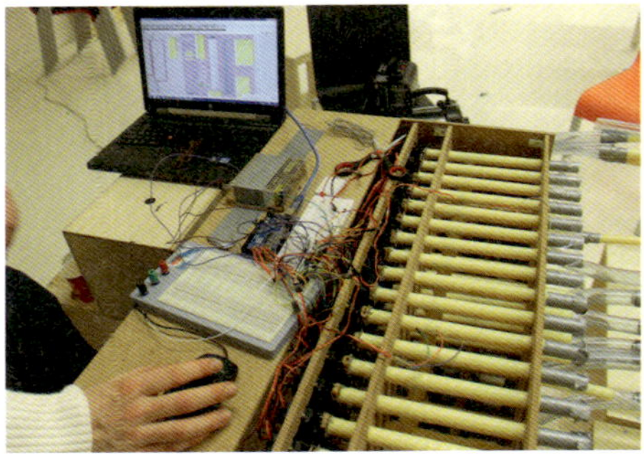

Adaptive City Thesis Studio

TU Delft, Faculty of Architecture

February 2013 – June 2014
Sarah Chebaro, Stavros Gargaretas, Jayson Johnstone, Albert Mark, Leo Stuckardt, Lara Tomholt, Teun Verkerk
Teachers: Ulf Hackauf, Adrien Ravon, Huib Plomp with Winy Maas

This graduation studio first arose independently from the Barba research. Adaptation and adaptive technologies were considered in a broader sense. The group produced collective research on adaptabilities in cities, from which a number of independent graduation projects developed. But because a lot of the work of The Why Factory is strongly linked, some of these projects eventually showed enough overlap with the Barba project to be presented in this context.

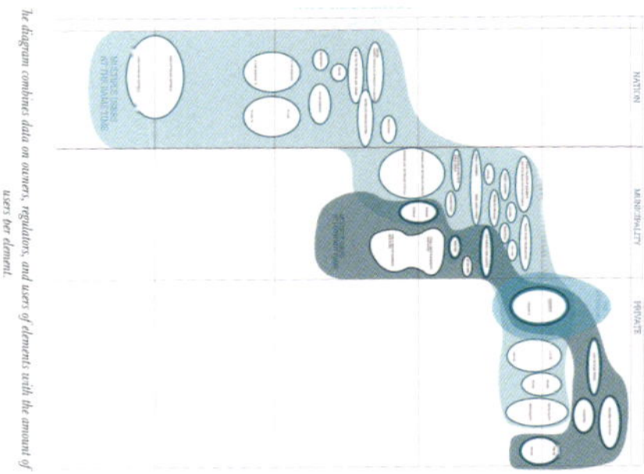

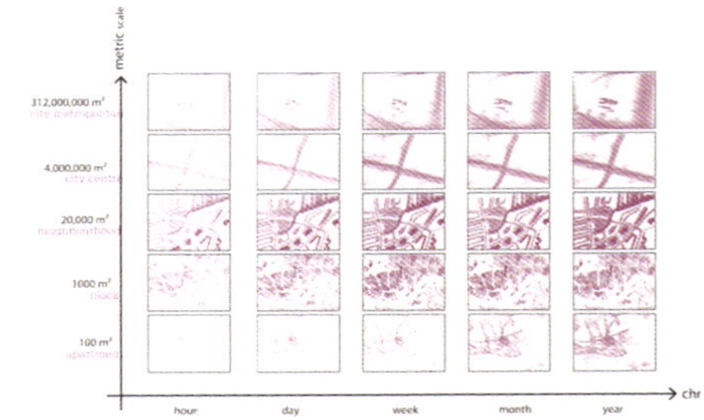

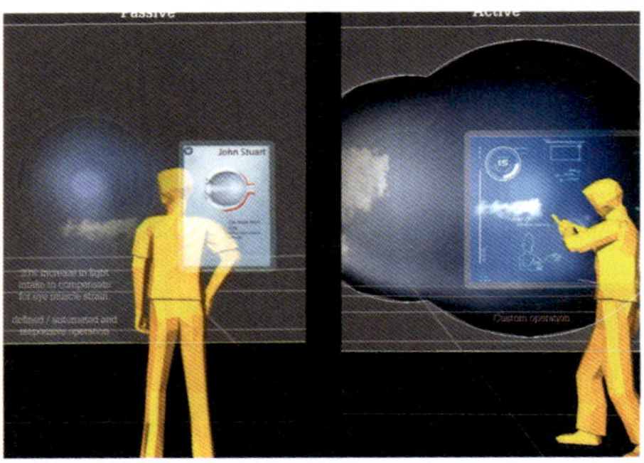

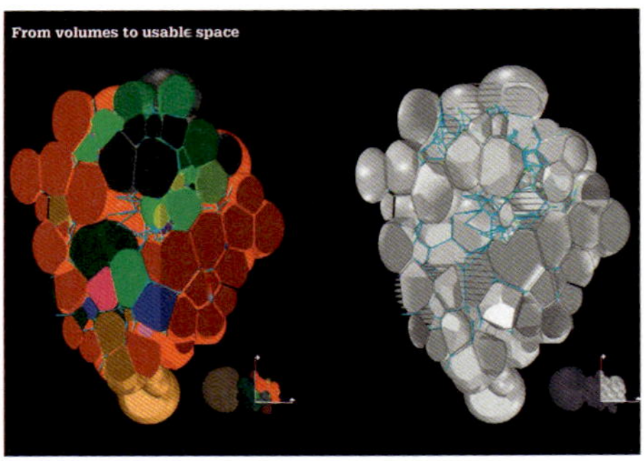

The Why Factory Research Team
TU Delft, Faculty of Architecture

Ulf Hackauf, Adrien Ravon, Arslane Benamar, Stavros Gargaretas, Alise Jeksabone, Lex te Loo, Jeronimo Mejia, Sander Mulders, Kaegan Walsh, Stefan Wülser

The work presented in this book is more than the direct result of the different studios and workshop. A team of teachers, researchers and research assistants has helped to edit studio material, fill gaps and make this book.

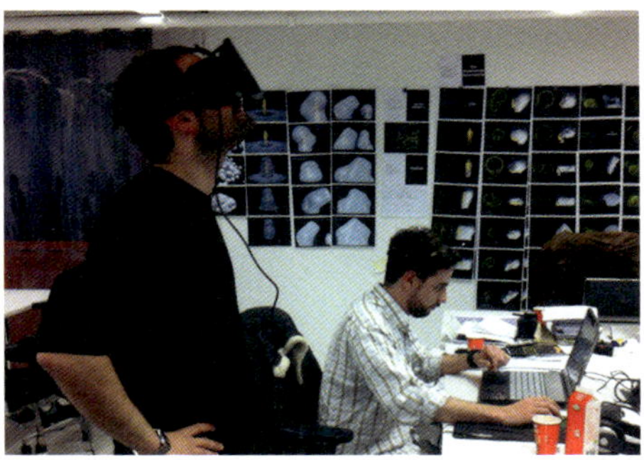

Mediating Presence
EIT ICT Lab contribution in collaboration with KTH Stockholm and TNO Nederland

March – September 2014
Ulf Hackauf and Adrien Ravon with Arslane Benamar, Sander Mulders, Lex te Loo and Stavros Gargaretas
EIT ICT Lab: Charlie Gullström (KTH Stockholm) and Joke Kort (TNO)

A presentation of the ongoing work on Barba led to this collaboration with the Mediating Presence research, an EIT ICT lab project led by Charlie Gullström. It allowed us to bring some of the interaction and material research of Barba into practice today. This served not as a tool for transforming buildings but as an adjustable, responsive interior element that becomes part of the multiple and innovative communication strategy developed in this framework. Barba goes out into the real world...

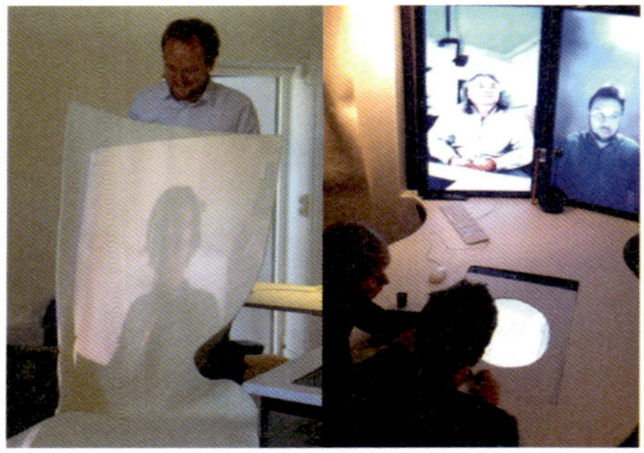

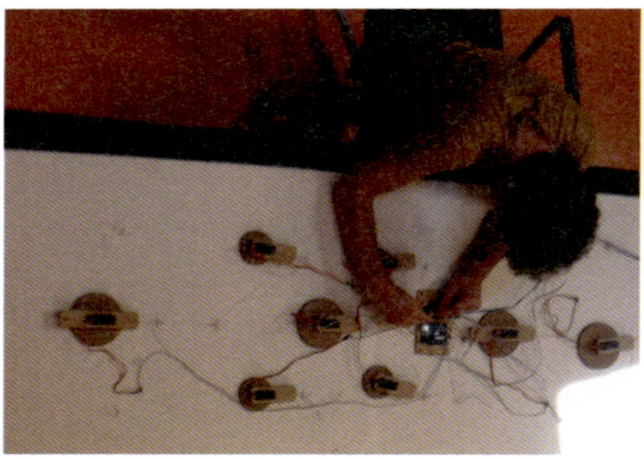

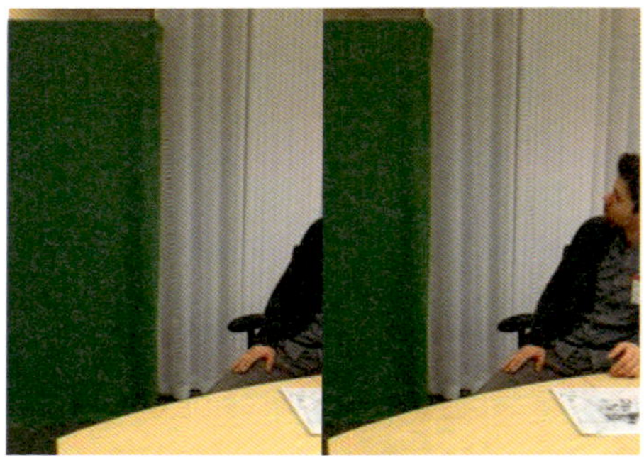

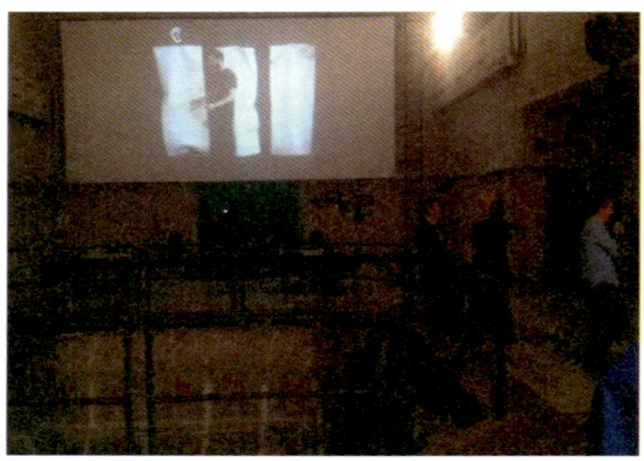

Ego City: The Next Step . . . ?

Ego City was the most recent design and research studio led by the Why Factory at Delft University of Technology. We began with the vision of an apartment slab in which 17 inhabitants with extremely large and diverse egos competed to accommodate their desires. While this project was not related to the Barba research and is based on fixed walls, created through complex battles, it could be the next test for Barba. The prompt led us to consider a space's capacity to negotiate conflicts and the ways in which it might reflect and accommodate the most extreme and diverse desires.
To be continued . . .

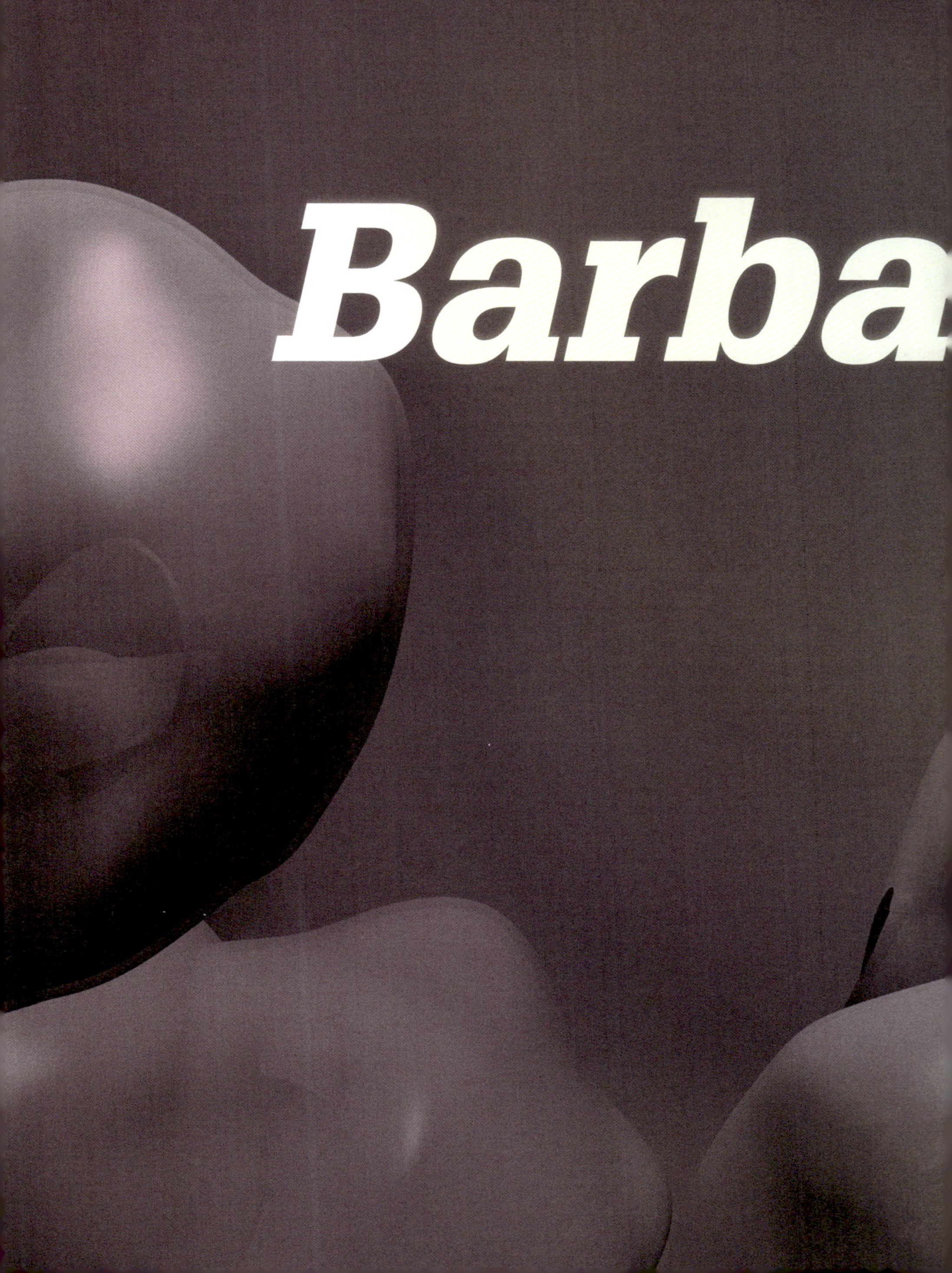
Barba

mama

The Why Factory: Winy Maas

The environment we created would act almost as if itself were alive. But can't – and shouldn't – what we once understood as the 'natural' world enter into this speculation? How might animals respond to these new materials? How would they inhabit them and how might they create shelters or hunting grounds, even?

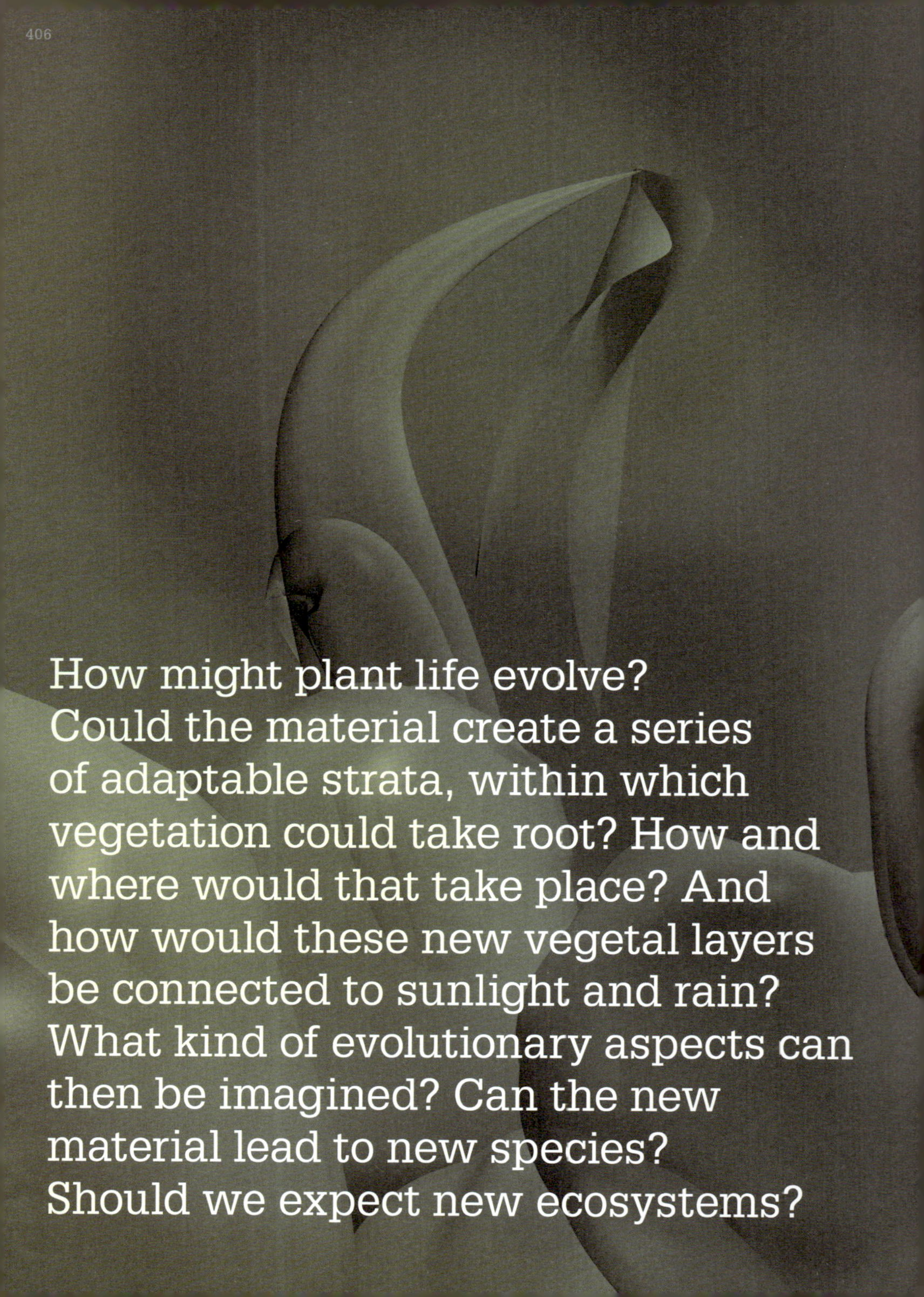

How might plant life evolve? Could the material create a series of adaptable strata, within which vegetation could take root? How and where would that take place? And how would these new vegetal layers be connected to sunlight and rain? What kind of evolutionary aspects can then be imagined? Can the new material lead to new species? Should we expect new ecosystems?

Could hydroponic systems, or even artificial photosynthesis, be incorporated to make sense of all of this? In a fully adaptable material world, where would we choose to position the natural realm? Everywhere? Or would we rather it adapt to follow our every human need?

What, then, is natural? Or – with the loss of authentically 'static' material and materialism – has all become essentially natural? Could Barba space become a new kind of nature that grows and evolves and gives birth?

Barba needs another dimension, a natural side. A Barbamama . . . Where would this story end? Or would it end at all? It could, eventually, continue across the planet. It could embrace everything, create a new surface on land and sea.

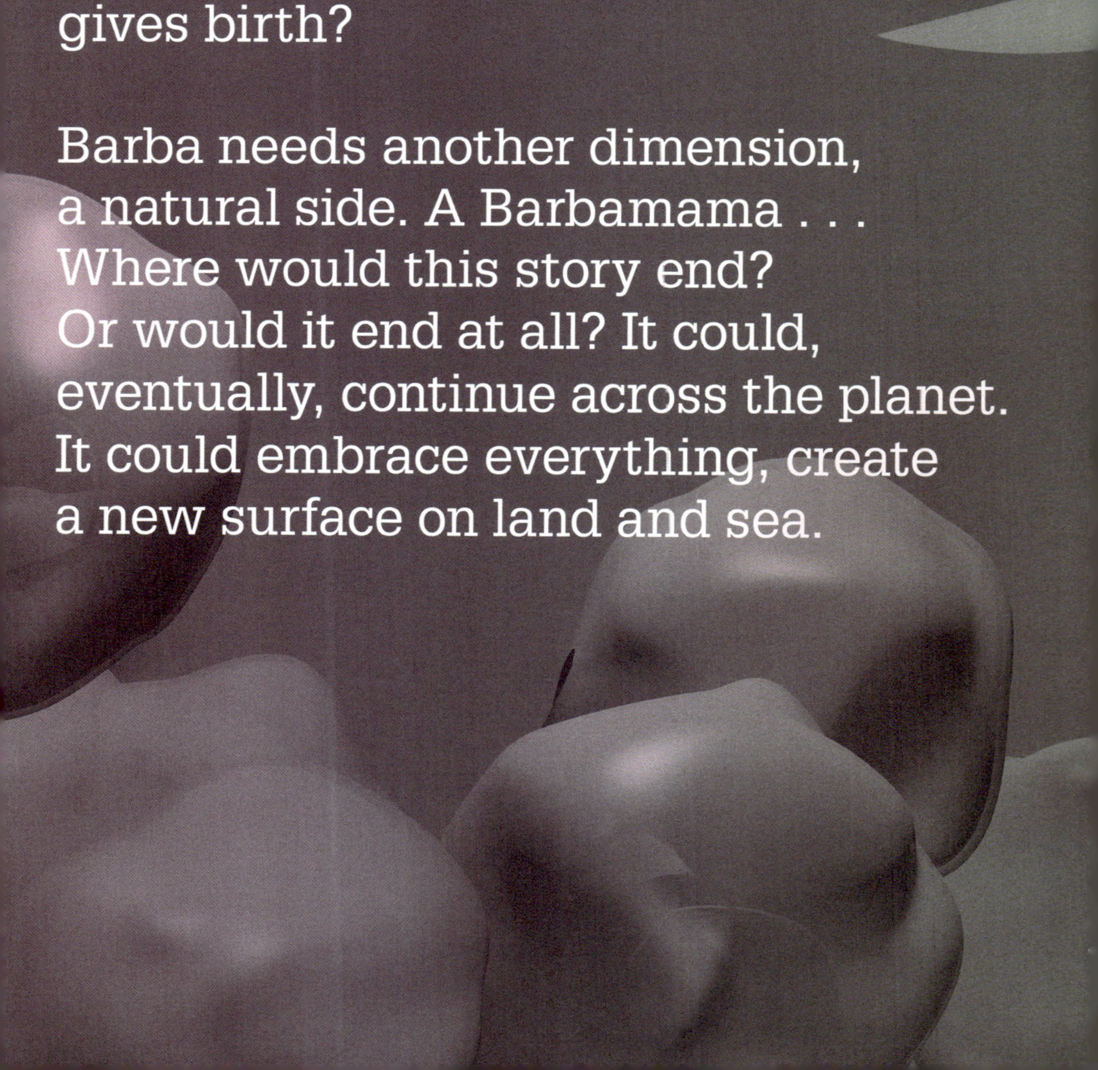

And where would it stop? Would it need to stop at all? Or could it eventually lift up and leave our planet? Could it create a new moon upon which we might live? Could it create a new planet? And might that new planet split in two? Might it divide itself even further, divide again and again and again?

Authors
Sources
Credits

The Why Factory

The Why Factory is a global think-tank and research institute, run by MVRDV and Delft University of Technology and led by professor Winy Maas. The Why Factory's Future Cities research programme explores possibilities for the development of our cities by focusing on the production of models and visualizations for cities of the future. The results of this research programme are being presented in a series of books – the *Future Cities* series – published in association with nai010 publishers in Rotterdam and designed by Bastiaan de Wolff of Thonik / BENG in Amsterdam. *Barba* is the eighth book in the Why Factory's *Future Cities* series, and follows *Visionary Cities*, *The Why Factor(y)*, *Green Dream*, *Vertical Village*, *Hong Kong Fantasies*, *City Shocks* and *We Want World Wonders*. For more information see: www.thewhyfactory.com

Authors

Winy Maas
Prof. Ir. Ing FRIBA HAIA (1959, Schijndel, the Netherlands) is one of the founding directors of the globally operating architecture and urban planning firm MVRDV, based in Rotterdam and known for projects such as the Expo 2000 and the vision for greater Paris, Grand Paris Plus Petit. Maas is also a professor at Delft University of Technology and the director of The Why Factory, a research institute for the future city, which he founded in 2008. Since 2013 he has been a visiting professor at the University of Hong Kong, and he has served as a professor at ETH Zürich, the Berlage Institute, MIT, Ohio State University and Yale University. In addition, he designs stage sets and objects and was the curator of Indesem 2007. He curates exhibitions, lectures throughout the world and takes part in international juries. Recently, Maas joined the Economic Development Board of Rotterdam (EDBR). In 2012, he was appointed urban supervisor for the city of Almere and, since 2003, he has been supervising the Bjorvika urban development in Oslo. With both MVRDV and The Why Factory he has published a series of research projects. For more information see: www.mvrdv.com

Ulf Hackauf
Dipl.-Ing. (1971, Braunschweig, Germany) was educated as an architect in Germany and Denmark.
After graduating from the Technische Universität Braunschweig, he worked as project architect at (EEA) Erick van Egeraat associated architects and at Neutelings Riedijk Architecten. In 2007, Ulf Hackauf joined The Why Factory, where his research includes work on topics of sustainability, bottom-up planning, technical innovation and interactive design.
He co-authored The Why Factory's publications *Visionary Cities*, *Green Dream* and *Vertical Village* and acted as The Why Factory's programme coordinator. Hackauf has taught at institutions including the Tamkang University, Taiwan; Tarumanagara University, Jakarta; Yonsei University, Seoul; Piet Zwart Institute, Rotterdam; Berlage Institute, Rotterdam; ETH Zürich; and the IIT in Chicago. Since 2015, Hackauf holds a research position at the chair of environmental technology and design, department of urbanism at Delft University of Technology. For more information see: www.hackauf.com

Adrien Ravon
Dipl.-Arch.(1986 Versailles, France) is an architect operating across the boundaries of different disciplines by combining both theoretical and technical approach. Ravon completed his studies at FADU-UBA and ENSAPM, Paris, where he defended his Master's thesis 'Read/Write Territories'. In parallel to his academic path in architecture, Ravon has studied free and open source programming, generativity and interactivity, multimedia and interactive games. After collaborating as an architect with Jacob+MacFarlane in Paris, Ravon joined The Why Factory in 2011 as a teacher and researcher. He has worked on various projects and had a key position in advancing the Future Models course, which provides specific computing support to The Why Factory's work. Adrien has worked with the Berlage Institute, Rotterdam; ETH Zürich; KTH Stockholm and the EIT ICT Labs.

Patrick Healy
(1955, Dublin, Ireland) studied philosophy, sociology and Near Eastern languages at the Pontifical University Maynooth, University College Dublin and has since been engaged in writing, research and teaching, mainly in the areas of aesthetics and contemporary art. His recent publications include works on aesthetics, the philosophy of science and various artists' biographies. He has served as Professor of Interdisciplinary Research at the Free International University, Amsterdam since 1997 and, at Delft University of Technology, works as a Senior Researcher and tutor in the Architecture Theory section, providing lecture series in aesthetics and the philosophy of science. For his publications and other activities, see: www.patrick-healy.com

Content Credits

Barba
The Why Factory: Winy Maas with Ulf Hackauf, Adrien Ravon, TU Delft MSc studio 2010 and Wieland and Grouwens
Text by Winy Maas

Wasted Space
The Why Factory
Study by Ulf Hackauf, based on the work of Khalid Ghobashi (Robotic City studio), the Transformer studio TU Delft and Stavros Gargaretas (Adaptive City Thesis studio)
Visualizations by Arslane Benamar

Life in Barba
Animation by Adrien Ravon with Arslane Benamar
Script by Adrien Ravon, Winy Maas and Ulf Hackauf, based on the movie 'Transformer', which Paolo Birrer produced in the framework of the Transformer at ETH Zürich studio and during the guest professorship of Winy Maas and The Why Factory at the department of architecture, ETH Zürich.
The renderings make use of the 3D model figure "Michael 6" by Daz3D. A user license has been purchased, copyright for the 3D model figure remain with Daz3D.
The movie is based on ideas by all the students who participated that studio:
Paolo Birrer, Tania Carl, Riccardo Caruso, Melina Cerfeda, Elena Chestnova, Nicholas Ganz, Marcel Hodel, Léone Kündig, Marisa Muscionico, Sebastian Oswald, Kai Peter, Pascal Ruckstuhl, Enrique Ruiz, Paula Schilliger, Vera Schmidt, Henry Stehli, Ying Yi Tan, Yangzom Wujohktsang

Space Makers
Concept and development by The Why Factory research team: Adrien Ravon with Ulf Hackauf, Sander Mulders, Stefan Wülser, Kaegan Walsh, Arslane Benamar, Lex te Loo
Based on the outcomes of the the Transformer at ETH Zürich studio during Winy Maas's guest professorship at the department of architecture, ETH Zürich:
Paolo Birrer, Tania Carl, Riccardo Caruso, Melina Cerfeda, Elena Chestnova, Nicholas Ganz, Marcel Hodel, Léone Kündig, Marisa Muscionico, Sebastian Oswald, Kai Peter, Pascal Ruckstuhl, Enrique Ruiz, Paula Schilliger, Vera Schmidt, Henry Stehli, Ying Yi Tan, Yangzom Wujohktsang
Visualizations by Adrien Ravon with Lex te Loo and Arslane Benamar
Project texts by Adrien Ravon and Ulf Hackauf

Materialization
Study by Ulf Hackauf, with support of Alise Jeksabone (The Why Factory research team)

Material Experiments
Developed and illustrated by the students of The Why Factory's 'Building Barbapapa' workshop at the Faculty of Architecture and the Built Environment at Delft University of Technology:
Callum Andrews, Laura Ospina Botero, Gloria Chen, Maosen Geng, Ching He, Aleksander Jensen, Mark de Klijn, Niek van Laere, Chen Li, Marie d'Oncieu de la Batie, Guendalina Rocchi, Ansis Sinke, Ugur Sütcü, Fei Wu
Teachers: Marcel Bilow (Buckylab, TU Delft), Ulf Hackauf, Adrien Ravon
Guest critics: Ulrich Knaack (façade research group, TU Delft), Winy Maas
Edited by Ulf Hackauf

Models
Developed and illustrated by the students of The Why Factory's 'Transformer Installations' design and research studio at the Faculty of Architecture and the Built Environment at Delft University of Technology:
Roxane Belot, Arslane Benamar, Nicola Campri, Sam Chia, Federico Gobbato, Nadia El Hakim, Ruth Hoogenraad, Cathrina Lee, Lex te Loo, Marine Manchon, Vincent Marchetto, Andrea Migotto, Rob Moors, Piet de Reuver, Dominik Saitl, Robbert Verheij, Hans Vlaskamp, Jordan Yerbury
Teachers: Ulf Hackauf, Adrien Ravon, Sander Mulders, Huib Plomp
Guest critics: Winy Maas, Marcel Bilow, Agata Jaworska
Editing and project description by Lex te Loo and Ulf Hackauf

Rietveld Revisited
Developed and illustrated by the students of The Why Factory's 'Transformer Installations' design and research studio at the Faculty of Architecture and the Built Environment at Delft University of Technology:
Roxane Belot, Arslane Benamar, Nicola Campri, Sam Chia, Federico Gobbato, Nadia El Hakim, Ruth Hoogenraad, Cathrina Lee, Lex te Loo, Marine Manchon, Vincent Marchetto, Andrea Migotto, Rob Moors, Piet de Reuver, Dominik Saitl, Robbert Verheij, Hans Vlaskamp, Jordan Yerbury
Teachers: Ulf Hackauf, Adrien Ravon, Sander Mulders, Huib Plomp
Guest critics: Winy Maas, Marcel Bilow, Agata Jaworska

Inhabiting Zero Wasted Space
Graduation project by Stavros Gargaretas, conceived as part of the Adaptive City thesis study by The Why Factory at the Faculty of Architecture and the Built Environment at Delft University of Technology
Teachers: Ulf Hackauf, Adrien Ravon, Huib Plomp with Winy Maas

Adaptive Dream Home Tower
Graduation project by Jayson Johnstone, conceived as part of the Adaptive City thesis study by The Why Factory at the Faculty of Architecture and the Built Environment at Delft University of Technology
Teachers: Ulf Hackauf, Adrien Ravon, Huib Plomp with Winy Maas

Twenty-four hour Block
Graduation project by Leo Stuckardt, conceived as part of the Adaptive City Thesis Study by The Why Factory at the Faculty of Architecture and the Built Environment at Delft University of Technology
Teachers: Ulf Hackauf, Adrien Ravon, Huib Plomp with

Winy Maas
Prototypes
Magic Carpet
Project by The Why Factory research team
Concept by Winy Maas, Ulf Hackauf and Adrien Ravon , based on an idea by Piet de Reuver and Robbert Verheij
Production: Arslane Benamar with Lex te Loo and Piet de Reuver
Text: Ulf Hackauf
Talking Napkin and Softwall
The Why Factory and EIT ICT Labs: The Why Factory research team with KTH Stockholm and TNO
The Softwall prototypes were made by The Why Factory research team:
Concept and supervision: Ulf Hackauf and Adrien Ravon
Development, prototyping and programming: Adrien Ravon, Arslane Benamar, Lex te Loo, Stavros Gargaretas and Sander Mulders
Report and text: Ulf Hackauf
The development was supported by Charlie Gullström (KTH Stockholm) Joke Kort (TNO) in the framework of the EIT ICT Lab 'Mediating Presence' 2014
Implications
Map of Questions
The Why Factory and ETH: ETH visiting studio 2012
The questions were posed by the students of the Transformer at ETH Zürich studio in the framework of Winy Maas's guest professorship:
Paolo Birrer, Tania Carl, Riccardo Caruso, Melina Cerfeda, Elena Chestnova, Nicholas Ganz, Marcel Hodel, Léone Kündig, Marisa Muscionico, Sebastian Oswald, Kai Peter, Pascal Ruckstuhl, Enrique Ruiz, Paula Schilliger, Vera Schmidt, Henry Stehli, Ying Yi Tan, Yangzom Wujohktsang
Editing, categorization and graphic layout of the questions: Stefan Wülser
Introduction: Ulf Hackauf
What if not?
The Why Factory: Ulf Hackauf
Through a Glass Darkly . . .
Text by Patrick Healy
Evolution
Edited by Ulf Hackauf
Barbamama
Text by Winy Maas
Graphics by Adrien Ravon

Image Credits

All images in this publication were produced and are copyright by The Why Factory, with the exception of the images used in the chapter 'Materialization':
Dynamic Building Parts
Image of the Rietveld Schröder House by basvb, available on flickr.com under creative commons
Isometric drawing of Rietveld Schröder House by Gerrit Rietveld
Image of the façade of the Institute du Monde d'Arab by Matito, available on flickr.com under creative commons
Image of a detail of the façade of the Institute du Monde d'Arab by Rory Hyde, available on flickr.com under creative commons
Images of the Muscle Tower II by Hyperbody, Delft University of Technology
Automated Assembly
Images by Gramazio&Kohler, Professorship for Architecture and Digital Fabrication, ETH Zurich and Raffaello D'Andrea Institute for Dynamic Systems and Control, ETH Zurich
Modular Robotics
Images 1 and 2 by Claytronics, Carnegie Mellon University
Images 3, 4, 5 and 5 by Distributed Robotics Lab, CSAIL, Massachusetts Institute of Technology
Organic Production
Image QuteMol03_Snapshot.png by snapshot, available on commons.wikipedia.org under creative commons.
Image foglet by J. Storrs Hall
Image from the "Grey Goo" video game published by Grey Box.

Credits

Concept and Editors
Winy Maas, Ulf Hackauf and Adrien Ravon

Associate Editor
Rachel Julia Engler

Copy Editor: D'Laine Camp
Graphic Design: www.BENG.nl
Design of The Why Factory logo: Thonik in collaboration with studio BENG
Lithography and Printing: Die Keure, Bruges (B)
Paper: Arctic Volume, 120 grs
Publisher: Marcel Witvoet, nai010 publishers, Rotterdam

We are grateful to the ETH Zürich, Delft University of Technology and the EIT ICT Labs for the financial support of this publication.

nai010 publishers is an internationally orientated publisher specialized in developing, producing and distributing books in the fields of architecture, urbanism, art and design.
www.nai010.com

nai010 books are available internationally at selected bookstores and from the following distribution partners:

North, Central and South America - Artbook | D.A.P., New York, USA, dap@dapinc.com

Rest of the world - Idea Books, Amsterdam, the Netherlands, idea@ideabooks.nl

For general questions, please contact nai010 publishers directly at sales@nai010.com or visit our website www.nai010.com for further information.

Printed and bound in Belgium
ISBN 978 94 6208 253 3

Barba is also available as e-book (pdf)
Barba e-book
ISBN 978 94 6208 256 4

Table of Contents

In the best tradition of science fiction, this book illustrates how a series of current technological innovations might change the way we build and live in cities in the future.
Our first inspiration came from 'Barbapapa', an illustrated cartoon character from the 1970s. Invented and drawn by Talus Taylor and Annette Tison, the friendly, blobby protagonist of the eponymous children's books and television programme could change his shape to resemble different objects. With Barbapapa's smooth morphosis in mind, we wondered how today's advancements in robotics, material science and computing might allow us to create environments that transform themselves as easily as Barbapapa could.
The project that emerged developed through a series of projects organized by The Why Factory and undertaken in collaboration between Delft University of Technology, ETH Zürich and the European Institute of Innovation and Technology. Neither Barbapapa's inventors nor anybody else from the team behind the cartoon were involved in this project, but we owe them absolute gratitude for the inspiration that their Barbapapa gave us.

Barba by The Why Factory:
Winy Maas, Ulf Hackauf and Adrien Ravon

— Dedicated to Talus Taylor (1933–2015) —